FAST NEUTRON PHYSICS

Beijing International Symposium on

FAST NEUTRON PHYSICS

Beijing, China 9 – 13 September 1991

Editors

SUN ZUXUN, TANG HONGQING, XU JINCHENG and ZHANG JINGSHANG

China Institute of Atomic Energy
Beijing

World Scientific
Singapore • New Jersey • London • Hong Kong

Published by

World Scientific Publishing Co Pte Ltd
5 Toh Tuck Link, Singapore 596224
USA office: 27 Warren Street, Suite 401-402, Hackensack, NJ 07601
UK office: 57 Shelton Street, Covent Garden, London WC2H 9HE

Proceedings of the Beijing International Symposium on
FAST NEUTRON PHYSICS

ISBN-13 9789810209223
ISBN-10 9810209223

Beijing International Symposium on Fast Neutron Physics, Sep. 9 - 13, 1991

Sponsors
China Institute of Atomic Energy
Beijing National Tandem Accelerator Laboratory
China Nuclear Physics Society

Organization
Chairman Sun Zuxun (President, CIAE)

PREFACE

These proceedings are a collection of the invited talks and contributed papers presented at the Beijing International Symposium on Fast Neutron Physics, held on September 9-13, 1991, Beijing, China and organized by China Institute of Atomic Energy.

This Symposium is the second one following the Beijing International Symposium on Physics at Tandem which was held in 1986. At that ime, the HI-13 tandem accelerator and its related experimental facilities were just before the formal operation. Since then, five years have passed and a lot of experiments at the tandem accelerator have been carried out. Some results on the studies of fast neutron physics can be found in these proceedings.

The proceedings focus on different topics of contemporary fast neutron physics including the recent progress and new achievements in fast neutron scattering and energy spectrum, nuclear fission, γ-ray spectroscopy and (n,γ) reaction mechanism, nuclear theory, activation cross sections, nuclear reaction and intermediate energy neutron physics.

The aim of the symposium was at reviewing the current status of fast neutron physics study by invited presentations and free discussions, exchanging experiences and physical ideas and promoting research work and collaboration. Owing to the efforts of more than 60 participants (including 15 foreigners from Belgium, Germany, Italy, Japan, USA and USSR), especially 20 invited speakers, the symposium was successful and reached its goal.

Here we would like to express our thanks to all attendees for their participating in the symposium, to all speakers for their interesting presentation, to the International Advisory Committee and Local Organizing Committee for their good suggestions, to the secretariat for its efficient work, to the sponsors (China Institute of Atomic Energy, Beijing National Tandem Accelerator Laboratory and Chinese Nuclear Physics Society) and ORTEC company for their financial support and to all people who made contributions to or concerned the symposium.

Editors

Opening Address

Wang Ganchang

Ladies and Gentlemen,

May I take the privilege to welcome you to the Beijing International Symposium on Fast Neutron Physics, on behalf of the China Institute of Atomic Energy, I'd like to congratulate this happy international gathering!

This symposium is about neutron. The February 17th of the next year will be the sixtieth anniversary of its discovery. In preparing this opening address, to my mind was recalled the situation when neutron was discovered. Would you be kind enough to allow me to recollect some pieces of the history and the reflexion on my own experience. As you might know, Professors Joliot-Curies missed the opportunity of neutron's discovery, here I would like to tell you another story.

From 1930 to 1934, it was a golden era, marked with many profound discoveries in nuclear physics, when I was in University of Berlin as a graduate student. Professor Lise Meitner was my advisor then. I remember that someone reported at a colloquium one day in 1930 the experiment just done by Professor Walther Bothe and his student Dr. H. Becker. They bombarded Beryllium with Polonium emitted alpha particles and found a strongly penetrating radiation, which, unfortunately, they interpreted as gamma radiation. I had learned that Professor Lise Meitner had done a series of experiments on the properties of gamma-rays. I doubted then that the so-called gamma radiation could have so high energy as to reveal so strong penetrating capability. Bothe used the counter in his experiment. I thought that if we used the cloud chamber as the detector, the nature of the radiation might be clarified more easily. I proposed my idea to Madam Meitner, but it was a pity that we did not put the idea into practice. After the discovery of neutron, she once said to me, regretfully: "We were out of luck." The story passed long ago, but I learned a lesson since then. That is, you should insist on your good idea, and then realize it with your full strength.

Five years ago, right before the HI-13 tandem accelerator of the China Institute of Atomic Energy put into operation, we convened the Beijing Intenational Symposium on Physics at Tandem. I'am satisfactory with that, since the operation of the tandem, it has provided good beams for the nuclear experiments, and based on these experiments done at the tandem, our relevant theoretical work has also advanced forward. The progress we have made on fast neutron physics will be presented at this symposium. we would like to hear comments and suggestions.

Although neutron physics has gone through for almost sixty years, it is still developing. On the one hand, new requirements from nuclear power development and nuclear technique development, continue to challenge neutron physics, particularly nuclear data measurements. On the other hand, our understanding of neutron and its interaction with materials is far from completion. The invited talks and presentations in this symposium will, I believe, greatly deepen our insight into this subject, and will help us to exchange the latest developments on fast neutron physics.

The advancement of science is not the enterprise of just one country or one nation, but it is the endeavour of the whole human being. Chinese nuclear physics community will do our best to promote the study of nuclear physics, including the topic of this symposium,fast neutron physics. Of course,we need to strengthen the scientific exchange and collaboration between China and other countries. I believe it will be benefitial to the physicists in this field. I hope this symposium will give impetus to this effort.

Again, welcome and best wishes for a successful symposium. For our guests from abroad, I hope you all enjoy your stay in Beijing and in China.

CONTENTS

5. Activation Cross Sections

6. Nuclear Reactions

7. Intermediate Energy and Others

SCATTERING AND ENERGY SPECTRUM

MEASUREMENT OF DOUBLE DIFFERENTIAL NEUTRON EMISSION CROSS SECTIONS AT 14 MeV

Akito TAKAHASHI
Department of Nuclear Engineering, Osaka University
Yamadaoka 2-1, Suita, Osaka, Japan

ABSTRACT

Using an 8 m long TOF spectrometer with 2 ns pulsed D-T neutron source, double differential neutron emission cross sections at 14 MeV have been measured for 30 nuclides (Li-6, Li-7, Be, B-10, B-11, C, O, F, Mg, Al, Si, Ca, Ti, V, Cr, Mn, Fe, Co, Ni, Cu, Zr, Nb, Mo, Sn, Sb, Ta, W, Pb, Bi, U-238). Data are taken for 16 LAB angles (15 to 160 deg), in 0.5-15 MeV range of secondary neutrons and with 0.2 MeV energy resolution.

Angle-differential cross sections (ADX) are obtained for resolved peaks of DDX data; elastic and discrete inelastic scatterings. Angle-integrated neutron emission spectra in CMS (EDX) are also deduced for 30 nuclides.

Systematic comparison study has been made for DDX, ADX and EDX data with evaluated nuclear data (JENDL-3 and ENDF/B-VI). There exist discrepancies between experiments and evaluations for some nuclides. To meet better agreements, additional calculational analyses are made using available nuclear theory codes of direct, precompound and compound processes.

1. Introduction

Double differential particle emission cross sections induced by fast neutrons are required for nuclear energy applications. Especially, double differential neutron emissiomn cross sections (DDX in this paper) at incident neutron energy of 14 MeV are of primary interest in fusion reactor analyses; e. g., tritium breeding in blankets, nuclear heating, radiation shielding and material damage. DDX data are also useful to study nuclear reaction mechanisms by fast neutrons which enrgies exceed nucleon binding energies of target nuclides; neutron-induced reactions with 14 MeV incident energy show the competing process between the direct (inelastic scattering and multibody breakup), the precompound and the compound processes, which are reflected in energy spectral shapes of DDX data. DDX data have been therefore utilized to the assessment of nuclear model codes for evaluation works of JENDL-3[1], ENDF/B-VI[2], EFF1[3], FENDL[4], etc..

DDX data at 14 MeV have been laboriously measured at Dresden[5], Sendai[6], Osaka[7] and so on in these ten years for most nuclides and elements which are or will be used in nuclear energy applications. Experiments are now extending to the "missing range" of 7-13 MeV and higher enrgy range than 20 MeV.

This paper reviews DDX measurements[8-16] at OKTAVIAN; experimental method, DDX data, comparisons with evaluated data and some additional theoretical calculations. We point out that a) multi-body direct breakup processes play dominant roles for light elements (Li, Be, C) and b) combinational analyses of direct, precompound plus compound theories are quite sucessful to reproduce DDX data for medium to heavier elements.

2. Experimental Method

A time-of-flight neutron spectrometer using a pulsed D-T neutron source is used. The OKTAVIAN accelerator[17] provides 1.5-2.0 ns pulsed deuteron beam with about 15 micro-amp. time-averaged current for 1 MHz repetition frequency, to generate 14 MeV neutrons with an air-cooled TiT target. An 8.2 m long TOF facility is installed in the 85 degree line against the deuteron beam line as shown in Fig.1. This TOF facility is fixed and not rotated as usual goniometer systems, since we employed the method of sample rotation around the tritium target as shown with the inserted figure (upper left) of Fig.1, utilizing that source neutron energy is constant (14.1 MeV) in the plane virtical to the deuteron beam line. Actually, the TOF facility is installed at 85 degree line due to the restriction of building. A cylindrical sample with 2-3.5 cm in diameter and 5-7 cm in length is set on the top of a thin paper pipe and the sample position is moved along the arc, to change scattering angle. Neutron collimation and shielding around the main neutron detector (10 inch diam. 10 cm thick NE213) was finally set with the material configuration shown in Fig.1, after many trial BG runs changing the configuration.

Blockdiagram of the measuring electronics is shown in Fig.2. Circuits in usual TOF technique are employed. To minimize background signals, the n-gamma pulse shape discrimination is essential. Two parallel n-gamma separation systems with low and high gain (gain ratio = 1 : 10) delay line amplifiers (DLA) can cover a neutron energy range of 0.5 to 15 MeV. Calibration of the detector efficiency is done by two methods; 1) measuring H(n,n) peak areas changing scattering angles with a 1.5 cm diam. 5 cm long polyethylene sample (6 to 14 MeV) and 2) Cf-252 neutron source (0.5 to 8 MeV). Calibration of absolute DDX values is also done by referring standard H(n,n) differential cross sections.

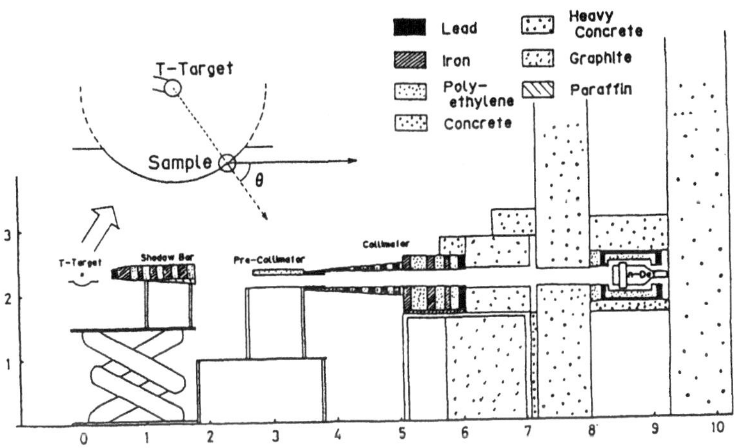

Fig. 1 Schematic view of neutron TOF facility (scale in meter)

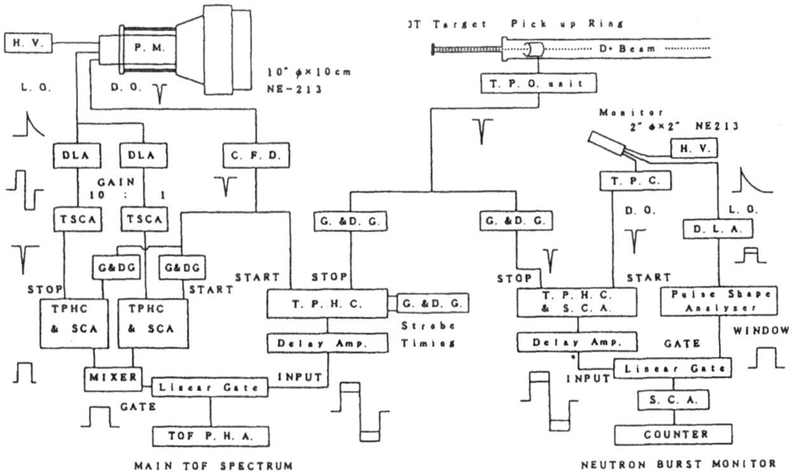

Fig.2 Block-diagram of TOF analysis system with n-gamma discrimination and source neutron yield monitor

Corrections for multiple-scattering in sample, lower energy contamination of source neutron spectrum and source yield variation around the tritium target are done all together using the MUSCC3 code[18]. To deduce EDX data, measured DDX data are once converted to the DDX data in CMS and integrated over scattering angle. To obtain SDX data, resolved peaks in DDX spectra are searched and values of peak-areas are obtained.

Other details of the experimental method is described in Ref.8.

3. Results and Discussions

Obtained DDX, EDX and SDX data are so numerous that only examples are shown in the following. More detailed information is given in Refs. 8 through 19.

3.1 Light Elements (Li-6, Li-7, Be, B-10, B-11, C, O, F)

Lithium: EDX data for Li-6 are shown in Figs. 4 and 5 in comparison with JENDL-3 and ENDF/B-VI. Lithium isotpes are key nuclides for tritium breeding. Evaluated data in JENDL-3 and ENDF/B-VI reproduce fairly well the measured DDX data for both of Li-6 and Li-7. Pseudo-level presentations for continuous neutron emission, which would be the products of direct multibody breakup reactions, appear to be successful. However, there remain still considerable disagreements (underestimations by both of JENDL-3 and ENDF/B-VI for Li-6, as shown in Figs.4 & 5) in the 0-5 MeV region. We need further improvements if precisions are required in fusion neutronics calculations.

Be-9: DDX data of $^{9}Be(n,2n)$ reaction are of primary importance for neutron multiplication in fusion blankets. 3-body direct breakup to n_1 + ^{5}He + ^{4}He and 4-body breakup to n_1 + n_2 + $2^{4}He$ play dominant role, as shown in Fig.6 where measured DDX data are compared with the Monte Carlo calculation with multibody break-ups. As is the case of $^{7}Li(n,nt)$ reaction, the alpha-clustering at the final state interaction will be characteristic in this reaction[20]. The evaluated data in JENDL-3 reproduce well the measued DDX data, while the ENDF/B-VI data show disagreement in low energy region.

EDX data are shown for B-10, B-11, C, O and F in comparison with evaluations in Fig.7 through Fig.11. For B-10 and B-11, disagreements with JENDL-3 exist in the energy region lower than 5 MeV. Agreement is excellent for C. Disagreement is large for F, due to the neglection of direct discrete inelastic scatterings. More detailed theoretical analyses are needed for these elements except carbon.

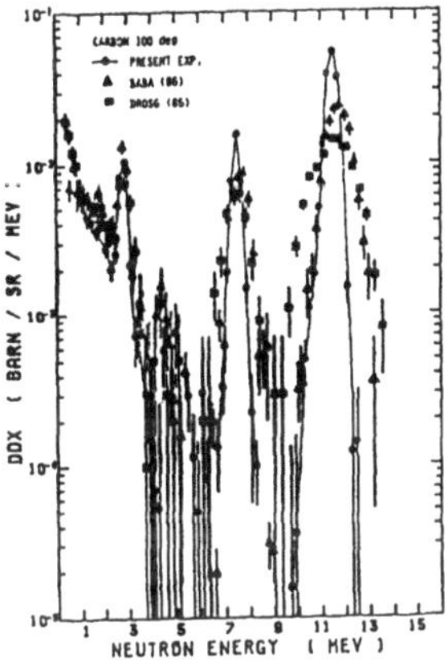

Fig.3 DDX data for C, showing energy resolutions

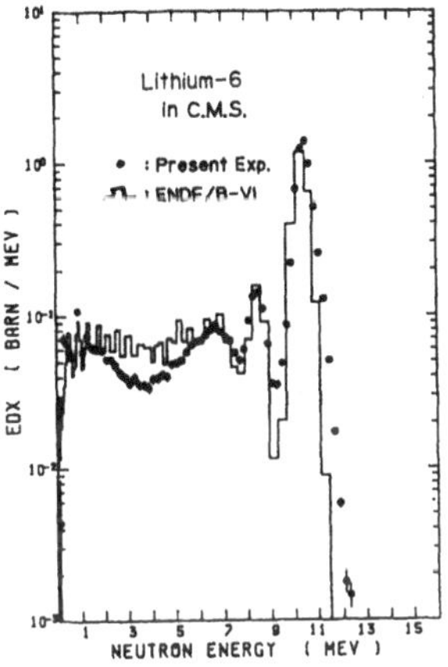

Fig.5 EDX for Li-6, compared with ENDF/B-VI

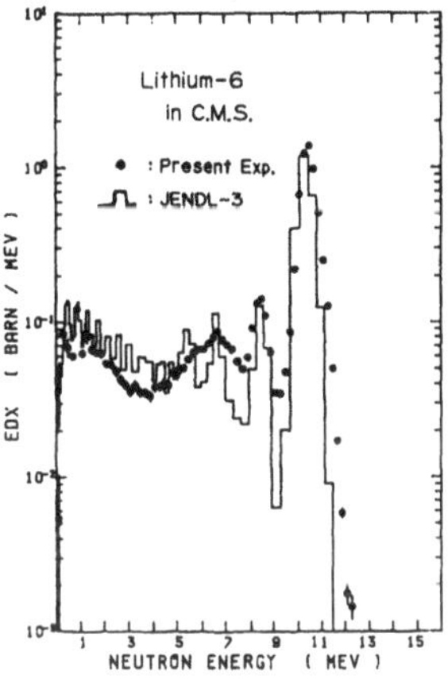

Fig.4 EDX for Li-6, compared with JENDL-3

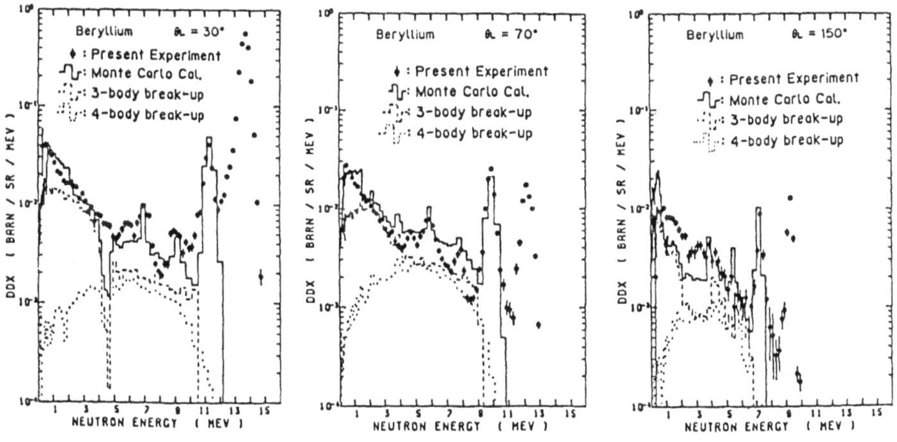

(a) (b) (c)

Fig.6 DDX data for Be, compared with Monte Carlo calculation

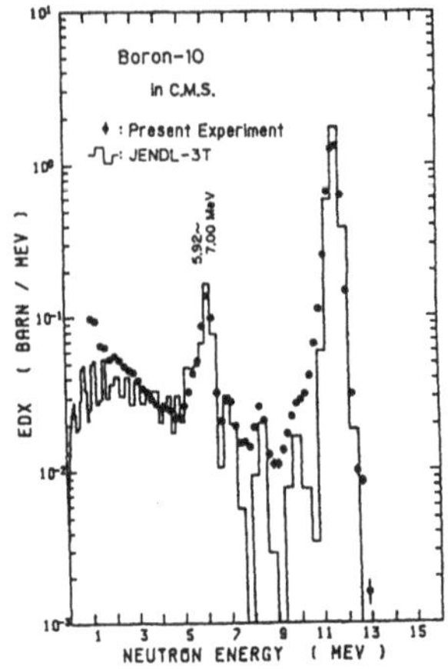

Fig.7 EDX for B-10, compared with JENDL-3

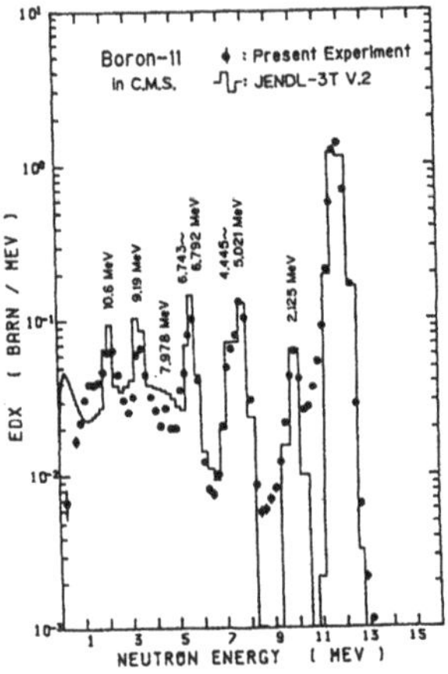

Fig.8 EDX for B-11, compared with JENDL-3

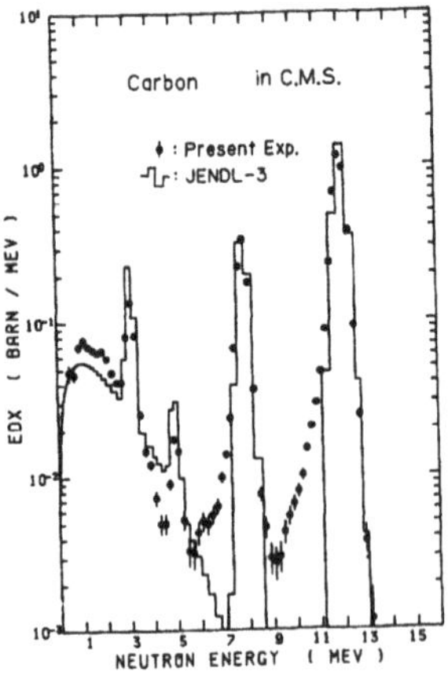

Fig.9 EDX for C-12, compared with JENDL-3

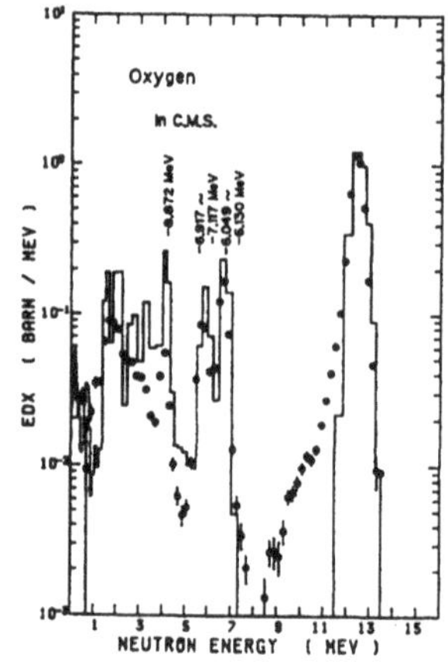

Fig.10 EDX for O-16, compared with JENDL-3

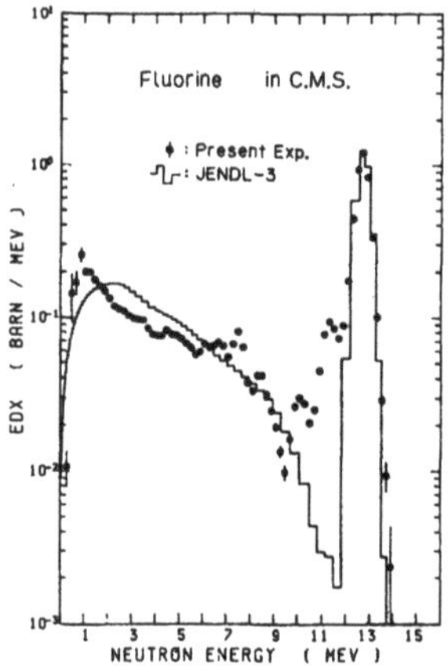

Fig.11 EDX for F-19, compared with JENDL-3

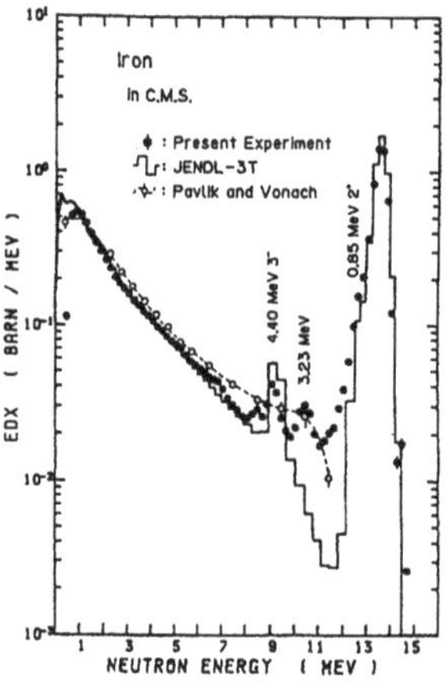

Fig.12 EDX for Fe, compared with JENDL-3

3.2 Medium Elements (Mg to Sb)

Medium weight metal elements are important as constituents of reactor structural materials, semiconductors, superconducting materials and first walls. Generally, good evaluations in JENDL-3 and ENDF/B-VI have been made based on the combined theoretical models of direct (DWBA), preequyilibrium (exciton model) and equlibrium (Hauser-Feshbach model) processes. Examples of the JENDL-3 evaluations are shown in Figs.12 and 13.

A sytematic analysis for these elements has been made in the present work using the EGNASH code[21] which treats the direct, preequilibrium and equilibrium processes. We did not calculate elastic scatterings (rightest peaks in measured DDX spectra). We treated discrete inelastic scatterings as many as possible, for example; 14 levels from 0.744 MeV to 1.9687 MeV for Nb, adjusting partial cross sections in DWUCKY calculations. We considered all isotopes for natural elements. We adjusted level density parameters (a-parameters) and Kalbach constants to fit measured EDX spectra in continuous emission regions. Some examples are shown in Fig.14 through 17, for V, Mn, Co and Cu. We obtained the result that measured EDX spectra were very well reproduced with the combinations of three components; direct inelastic scatterings, preequilibrium neutron emission and equilibrium neutron emission. Used level density parameters are shown in Fig.18 as a function of mass number; there is a trend that a-parameters increase in higher mass nuclides.

Angular dependences of DDX data can not well be reproduced by the EGNASH calculations, which give isotropic neutron emissions for the continuous components. Using the Kalbach-Mann[22] and the Kumabe[23] systematics, we can almost completely reproduce measured DDX data as shown in examples of Figs.18 and 19.

3.3 Heavier Elements (Ta to Bi)

We are trying EGNASH calculations for these heavier nuclides, but final results are not yet obtained. An interim result for Ta is shown in Fig.20 which does not include DWBA calculations because there are too many excited levels to be considered. It seems that the calculation for continuous neutron emission agrees very well with the measured EDX spectra though we have to employ high a-parameters compared with Fig.18; situations for W, Pb and Bi are the same. For other example, EDX data for Bi are shown in Fig.21, compared with ENDL86 which gives apparently harder spectrum in the energy range lower than 6 MeV.
Further analyses are needed for heavier elements. DDX data for U-238 are preliminary since we used an oxide sample.

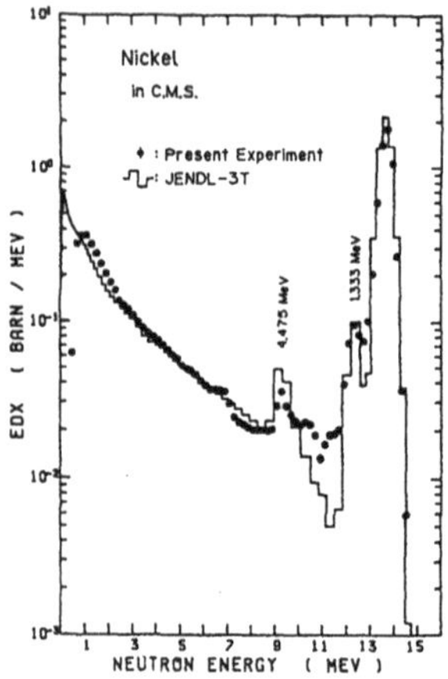

Fig.13 EDX for Ni, compared with JENDL-3

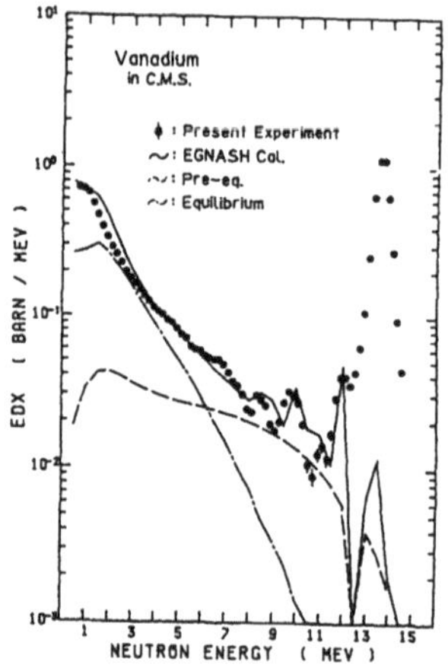

Fig.14 EDX for V, compared with calculation

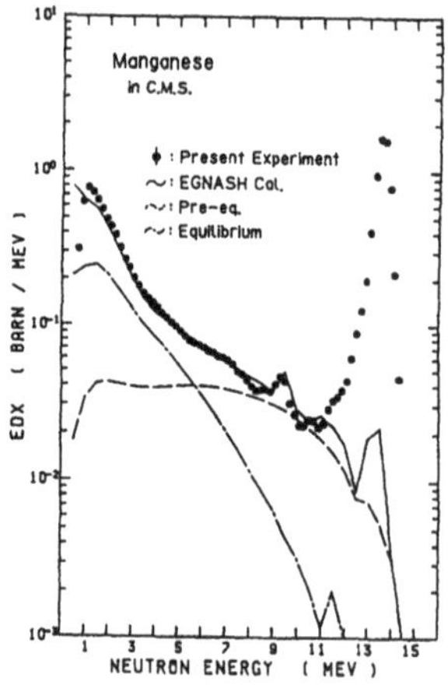

Fig.15 EDX for Mn, compared with calculation

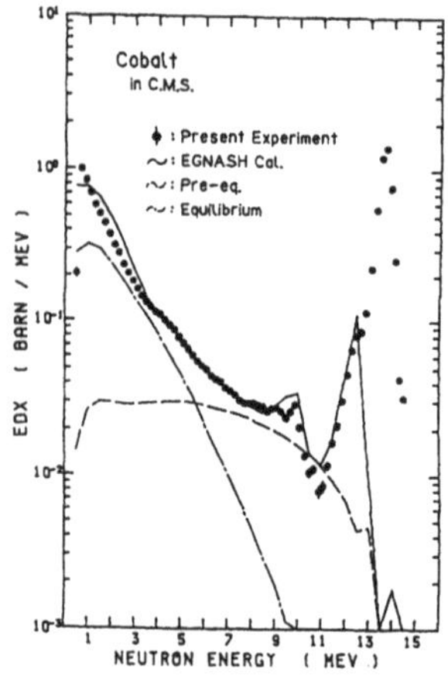

Fig.16 EDX for Co, compared with calculation

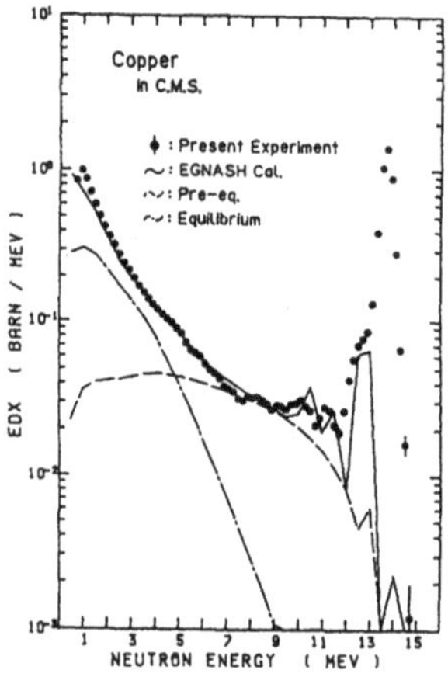

Fig.17 EDX for Cu, compared with calculation

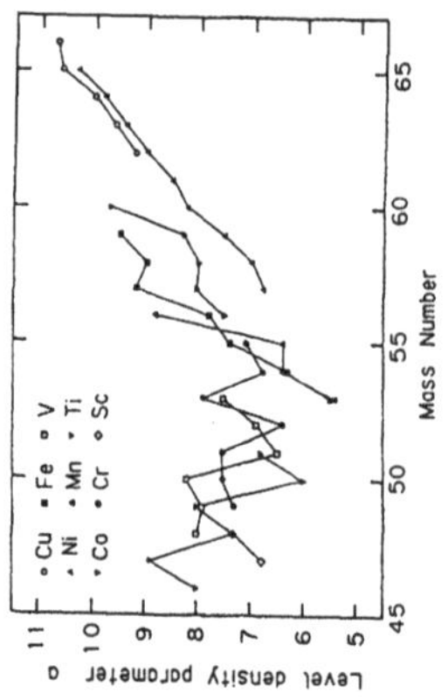

Fig.18 Used level density parameters

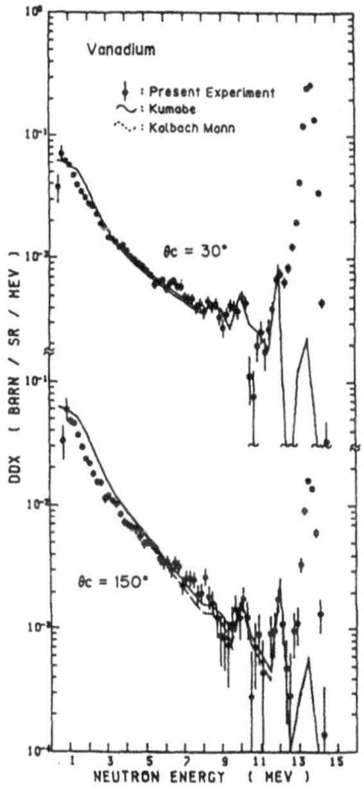

Fig.19 DDX data for V, compared with calcul.

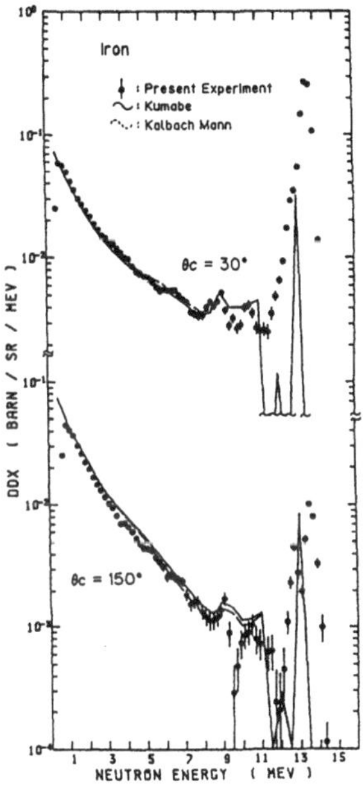

Fig.20 DDX data for Fe, compared with calcul.

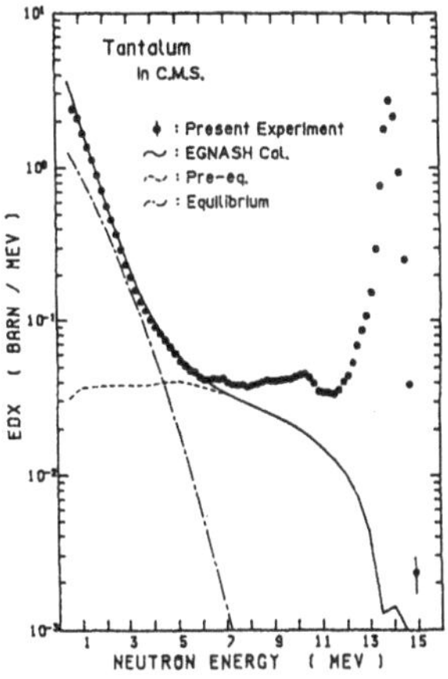

Fig.21 EDX data for Ta, compared with EGNASH calcul.

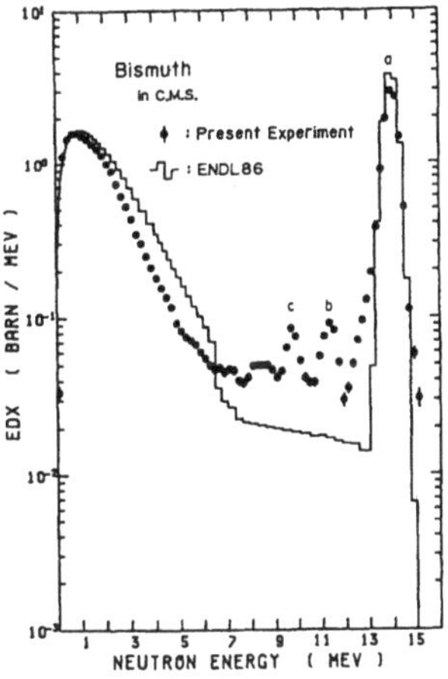

Fig.22 EDX for Bi, compared with ENDL86

References

1) Shibata, K., et al.: "JENDL-3 General Purpose File", JAERI 1319, (1990)
2) ENDF/B-VI, American Nuclear Data Library, BNL (1990)
3) EFF1, Europian Fusion File, Petten, (1990)
4) Kopecky, J.: INDC(NDS)-223/GF, 62, (1989)
5) for example, Obiesi, H.A., et al.: "Double-differential neutron emission cross sections of V, Ta, W and U...", Proc. Conf. Nucl. Data, May 1991, Juelich
6) for example, Baba, M., et al.: Proc. Conf. Nucl. Data, 1988, Mito, p.209
7) Takahashi, A.: "Measurements of Double Differential Cross Sections at OKTAVIAN" (review), J. Nucl. Sci. Technol., 26, 15-20, (1989)
8) Takahashi, A., et al.: J. Nucl. Sci. Technol., 25, 215-232, (1988)
9) Takahashi, A., et al.: Proc. Conf. Nucl. Data, Antwerp 1982, pp.360-367, (1983)
10) Takahashi, A., et al.: JAERI-M 88-065, pp.279-318, (1988)
11) Takahashi, A., et al.: JAERI-M 88-102, (1988)
12) Takahashi, A., et al.: Proc. Conf. Nucl. Data, Mito 1988, pp.205-208, (1988)
13) Takahashi, A., et al/: JAERI-M 89-214, (1989)
14) Takahashi, A., et al.: JAERI-M 90-220, (1990)
15) Gotoh, M. and Takahashi, A.: JAERI-M 91-032, pp.227-236, (1991)
16) Takahashi, A., et al.: INDC(JPN)-118/L, (1989)
17) Takahashi, A., et al.: J. Nucl. Sci. Technol., 21, 577-598, (1984)
18) Ichimura, E., Takahashi, A.: Oktavian Rep. A-87-02, Osaka Univ., (1987)
19) Takahashi, A., et al.: ibid., A-87-03, (1987)
20) Chiba, S., et al.: "Neutron scattering: technological achievements and illustrative results", Proc. Conf. Nucl. Data, Juelich 1991
21) Yamamuro, N.: Proc. Conf. Nucl. Data, Mito 1988, pp.489, (1988)
22) Kalbach, C., Mann, F.: Phys. Rev., C23, 112, (1981)
23) Kumabe, I.: private communication

FAST NEUTRON SCATTERING CROSS SECTIONS AT SMALL ANGLES AND OPTICAL MODEL PARAMETERS*

QI HUIQUAN, CHEN QIANKUN, CHEN YINGTANG,
CHEN ZHENPENG and CHEN ZEMIN
Department of Physics, Tsinghua University
Beijing, P. R. China

ABSTRACT

Differential cross sections of 14.8 MeV (or 14.7 MeV) neutrons scattered from ^{238}U, Pb, Ni, Fe, Cr, Al and O, and of 12 MeV neutrons from ^{238}U and Fe between 2.5° and 15° were measured. The angular distributions are analyzed using the spherical optical model or coupled channel model (code ECIS79).

1. Introduction

There is a drastic increase at forward angles in the angular distribution of elastic scattering of 14 MeV neutrons on medium and heavy nuclei. The precise knowledge of the differential cross sections at small angles is important. However, the data for the angles smaller than 15° are very scarce because of the difficulties in geometrical arrangement in the measurement. By using the optical model, the differential cross sections at small angles can be calculated. But most of the optical model parameters published in the past were obtained without experimental data in the small angular region. It is not assured that they can fit the small angle scattering data very well. Hence it is necessary to measure elastic differential cross sections at small angles. ^{238}U, Pb, Ni, Fe, Cr and Al are important materials for nuclear fusion and fission devices. But no data of small angle scattering have been published for Cr and O, and only a few experimental data were measured for other nuclei. In view of the importance in the application and to obtain the optimum optical model parameters, the measurements of the differential cross sections of neutron scattering at small angles have been performed for these nuclei.

The measurements were carried out at Cockcroft-Walton accelerator, where ^{238}U, Pb, Ni, Fe, Cr, Al and O were measured at E_n = 14.8 MeV or 14.7 MeV using the T(d, n)^{4}He reaction, and HI-13 tandem Van de Graaff where experiment

*Supported in part by the International Atomic Energy Agency, National Natural Science Foundation of China and China National Nuclear Corporation.

on ^{238}U and Fe was carried out at E_n = 14 and 12 MeV using the D(d, n)^{3}He reaction.

2. Experimental Arrangement

The measurements of differential scattering cross sections at small angles in the past several years at Tsinghua University have been reported in Refs. [1-3]. Only a brief description about the experimental arrangement is given here, which is shown schematically in Fig. 1.

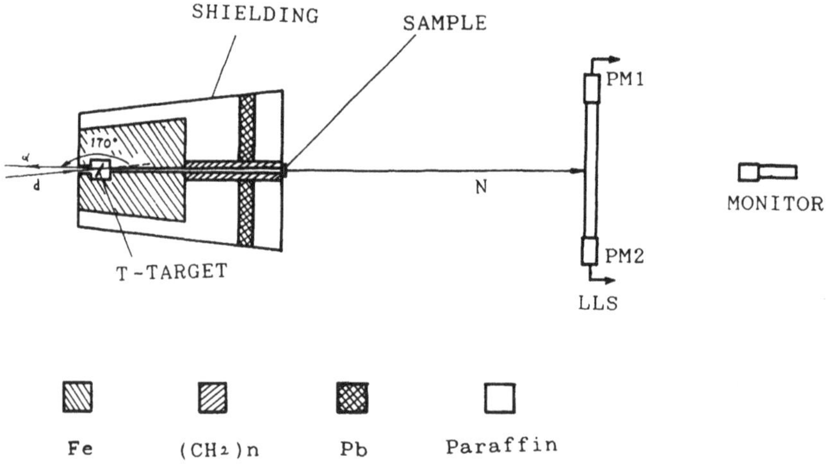

Fig. 1. The Experimental Arrangement

TABLE 1. The parameters of the samples

sample	thickness (cm)	diameter (cm)	number of nuclei per cm^3	trans. %	purity %
U	0.8	6.0	3.67×10^{22}	80.0	
Pb	2.0	2.8	9.11×10^{22}	62.3	99.9
Ni	1.6	4.0	1.45×0^{23}	68.2	00.5
Fe	2.0	4.0	1.69×10^{23}	64.5	99.99
Cr	2.1	4.0	1.14×10^{23}	70.0	99.5
Al	3.0	4.0	1.81×10^{23}	73.0	99.5

The neutron beam is collimated in 0.5° by a double-truncated conical collimator located in the centre of the shield consisting of 70 cm iron, 60 cm paraffin and 10 cm lead in thickness. The T-Ti target is in the shield. The main detector is a position sensitive neutron detector which consists of a long cylindric liquid scintillator tube (LLS) of 100 cm in length and 5.3 cm in inner diameter with two photomultipliers at both ends of the tube. The position information of the incident neutrons is extracted from the time difference between the output signals of the two photomultipliers.

The d-D neutron source is produced by bombarding a deuterium gas target with a deuteron beam. The gas target is 2 cm long and 1 cm in diameter with a gas pressure of 6 atm. The repetition rate of the pulsed deuteron beam is 4 MHz and the full width at half maximum for primary neutrons is 2 ns. The time of flight spectrum of 12 MeV neutrons by the D(d, n) reaction is shown in Fig. 2. As shown in the figure, the break up neutrons are separated from monoenergetic neutrons clearly.

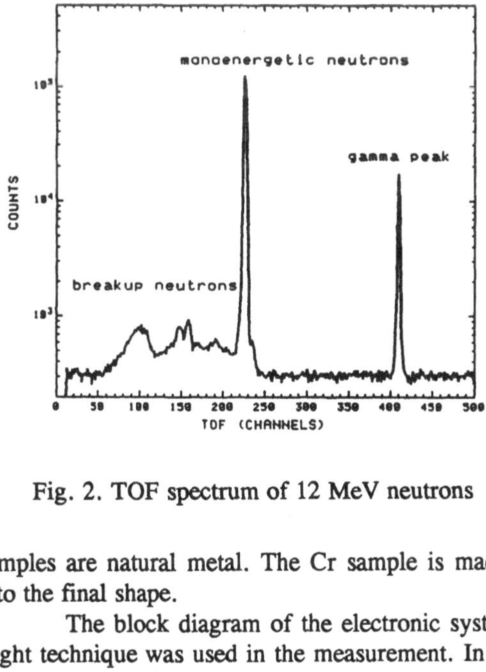

Fig. 2. TOF spectrum of 12 MeV neutrons

The samples are in disk shape, their parameters are listed in Table 1. All samples are natural metal. The Cr sample is made by pressing metal Cr powder into the final shape.

The block diagram of the electronic system is shown in Fig. 3. Time of flight technique was used in the measurement. In order to reduce the background, n-gamma discrimination was used.

3. Data Analyses

The experimental results were corrected for neutron flux attenuation, multiple scattering and finite geometry of samples and detector by Monte-Carlo simulation. The Schwinger scattering cross sections have been corrected for ^{238}U and Pb and can be neglected for other nuclei. After these corrections, the differential elastic scattering cross sections including the data in the larger angular regions from other laboratories[4,5] were analyzed for each nucleus using code ECIS79. Based on the analyses of individual nucleus, the global optical model parameters were searched for. At first the average geometry parameters were kept fixed and the relations of the potentials to the mass number A and asymmetry terms (N-Z)/A were investigated. Through adjusting the potentials, a set of optimum potential parameters for Pb, Ni, Fe and Cr has been obtained. The values are as follows.

The geometrical parameters (in fm) are:

$r_o = 1.15$	$r_v = 1.19$	$r_s = 1.32$	$r_{so} = 1.20$
$a_o = 0.73$	$a_v = 0.60$	$a_s = 0.56$	$a_{so} = 0.74$

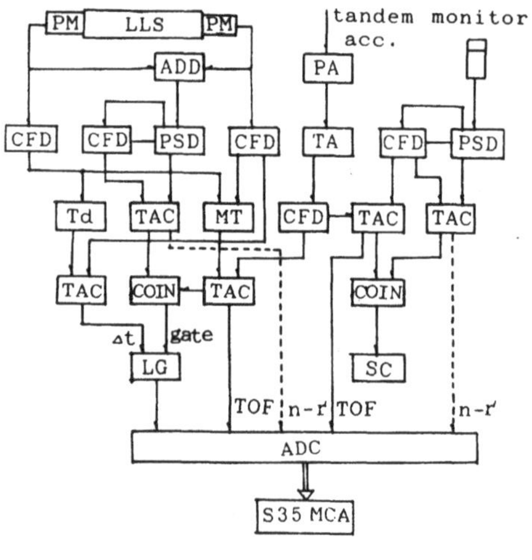

Fig. 3 Block diagram of electronic system

ADC:	Analog to Digital Converter	ADD:	Add Circuit
CFD:	Constant Fraction Discriminator	COIN:	Fast Coincidence
LLS:	Long Liquid Scintillator Detector	LG:	Linear Gate
MCA:	Multichannel Analyser	MT:	Mean Timer
PSD:	Pulse Shape Discriminator	PA:	Preamplifier
SC:	Scaler	TA:	Timing Amplifier
TAC:	Time to Amplitude Analyser with SCA	TD:	Time Delay

The subscripts o, v, s and so represent real, volume imaginary, surface imaginary and spin-orbit respectively. The second order deformation parameters.

β is equal to 0.19, 0.24, and 0.16 for Ni, Fe and Cr respectively.

The depth of the potentials (in MeV) is as follows.

$$V_o = 51.0 - 16(N - Z)/A, \ W_v = 1.41, \ W_s = 6.8 - 0.0093A, \text{ and } V_{so} = 7.0.$$

4. Results

The comparisons of the experiment and calculation of the angular distributions at small angles for Pb, Ni, Fe and Cr are shown in Fig.4 and those between 0° and 180° shown in Fig. 5. The curves represent calculated data using our optimum parameters. C. S. is the abbreviation of cross section. Arrows indicate the Wick limits. All of the differential scattering cross sections of Pb, Ni, Fe and Cr at zero degree are larger than their Wick limits respectively. As the figures show, the present data are in agreement with other measurements.

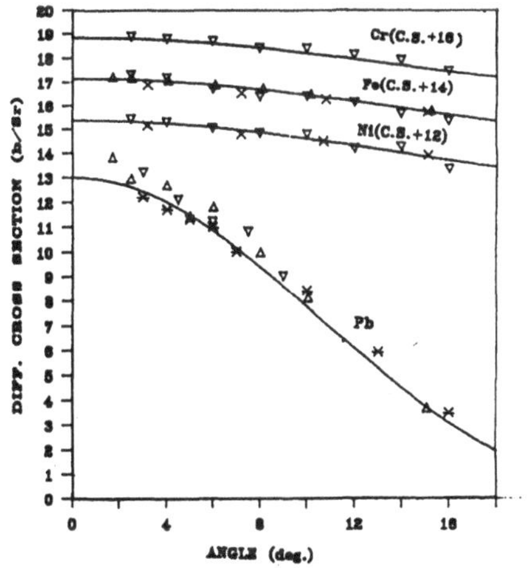

Fig. 4 Differential cross sections at small angles
∇ present work, — cal., Δ ref. 6, × ref. 7, ⋇ref. 8

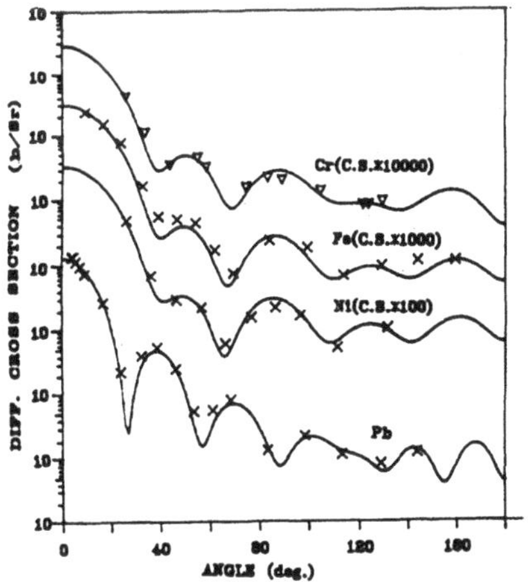

Fig. 5 Differential cross sections between 0° and 180°
× ref. 4, Δ ref. 5, — cal.

5. Summary

The differential scattering cross sections of 14.8 MeV or 14.7 MeV neutrons from ^{238}U, Pb, Ni, Fe, Cr, Al and O between 2.5° and 16° have been measured absolutely and those of 12 MeV neutrons from ^{238}U and Fe have also been measured. The data for Cr and O have not been seen in previous publications. The data for other nuclei are in agreement with those of Refs. [6-8] which were obtained by completely different experimental method. The optimum O.M. parameters were obtained which can fit the experimental data very well both in the small angular region and at large angles.

References

1. Qi Huiquan et al., *Proc. of the International Conf. on Nucl. Data for Basic and Applied Sci.* (Santa Fe, U.S.A.), **Vol. 2** (1985) 1355
2. Qi Huiquan et al., *INDC(CPR)-005/G,* (1985); *INDC(CPR)-006/L,* (1985)
3. Qi Huiquan et al., *Proc. of the International Conf. on Nuclear Data for Science and Technology,* (MITO, Japan), (1988) P. 799
4. L. F. Hansen et al., *Phys. Rev.* **C31** (1985) 111
5. P. H. Stelson et al., *Nucl. Phys.* **68** (1065) 97
6. Li Jingde et al., *Chinese J. of Nucl. Phys.* **11** (1989) 19
7. W. P. Bucher et al., *Phys. Lett.* **58B** (1975) 277
8. J. V. Dukarevich et al., *J. ZET,* **44** (1961) 1

Hadron Dependence of Collective Excitations and Neutron Scattering to Collective Levels

M. T. McEllistrem
Physics and Astronomy, University of Kentucky,
Lexington, Kentucky 40506-0055

October 29, 1991

Abstract

The large difference of the effective interaction for unlike nucleons from that for like nucleons in complex nuclei provides a powerful means of separating target proton from target neutron roles in collective excitations in nuclei when neutron scattering is combined with electromagnetic excitations or decay strengths. Results obtained for ^{48}Ca excitations are presented to illustrate the power of this separation method. Combining γ-ray detection with neutron detection in neutron inelastic scattering enhances the ability to separate levels and identify varieties of collective excitations; lifetime measurements made with DSAM techniques in neutron scattering are a powerful means of fixing the needed electromagnetic strengths. The power of γ-ray methods is illustrated by presenting multi-phonon multiplets for ^{96}Zr and for ^{144}Sm.

1 Introduction

This paper presents neutron scattering in its use as a probe of low-lying excited levels and their properties, particularly collective levels, rather than as a means of examining scattering mechanisms or neutron-induced reactions. There are many different goals associated with neutron-induced reaction studies, among them:

1) Mean field descriptions of scattering 2) Mechanisms of non-elastic reactions 3) Level-densities and systematics of levels 4) Activation reactions – analysis applications of neutron cross sections 5) Nuclear rates for astrophysical problems 6) Fundamental nucleon-nucleon interaction studies 7) Structure questions in complex nuclei.

The focus of this paper is only part of the last topic in this list. Collective excited levels in nuclei are studied at Kentucky using both neutron detection experiments and γ-ray detection in neutron inelastic scattering experiments.

Work of the decades of the 1970s and 80s, and to the present, yielded several developments which gave us a richly detailed and fairly confident understanding of neutron scattering, at least at incident neutron energies near and below 25 MeV. The most recent scattering field developments have been the dispersion corrections[1, 2,3] which have shown the connection between scattering fields at positive neutron energies and bound state potentials at negative energies. Knowing that a scattering potential can lead directly to a bound state potential, which correctly describes the distribution of bound single particle strengths, provides great confidence in the reality and accuracy of the scattering potentials. This is important for using scattering to study structure properties, since such studies depend on analyses of nonelastic reactions, which in turn depend on the accuracy of the fields.

Confidence in scattering fields and extracted structure properties is also greatly reinforced by two practices: 1) the measurement and description of many neutron scattering observables for each nucleus studied, and observables spanning a wide range of neutron energies,[4,5] and 2) consistently extracting structure properties from separate experiments conducted at several different neutron energies.[5,6]

Strong coupling models invoked to describe inelastic scattering excitations of levels of collective nuclei[7,4,6] show that collective levels are very strongly excited in neutron scattering, and that scattering can be used with these models to quantify properties proper to the target nuclei in a way that does not depend on specific structure models.[6]

1.1 Three Topics to be illustrated

This paper deals with three illustrations of scattering as a probe of nuclear structure. The first is of scattering from ^{48}Ca, to show that the separate roles of target neutron and proton excitations in low-lying levels can be confidently fixed. The second topic is the relation of hadron dependence in scattering to excited levels of collective heavy nuclei, and the qualitative relation of hadron scattering amplitudes to model descriptions of the bound levels themselves, particularly the effective charges of boson-based models. Both of these studies rely principally on neutron detection data from neutron time-of-flight (TOF) experiments.

The third topic of this work illustrates the use of γ-ray detection in scattering to do two things: 1) separate, with its high resolution capabilities, levels which cannot be separated in neutron detection, and 2) detect and identify double-phonon excitations in heavy nuclei. This third type of study enables the identification and study of the low spin parts of band structures which would be quite difficult to study with heavy-ion reactions.

2 Excitations in ^{48}Ca

Doorway or intermediate structure resonances were known for a long time to be directly identifiable in scattering in only a few favorable cases,[8] in which the compound nucleus was produced at relatively low excitation energy. This occurs for magic or semi-magic target nuclei, or in light nuclei, where nuclear level densities are particularly low.

A well known example, known for more than 25 years,[9] was that of neutron scattering from ^{48}Ca. It was clear that at least one doorway state had been identified near 0.96 MeV incident energy; but whether this was a result of a single particle-hole (p-h) configuration or the result of a core collective excitation was unresolved.[9] Rather recent high resolution measurements of total cross sections[10,11] showed that there were at least three visible doorways, and possibly four,[11] and further that all of them occur in $d_{5/2}$ scattering from the ground state at incident energies below 3.6 MeV. The apparent presence of three or four strong doorways in the same scattering channel led to the suggestion[11] that they were the result of coupling of $p_{3/2}$, $f_{5/2}$, and $p_{1/2}$ amplitudes of the inelastically scattered neutrons to the strong, collective 3_1^- core of ^{48}Ca.

Since the resonances are at energies below the excitation energy of the first excited level, the neutrons coupled to the core excited 3^- level must be at negative energies; that is, they are bound levels of the A = 49 system. This is illustrated in Fig. 1, where we provide the core excitations of ^{48}Ca. Only the coupling of these configurations to the $d_{5/2}$ ground state scattering gave them escape widths, producing intermediate-structure resonances as observed.[10,11]

It has been known for a long time that the 3_1^- level of ^{40}Ca is a very strong excitation; it was natural to assume that the same level of ^{48}Ca would dominate. Of course, this sort of description led to several subsidiary questions - why wouldn't there be resonances from the coupling to other core states, and why wouldn't there be resonances in channels other than the ground state $d_{5/2}$ channel? These questions were all successfully dealt with in the recent study of differential elastic and inelastic scattering cross sections.[12]

The new scattering study of differential elastic and inelastic scattering cross sections led, among other things, to the analyses of total cross sections shown in the two panels of Fig. 2.

The fine curves of that figure trace the experimentally measured total cross sections. The heavy curves in both panels are strong-coupling model calculations, and comparing them reveals something interesting about ^{48}Ca structure. The most prominent resonance, that which had been discovered about 25 years ago,[9] is the isolated resonance at 0.96 MeV. This feature is completely missing in the analysis shown in the lower panel. The only difference between the analyses of the two panels is that the second 3^- level of Fig. 1, at an excitation energy of 5.37 MeV, was left out of the strong coupling model space, following the suggestions of Ref. 13. The model of the upper panel includes *both* 3^- core excitations. The small shift of the calculated, large resonance from that measured is associated with a 4% shift in the strength of the coupling potential, coupling ground state scattering to excited level scattering. The shift of coupling strength was left unadjusted in the figure so that the calculated and measured strengths and widths could be directly seen and compared.

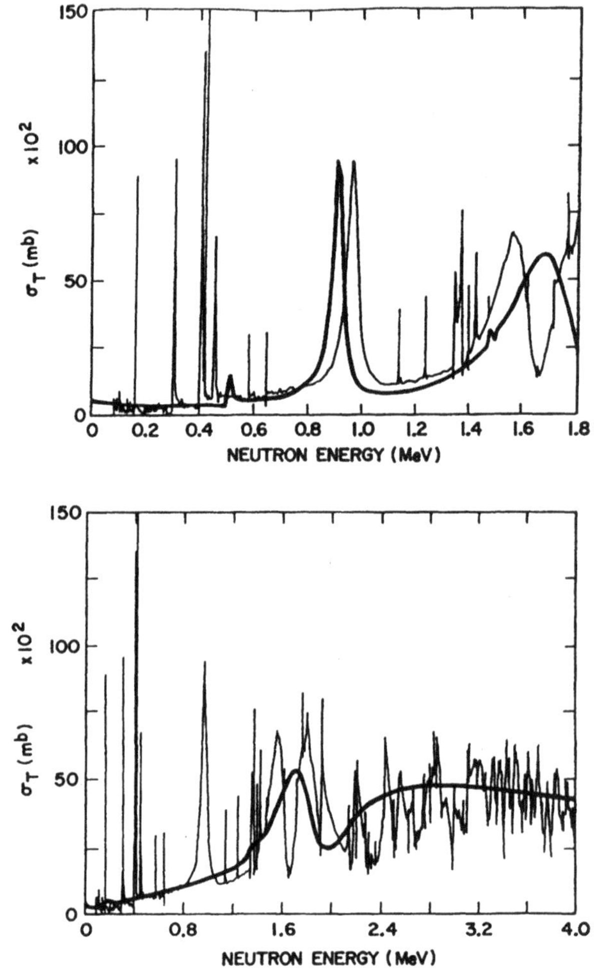

New results which resulted from the differential scattering cross section study[12] were then that the 0.96 MeV resonance resulted primarily from electric octupole (E3) coupling between the ground state and the second 3^- level, rather than the lowest 3^- as previously thought. Secondly, except for a a small, 4% shift in coupling strength, the electric quadrupole (E2) and E3 couplings which account for the decay of the core excited states in ^{48}Ca are similar to the couplings which account for the strengths and widths of the intermediate structure resonances. The collective couplings connect the bound excited core configurations to the ground state scattering channel. The large, wide calculated resonance near 1.65 MeV is actually split into two $d_{5/2}$ fragments by residual couplings not in our model.

It turns out that $f_{5/2}$ excited state neutron scattering does not contribute at all, as had been suggested;[11] penetrability effects damp it out completely. The important model components are $p_{3/2}$ and $p_{1/2}$ amplitudes coupled to *both* 3^- core excitations; this produces four doorway resonances.

Why had not the 3_2^- level been included in the first proposed model, or other collective excitation studies of ^{48}Ca ? The answer is tied to probe or hadron dependence of scattering. Core excitations had been most extensively examined previously in intermediate energy electron scattering.[13,14] Those studies, separated by 16 years, both found that the second 3^- level was very weakly excited, less than 1/6 the strength of that for the 3_1^- level. Hence its collectivity was not regarded as very important. However, our neutron scattering study[12] found it to be excited with 40% the strength of the lower energy 3^- level. In fact, if the 3^- excitation amplitudes of the two levels are added together, the sum comes very close[15] to the collective excitation amplitude of neutron scattering to the single 3^- level of ^{40}Ca.

This low energy resonance study illustrates several structure properties which could only be ascertained in neutron scattering. First, the resonances give us important insight into configurations of levels at about 5 MeV excitation energy in ^{49}Ca, about 1 to 3 MeV above the neutron separation energy. They show that core-excited levels actually dominate in certain energy regions. Further, the striking difference between excitations of 3^- levels of ^{48}Ca by electron scattering and neutron scattering gives us valued information about the p-h excitations of these two excited core levels. The information about p-h roles comes from examination of hadron dependence in excitation strength.

As noted in the study of Gamma-Soft heavy nuclei,[16] and following the work of V. R. Brown, V. A. Madsen, A. M. Bernstein and collaborators,[17] one can characterize excitations in terms of target neutron and proton structure amplitudes separately:

$$\frac{M_{nn'}}{M_{em}} = \left[\frac{(\chi_{np}M_p + \chi_{nn}M_n)/(\chi_{np}N_p + \chi_{nn}N_n)}{M_p/N_p}\right]$$

where χ_{np} is a weight factor for the effective neutron-proton interaction strength, χ_{nn} is a similar factor for the neutron-neutron interaction strength, and M_p and M_n are matrix elements for excitation of target protons and neutrons respectively. The other weight factors are N_p and N_n, the numbers of valence protons and valence neutrons respectively.

The $M_{n,n'}$ and M_{em} are respectively neutron scattering and electromagnetic excitation amplitudes for scattering to collective levels. Clearly similar expressions can be written involving proton and α-particle scattering, and various ratios of hadron scattering amplitudes constructed. The normalization of these ratios is such that if all *valence* particles participate equally in a particular excitation, then the ratio above will be 1, and the E2 matrix element per valence particle would be the same for neutrons as for protons. This is what is found always for well deformed rotors as target nuclei, for example.[4,7]

The very strong difference between electromagnetic (EM) and neutron excitation amplitudes for the 3_2^- level of ^{48}Ca, then, tells us that *neutron* particle-hole amplitudes dominate that particular excitation. The intermediate structure resonances, together with an elastic and inelastic neutron scattering experiment at 8 MeV incident,[12] have told us that core collective levels dominate near 6 – 9 MeV excitation energy in ^{49}Ca, and that the 3^- strength of ^{40}Ca is reproduced in ^{48}Ca, but split into two levels, the lower one having more strength in proton p-h excitations and the upper one quite dominated by valence neutron excitations.

Comparison of neutron and EM excitations of levels requires knowledge of the EM strengths as well as neutron scattering amplitudes; electron scattering and lifetime measurements provide EM strengths. The power of the lifetime measurements is illustrated in Fig. 3, drawn from the (n,n'γ) study[18] of ^{48}Ca of J. F. Vanhoy et al. This shows the 3832.2-keV γ ray de-exciting the first 2^+ level; the direct excitation component is shifted sharply in energy between 30° and 150°, while the portion coming from cascade contributions is unshifted. The actual lifetime measurements were made at a lower incident energy, where the cascade contribution was much weaker than shown in Fig. 3. Thus the EM strength of that 2^+ level was confirmed to be about 1.7 Weisskopf units (W.u.), whereas our neutron scattering strength was the equivalent of only about 0.7 W.u.

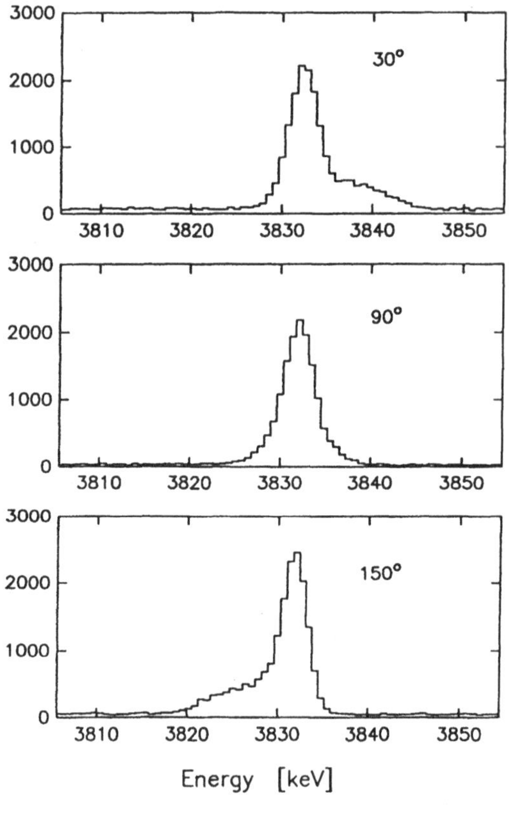

At the same time, α-particle scattering showed[18] an excitation strength of 4.9 W.u., quite a bit larger than either neutron scattering or EM strengths.

Any two of these strengths can be used to fix the relative roles of neutron and proton particle-hole excitations; we used the neutron excitation and EM strengths to determine that $M_n/M_p = 3.8$ for this 2^+ excitation; as shown in Table 1, this implies rather complete dominance by target neutron p-h excitations. This turned out to be in excellent agreement with that extraordinary ratio as determined ten years ago, without benefit of the neutron scattering results.[17]

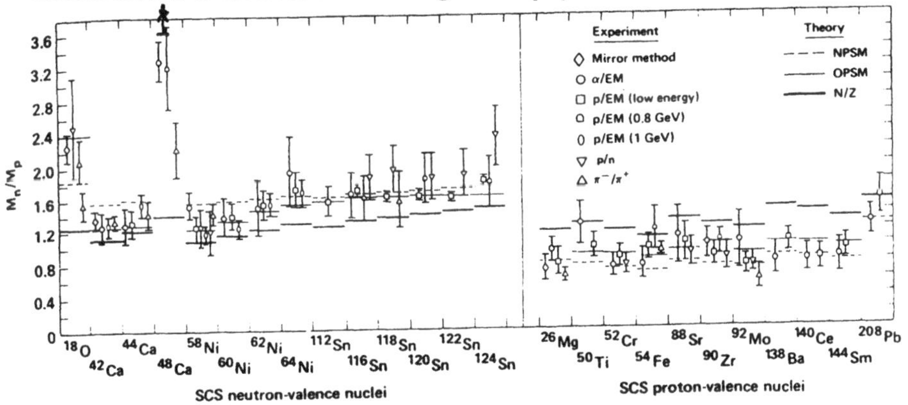

The ratios plotted in Fig. 4 show the compilation of Bernstein et al.[17] of these ratios for many nuclei, and show that only ^{48}Ca has such an anomalously large ratio of neutron to proton p-h excitations for first excited 2^+ levels. This remarkable ratio of neutron to proton excitations, unique in the periodic table, is thus well fixed by two independent comparisons, one ten years old, and this new one, developed from the neutron scattering and EM amplitudes.

Ratios of the target nucleus neutron and proton excitation matrix elements, M_n/M_p, for different levels of ^{48}Ca are determined from EM and neutron excitation amplitudes for several levels,[18] from combining electron scattering, lifetime measurements, and neutron excitation strengths.

Recognizing that the neutron to proton ratio, N/Z, is 1.4 for ^{48}Ca, one can convert the matrix element ratios into ratios of neutron matrix elements per neutron vs. proton matrix elements per proton. Thus we find the ratios normalized per nucleon shown in Table 1 for the lowest 2^+ and 3^- levels. The lowest 3^- level has a proton p-h component ~ 2.5 times the neutron component. Almost the reverse ratio applies to the second 3^- level; its neutron amplitude is about 2 times the proton component.

These ratios can be readily understood if one thinks of the nuclear shells involved. For the odd parity 3^- excitations the protons must excite from the closed $(1d, 2s)$ shell just to the next, $1f_{7/2}$ shell. The neutrons, on the other hand, must jump from the $1f_{7/2}$ to the $1g_{9/2}$ shell; the $(2p, 1f)$ shell must be skipped. Thus we would expect the lowest 3^- to be dominated by proton excitations and the second 3^- to be dominated by neutron excitations, so that the two levels would be orthogonal combinations of the two particle excitations. This is supported by the observation that if we sum our E3 excitation amplitudes for the two levels, we reach an amplitude very close to that determined[15] for the single 3^- excitation of ^{40}Ca. Neutron and EM amplitudes are shown[15] to be the same for the 3^- level of ^{40}Ca.

The even-parity excitations would be even more dominated by target neutron excitations, since for even parity excitations neutrons need only excite from the $1f_{7/2}$ to the $(1f, 2p)$ shell, whereas protons must excite from the $(1d, 2s)$ shell across *two* shells to the $1g_{9/2}$. The ratios of Table I should be quite useful to theorists interpreting the structure of ^{48}Ca.

3 Collective Excitations in Heavy Nuclei

The information which can be extracted from comparisons of neutron scattering matrix elements, $M_{n,n'}$, to EM amplitudes, M_{em} or M_p, for shape-transitional nuclei was illustrated with a recent comparison for neutron scattering[6] and Coulomb excitation results[19] for E2 excitations in the Pt and Os isotopes. We found the singular result[16] that the quasi-gamma bands in ^{190,192}Os showed strong dominance of target neutron excitations over proton excitations, while the ground bands seemed to show the usual result for deformed nuclei, that target neutrons and protons participate in the excitations equally. Thus direct excitation of collective levels in these nuclei reveals properties of nuclei very difficult to determine by other means.

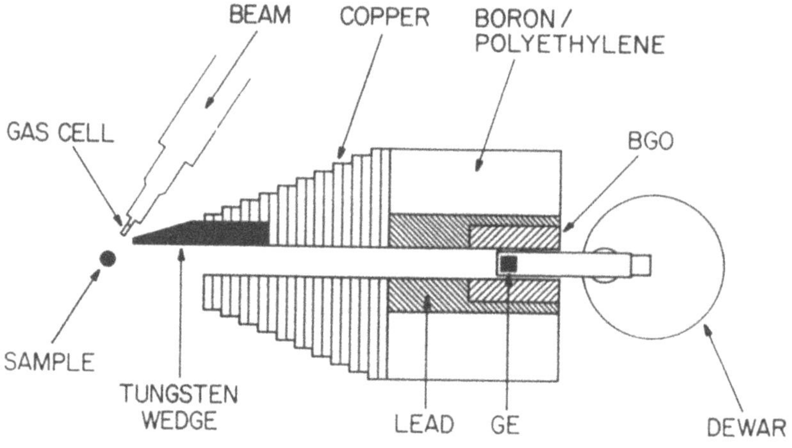

3.1 Multi-phonon E3 Excitations

Recently a great deal of attention has been turned to the topic of electric octupole, or E3, excitations. Many searches have been made for examples of multi-phonon level multiplets involving E3 as at least one of the phonons. These searches are best done with γ-ray detection, taking advantage of the high resolution and good efficiency available with modern HpGe detectors. The system shown in Fig. 5 shows a shielded Ge detector surrounded by an active, anticompton BGO shield, all enclosed in a large assembly designed to operate in a neutron environment, for neutrons up to 9 MeV. The present Kentucky system uses a 52% efficient Ge detector; the anticompton shield reduces the compton level by a factor between 4 for γ-ray energies above 3 MeV to between 6 and 8 for energies below 1 MeV. Thus the system has high sensitivity, and detects inelastic scattering for cross sections which range from 0.5 mb to 600 mb, or over a dynamic range in cross section larger than 1000.

3.1.1 Multi-phonon levels in ^{96}Zr

That large dynamic range is excellent for seeking the weakly excited multi-phonon levels, particularly those for which one phonon is an E3 excitation, as well as seeing the one phonon excitations themselves. A partial level scheme was presented[20] for ^{96}Zr, which shows a set of even-parity levels whose decays are dominated by E1 transitions to the one phonon 3^- level. These levels and their decays were ascertained in an (n,n'γ) experiment[21] carried out at Kentucky. An important part of the identification of E3 multiplets, like that of ^{96}Zr, is identifying the fast E1's of their decays. The speeds are obtained through lifetime measurements using DSAM techniques.[22] The usefulness of this lifetime technique is illustrated in Fig. 6, which shows the large shifts, almost 3 keV from the unshifted 90° line, found for forward and backward γ-ray detection. These shifts are for one of the E1 decays of ^{96}Zr.

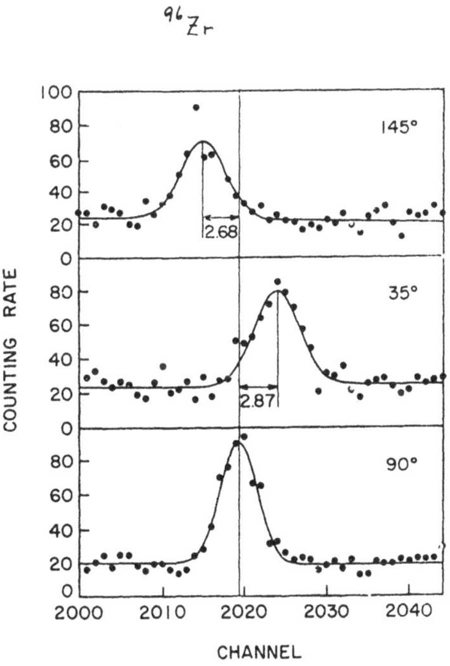

The E1 transition speeds are approximately an order of magnitude faster than those usually found[22] for decays not involving 3^- levels. An E1 decay from a collective 3^- level would not have to be abnormally fast. However, the phase coherences between p-h amplitudes which produce collective enhancements for E3 amplitudes would probably exist also for E1 amplitudes; it would be reasonable to expect them also to be enhanced. It is not easy, otherwise, to imagine a reason why an entire set of E1's from multi-phonon candidates should all be enhanced by a factor of ten.

3.1.2 Multi-phonon levels in ^{144}Sm

R. H. Spear had noted[23] that E3 excitations seem to be enhanced for nuclei not showing strong E2 enhancements. The nucleus ^{96}Zr had shown very strong E3 excitations, although far from being deformed. Another region where enhanced E3's might be found would be near[23] ^{146}Gd. Thus levels and transition rates were examined for ^{144}Sm. Here more complete multiplets were found[24] than in ^{96}Zr, as illustrated in Fig. 8. The double E3 multiplet, just the four levels expected, is shown on the left in the figure, and the mixed E2, E3 multiplet is shown on the right. Once again, the E1 transitions are more than an order of magnitude faster than normal[20] for transitions in this mass region. This is particularly evident for the $1^- \rightarrow 0^+$ decay of the 3235.3 keV level. This E1 transition has just the greatly enhanced speed expected of a mixed E2-E3 multiplet, interpreting the E1 as the result of E2 and E3 interference.

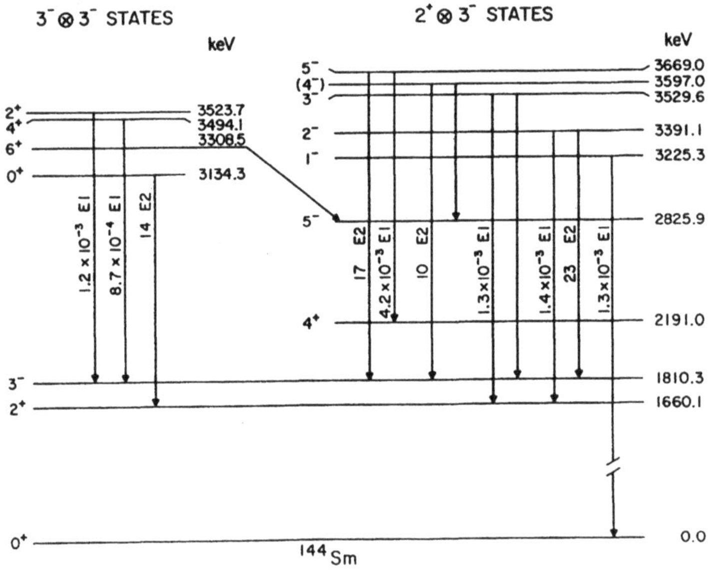

Table 1: Matrix Elements from Lifetimes and Hadron Excitation Cross Sections

J^π	β_{em}	$\beta_{p,p'}$	$\beta_{n,n'}$	β_α	M_n/M_p	M_nZ/M_pN
2_1^+		0.14	0.09	0.25	3.8	2.7
3_1^-	0.16 – 0.17	0.17	0.15		0.6	1/2.5
3_2^-	0.067		0.10		3.0	2.1

4 Summary

Neutron detection experiments in neutron scattering allow us to determine the separate roles played by target neutrons and target protons in collective excitations, when combined with either lifetime measurements, Coulomb excitation experiments, or electron scattering intensities. Neutron scattering is particularly sensitive to excitation differences, because the neutron-proton effective interaction in nuclei is 3 times as strong as the proton-proton or neutron-neutron interaction. Taking advantage of this fact, we find that the low-lying collective 3^- level of ^{40}Ca is split into two 3^- levels in ^{48}Ca, the lower one dominated by proton p-h excitations and the upper correspondingly dominated by neutron p-h excitations. The even parity levels, at least the first 2^+, are dominated by neutron p-h excitations.

Gamma-ray detection experiments are sensitive to a dynamic range of excitation strengths which range over more than three orders of magnitude in neutron scattering. This large range, together with the superb resolution and high efficiency available with modern Ge detectors makes it possible to find multi-phonon excitations through associating them with the one phonon levels to which they dominantly decay. Using these capacities, double E3 and E3 coupled to E2 multiplets have been clearly identified in ^{144}Sm, and probably also in ^{96}Zr. It appears further that in a nucleus with strong octupole correlations the E1 decays of many levels are strongly enhanced,[24] not just the phonon excitations. The E3 correlations affect the entire level scheme. The combination of both neutron and gamma-ray detection allows neutron inelastic scattering to be a powerful probe[18] of the properties of collective excited levels.

References

[1] C. Mahaux and N. Ngo, Nucl. Phys. A378, 205 (9182); *loc cit.*, Phys. Letters 126B, 1; C. Mahaux and R. Sartor, Nucl. Phys. A468, 193 (1987).

[2] C. H. Johnson, D. J. Horen, and C. Mahaux, Phys. Rev. C36, 2252 (1987); C. H. Johnson and C. Mahaux, Phys. Rev. C38, 2589 (1988).

[3] S. E. Hicks and M. T. McEllistrem, Phys. Rev. C37, 1787 (1988).

[4] J. P. Delaroche, G. Haouat, J. Lachkar, Y. Patin, J. Sigaud, and J. Chardine, Phys. Rev. C23, 136 (1982).

[5] S. E. Hicks, Z. Cao, M. C. Mirzaa, J. L. Weil, J. M. Hanly, J. Sa, and M. T. McEllistrem, Phys. Rev. C40, 509 (1989)

[6] T. B. Clegg, G. Haouat, J. P. Delaroche, Ch. Lagrange, J. Chardine, S. E. Hicks, G. R. Shen, and M. T. McEllistrem, Phys. Rev. C40, 2527 (1989).

[7] J. Lachkar, M. T. McEllistrem, G. Haouat, Y. Patin, J. Sigaud, and F. Cocu, Phys. Rev. C14, 933 (1976).

[8] *"Intermediate Structure in Nuclear Reactions"*, ed. by Hugh P. Kennedy and Rudolph Schrils (University of Kentucky Press, (1968)).

[9] William P. Beres and M. Divadeenam, Phys. Rev. Lett. 25, 596 (1987); M. Divadeenam, W. P. Beres, and H. W. Newson, Ann. of Phys. 69, 428 (1972).

[10] J. A. Harvey, C. H. Johnson, R. F. Carlton, and B. Castel, Phys. Rev. C32, 1114 (1985).

[11] R. F. Carlton, J. A. Harvey, R. L. Macklin, C. H. Johnson, and Boris Castel, Nucl. Phys. A465, 274 (1987).

[12] Sally F. Hicks, S. E. Hicks, G. R. Shen, and M. T. McEllistrem, Phys. Rev. C41, 2560 (1990).

[13] R. A. Eisenstein, D. W. Madsen, H. Theissen, L. S. Cardman, and C. K. Bockelman, Phys. Rev. 188, 1815 (1969).

[14] J. E. Wise, J. S. McCarthy, R. Altemus, B. E. Norum, R. R. Whitney, J. Heisenberg, J. Dawson, and O. Schwentker, Phys. Rev. C31, 1699 (1985).

[15] G. M. Honoré, W. Tornow, C. R. Howell, R. S. Pedroni, R. C. Byrd, R. L. Walter, and J. P. Delaroche, Phys. Rev. C33, 1129 (1986).

[16] M. T. McEllistrem, "Gamma-softness in shape-transitional nuclei and separation of neutron and proton dynamics", M. T. McEllistrem, in *Proc. International Conference on High Spin Physics and Gamma-Soft Nuclei*, ed. by R. A. Sorenson and J. X. Saladin, Pittsburgh/Carnegie Mellon University, Pittsburgh, PA., Sept. 17-21, 1990.

[17] V. A. Madsen, V. R. Brown, and J. D. Anderson, Phys Rev. C12, 1205 (1975); A. M. Bernstein, V. R. Brown, and V. A. Madsen, Phys. Letters 103B, 255 (1981).

[18] J. R. Vanhoy, Sally F. Hicks, M. T. McEllistrem, R. A. Gatenby, E. M. Baum, E. L. Johnson, G. Molnár, and S. W. Yates, to be published in The Physical Review C.

[19] Ching-Yen Wu, Ph. D. Dissertation, The University of Rochester (unpublished), 1983; D. Cline, Ann. Rev. Nucl. Part. Sci. 36, 683 (1986).

[20] T. Belgya, B. Fazekas, G. Molnár, A. Veres, R. A. Gatenby, J. R. Vanhoy, E. M. Baum, E. L. Johnson, and S. W. Yates, in *"Understanding the Variety of Nuclear Excitations"*, (ed. by Aldo Covello, World Scientific Press (1991)), p. 191.

[21] G. Molnár, T. Belgya, B. Fazekas, A. Veres, S. W. Yates, E. W. Kleppinger, R. A. Gatenby, R. Julin, J. Kumpulainen, A. Passoja and E. Verho, Nucl. Phys. A500, 43 (1989).

[22] T. Belgya, G. Molnár, B. Fazekas, A. Veres, R. A. Gatenby, and S. W. Yates, Nucl. Phys. A500, 77 (1989).

[23] R. H. Spear, Atomic Data and Nucl. Data Tables 42, 55 (1989).

[24] R. A. Gatenby, J. R. Vanhoy, E. M. Baum, E. L. Johnson, S. W. Yates, T. Belgya, B. Fazekas, A. Veres, and G. Molnár, Phys. Rev. C41, R414 (1990).

MEASUREMENT OF DOUBLE DIFFERENTIAL CROSS SECTIONS OF SECONDARY NEUTRONS IN THE INCIDENT ENERGY RANGE 9-13 MeV*

TANG HONGQING, QI BUJIA, ZHOU ZUYING, SA JUN, KE ZUNJIAN, SUI QINGCHANG, XIA HAIHONG AND SHEN GUANREN

China Institute of Atomic Energy

P. O. Box 275-46, Beijing 102413, P. R. China

ABSTRACT

The status and technique of double differential cross section measurement of secondary neutrons in the incident neutron energy range 9 to 13 MeV is reviewed with emphasis on the work done at CIAE.

There are scarce measurements of secondary neutron double differential cross sections in this energy region upto now. A main difficulty for this is lack of an applicable monoenergetic neutron source. When monoenergetic neutron energy reaches 8 MeV, the break-up neutrons from the d+D or p+T reaction starts to become significant. It is difficult to get a pure secondary neutron spectrum induced only by monoenergetic neutrons. To solve this problem an abnormal fast neutron TOF facility was designed and tested.

Double differential neutron emission cross sections of ^{238}U and ^{209}Bi at 10 MeV were obtained by combining the data measured by both normal and abnormal TOF spectrometers and a good agreement between measurement and calculation was achieved.

1. Introduction

Measurement of double differential cross sections of secondary neutrons have been intensively carried out around 14 MeV[1, 2, 3] and there are some measurements below 8 MeV. However, measurement of double differential cross sections in the incident neutron energy range 9 to 13 MeV is scarce. So far only 4 elements which are Li, Be, B and C have been measured at 10 MeV at LANL using an H(t, n) reaction neutron source[4, 5]. ^{7}Li was also measured in Geel laboratory in the incident neutron energy bins of 9.6-11.7 and 11.7-13.8MeV[6].

Why is the measurement of double differential cross sections in this energy region so few? The answer comes from monoenergetic neutron source. As we know, for the conventional neutron source, like d+D and p+T reactions, when the monoenergetic neutron energy reaches 8 MeV, the break-up neutrons will

*This work is partly supported by the National Natural Science Foundation of China and the International Atomic Energy Agency

occur. The ratio of the low energy neutrons to the monoenergetic ones increases with increasing of neutron energy. The break-up neutrons will interfere secondary neutron spectrum measurement. To solve this problem, a lot of efforts have been made in different laboratories. Besides the most promising neutron source i.e. accelerating triton and using the H(t, n) reaction, following methods have been explored:

(1). Production of a continuous spectrum neutrons which simulate the source break-up ones. In this case , two runs of measurements respectively with d+D (or p+T) neutron source and the simulation neutrons should be performed. The break-up neutron interference is eliminated by subtracting the spectrum induced by simulation neutrons from the one induced by d+D reaction neutrons.

(2). Searching for new monoenergetic neutron sources.
The following heavy ion reaction neutron sources have been studied. The H(^{7}Li, n) reaction was investigated at TUNL[7]. Our laboratory tested the ^{3}He(^{12}C, n) reaction[8]. The H(^{11}B, n) reaction was studied by scientists at JAERI[9]. Because of their low intensity in neutron flux and large energy spread, they may not be suitable for the secondary neutron spectrum measurement.

(3). Using pulsed white neutron source and unfolding pulse height distribution. This kind of technique was developed in Geel laboratory[6], The incident neutron energy was determined by TOF technique with a long flight path between neutron source and sample, and a negligible distance between sample and neutron detectors. The secondary neutron spectrum was obtained by unfolding pulse height spectrum of the neutron detector.

(4). CIAE Method[10]. The detail is described in the following section.

2. CIAE Method

Our method is realized in two steps.

(1). Double differential cross sections in higher energy part of the secondary neutron spectrum (En > 4MeV) was obtained by using the normal fast neutron TOF spectrometer.

(2). To eliminate source break-up neutron interference, double differential cross sections in lower energy part of the secondary neutron spectrum (En < 4MeV) was obtained using an unusual geometry (we call it abnormal TOF spectrometer). The double differential cross sections in the whole energy region was given by combination of these two datum sets.

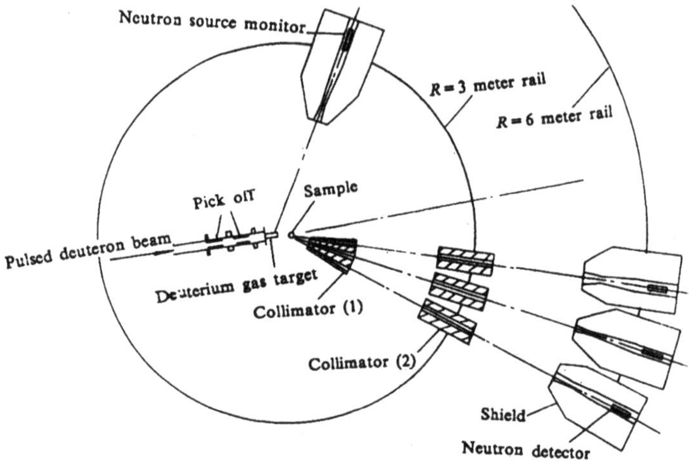

Fig. 1. Normal fast neutron TOF spectrometer

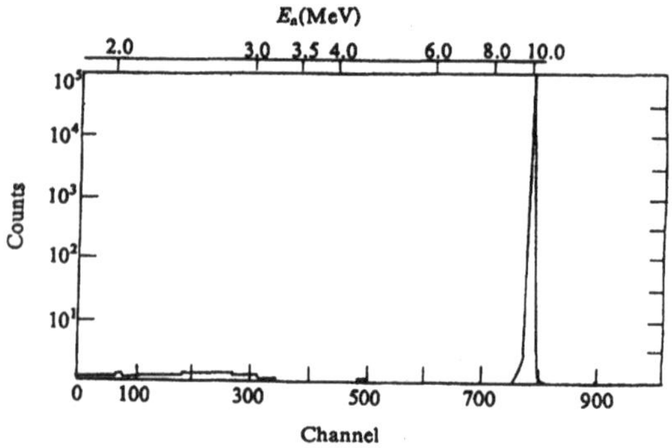

Fig. 2. TOF spectrum of the d+D reaction neutrons at 0° at bombarding energy E_d=7.0 MeV

The normal fast neutron TOF spectrometer is shown in Fig. 1. The detailed description can be found in ref. [11]. Fig. 2 shows the incident neutron TOF spectrum of d+D reaction at the bombarding energy E_d = 7.0 MeV (the monoenergetic neutron energy is about 10 MeV), measured by the normal TOF spectrometer. It can be seen from the figure that the break-up neutron with energy below 3.5 MeV is significant. It is the major trouble for secondary neutron spectrum measurement. Because of its interference, the lower energy part below 3.5 MeV of the spectrum is contaminated, as seen in Fig. 3. It is hard to get a pure secondary neutron spectrum for 10 MeV incident monoenergetic neutrons.

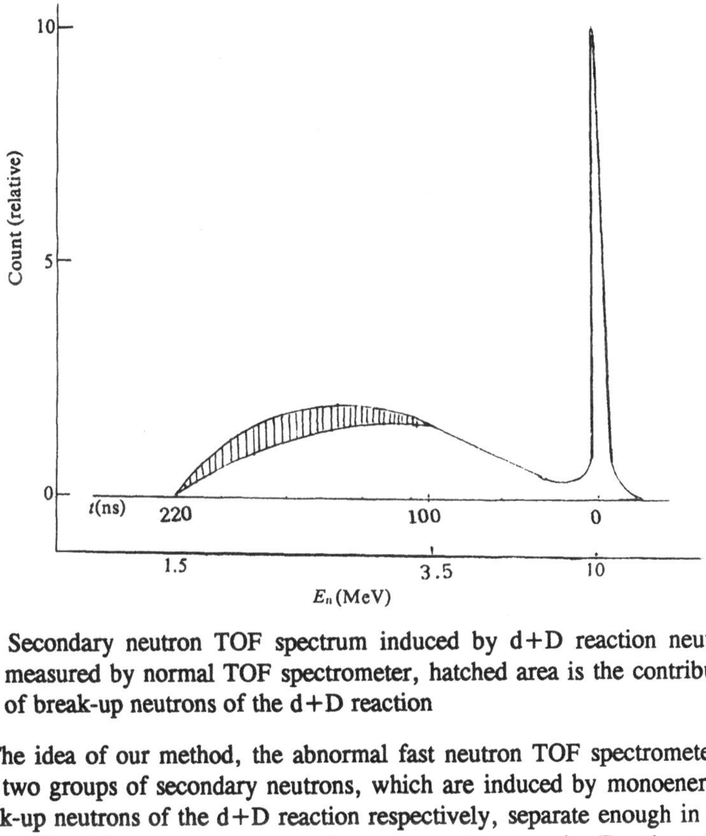

Fig. 3. Secondary neutron TOF spectrum induced by d+D reaction neutrons measured by normal TOF spectrometer, hatched area is the contribution of break-up neutrons of the d+D reaction

The idea of our method, the abnormal fast neutron TOF spectrometer, is to make two groups of secondary neutrons, which are induced by monoenergetic and break-up neutrons of the d+D reaction respectively, separate enough in their flight time. For this purpose, an unusual arrangement was made. For the normal TOF spectrometer, the distance between neutron source and sample is about 10-20 cm and that between sample and neutron detector is several meters (In our case, it is about 6 meters). However, for the abnormal TOF facility the distance between neutron source and sample is much longer than that used in the normal TOF spectrometer. In our case, it is 2.2 meters. So that the difference of the flight time between monoenergetic and break-up neutrons of the d+D reaction over this distance is much larger than usual (usually the separation is 3 ns, but for the abnormal TOF spectrometer it is 35 ns assuming that the monoenergetic neutron energy is 10 MeV and the maximum break-up neutron energy is 3.5 MeV). Meanwhile, the distance between sample and neutron detector is shortened. In our arrangement, it is 70 cm. As a result the flight time of secondary neutrons with the lowest energy (say 1.5 MeV, cut by neutron detector bias) induced by monoenergetic neutrons is less than that of secondary neutrons with the highest energy (about 3.5 MeV) induced by break-up neutrons. Therefore two groups of neutrons which are induced respectively by monoenergetic and break-up neutrons of the d+D reaction are well separated in the time of flight spectrum. The flight time of two groups of secondary neutrons induced by 10 and 3.5 MeV neutrons under the above arrangement is listed in table 1. As the table indicates, the separation of these two groups of neutrons is 20.5 ns for the detector bias of 1.5 MeV and 11.2 ns for 1.0 MeV bias. The ideal TOF spectrum of secondary neutrons is shown in Fig. 4.

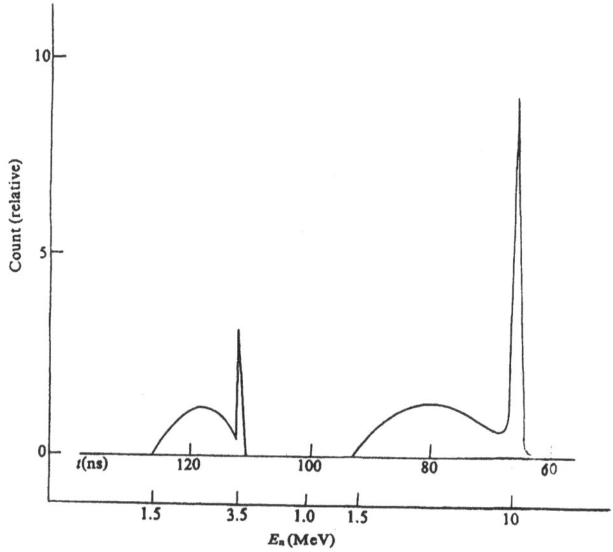

Fig. 4. Ideal time of flight spectrum of secondary neutrons induced by 10 MeV and 3.5 MeV neutrons.
Distance between neutron source and sample is 220 cm
Distance between sample and neutron detector is 70 cm

The abnormal fast neutron TOF spectrometer is shown in Fig. 5. The beam pick-off, gas target, neutron detector, electronics and data acquisition system are the same as those used in the normal TOF spectrometer[11]. Fig. 6 shows the secondary neutron TOF spectrum of ^{238}U induced by 10 MeV monoenergetic and break-up neutrons of the d+D reaction measured by the abnormal TOF spec-trometer. It is obvious that secondary neutrons are distributed in two groups which are induced by monoenergetic and break-up neutrons of the

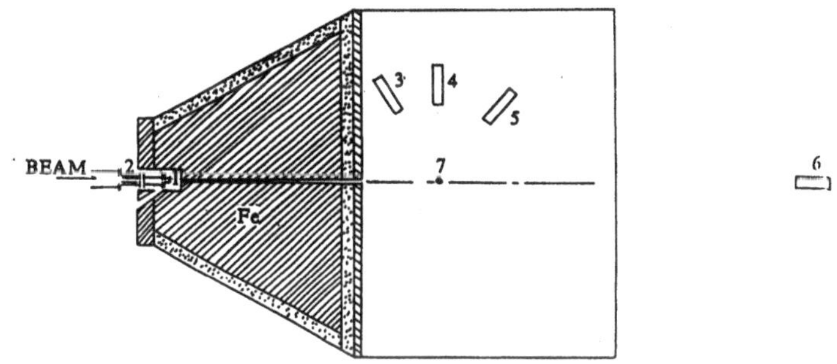

Fig. 5 Schematic diagram of the abnormal fast neutron TOF spectruometer
1 - neutron source 2 - beam pick-off
3, 4 and 5 - neutron detectors, 6 - source neutron monitor , 7 - sample
- paraffin + lithium carbonate
- lead - iron

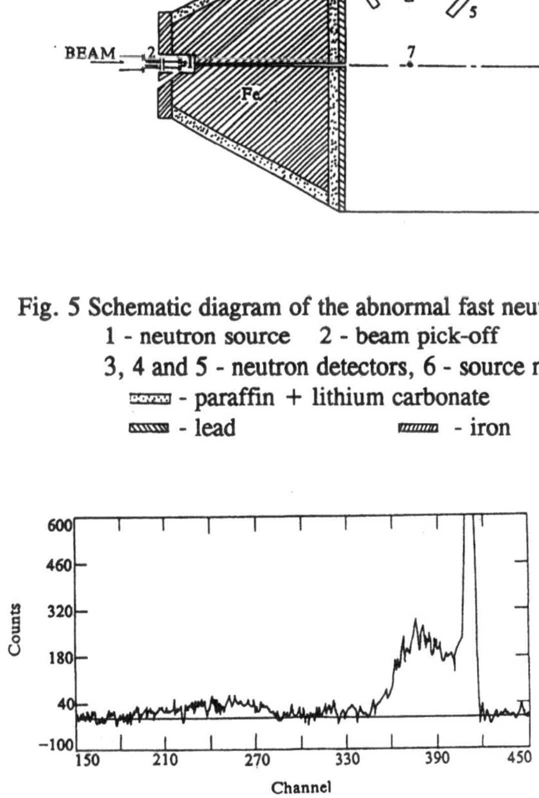

Fig. 6. TOF spectrum of secondary neutrons of ^{238}U induced by d+D reaction neutrons with monoenergy of 10 MeV measured by the abnormal TOF spectrometer, $\theta_1=45°$, random background subtracted

d+D reaction. Therefore, the contamination of break-up neutrons from the neutron source to the wanted secondary neutron spectrum can be fully eliminated and a pure secondary neutron spectrum induced only by monoenergetic neutrons can be obtained.

To get a net secondary neutron spectrum at an angle four runs of measurements are needed. The first run is performed with "sample in" and "target gas in". The second is done with "sample out" but "target gas in". The third is made with "sample in" but "target gas out". The fourth is for "sample out" and "target gas out". The net TOF spectrum of secondary neutrons is obtained by the following formula:

run 1 - run 2 - run 3 + run 4

Fig. 7 shows the individual and net neutron TOF spectra of ^{238}U induced by 10 MeV neutrons.

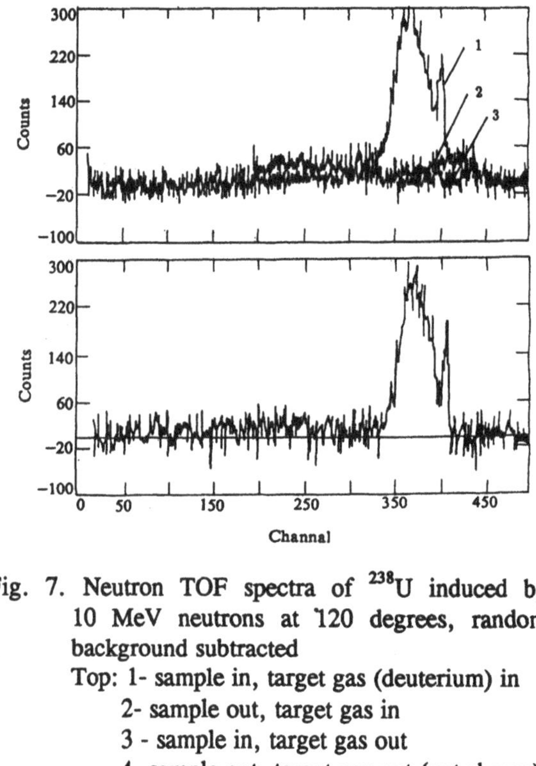

Fig. 7. Neutron TOF spectra of ^{238}U induced by 10 MeV neutrons at 120 degrees, random background subtracted
Top: 1- sample in, target gas (deuterium) in
2- sample out, target gas in
3 - sample in, target gas out
4- sample out, target gas out (not shown)
Bottom: [1] - [2] - [3] + [4]

The secondary neutron TOF spectrum was converted into energy spectrum. Double differential cross sections of ^{238}U at 45° at the incident neutron energy of 10 MeV measured by both normal and abnormal TOF spectrometers are shown in Fig. 8, symbolized with circles and crosses respectively. The shadow area is the source break-up neutron contribution which is moved away in the spectrum measured by the abnormal TOF facility.

3. Results

The double differential cross sections of secondary neutrons of ^{238}U and ^{209}Bi at 10 MeV have been measured so far. The final result was given in the following way. The double differential cross sections above 4 MeV were taken from the data measured by the normal TOF spectrometer to keep good energy resolution. The double differential cross sections below 4 MeV were given by the result obtained by the abnormal TOF device. Figs. 9-11 show the double differential cross sections of ^{238}U at 10 MeV, which have not yet been corrected for neutron flux attenuation and multiple scattering in the sample. The secondary neutron double differential cross sections of ^{209}Bi at 10 MeV incident

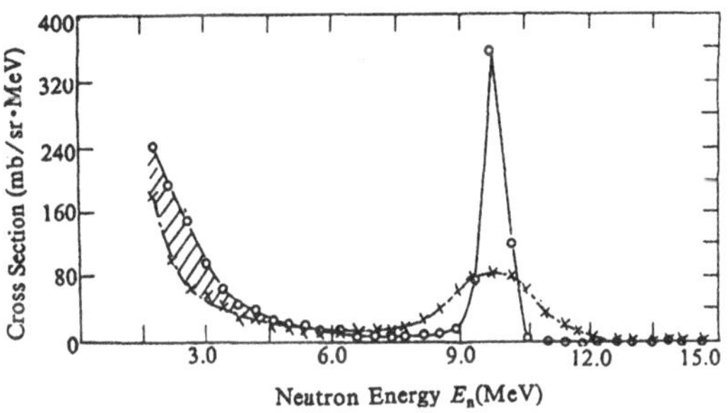

Fig. 8. Double differential neutron emission spectra of ^{238}U induced by 10 MeV neutrons at 45°
∘ using the normal TOF spectrometer
× using the abnormal TOF spectrometer

neutron energy at 45°, 70°, and 110° are shown in Figs. 12-14. The data were corrected for neutron flux attenuation, multiple scattering and finite geometry with a Monte-Carlo code and analysed with a semiclassical theory of multi-step reaction processes[12]. A good agreement was obtained between the measurement and calculation, as can be seen in the figures.

4. Discussion and Conclusion

As we know, the most promising method for measuring double differential cross sections of secondary neutrons in the incident neutron energy range 9 to 13 MeV is accelerating triton and using the H(t, n) reaction as a monoenergetic neutron source. However, this is not an easy way for most of laboratories. People have explored different approaches as mentioned in section 1, but few effective methods have been found in real use. The method developed in CIAE can be used to solve the problem with conventional neutron sources at the price of losing counting rate and energy resolution.

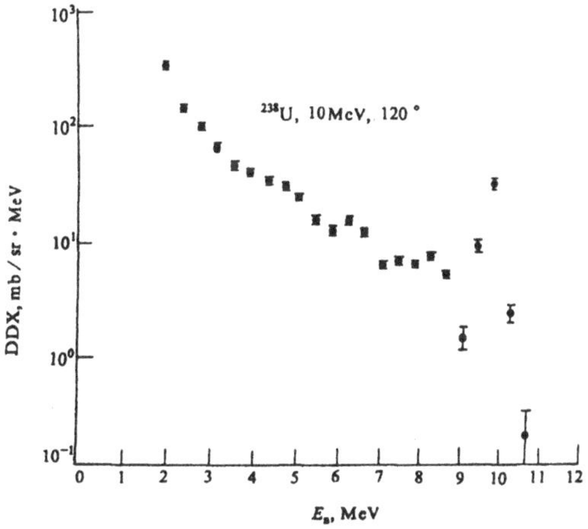

Fig. 9. Double differential cross sections of ^{238}U at 10 MeV, $\theta_1=120°$

The counting rate is a factor of 0.6, compared with that of the normal arrangement (supposing the distance between neutron source and sample is 20 cm and that between sample and neutron detector is 6 m for the normal TOF spectrometer). At present three neutron detectors are used in the measurement simultaneously. To increase the datum acquisition rate we plan to add three more neutron detectors. Because the neutron source is massively shielded in the forward semisphere, it is easy to put more neutron detectors without additional shielding.

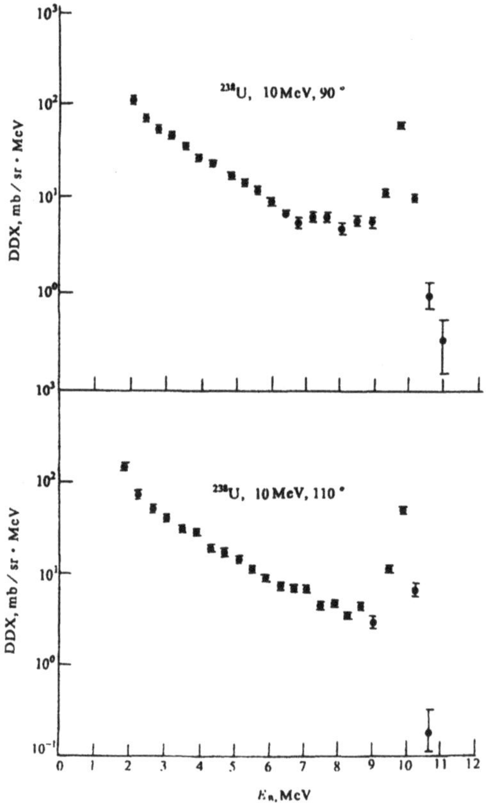

Fig. 10. Double differential cross sections of ^{238}U at 10 MeV, θ_1=90°, 110°

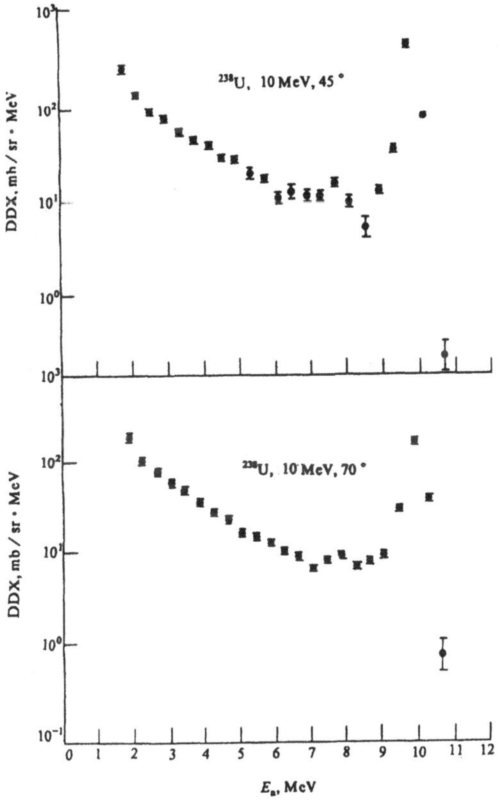

Fig. 11. Double differential cross sections of ^{238}U at 10 MeV, θ_1=45°, 70°

Now the threshold of the spectrometer is set at 1.5 MeV and the time separation in the spectrum between the secondary neutrons of the lowest energy (1.5 MeV) induced by monoenergetic neutrons and the one of highest energy (3.5 MeV) induced by break-up neutrons of the d+D reaction is about 20 ns. If the threshold is 1 MeV, the separation will be 11.2 ns,which is much larger than the time resolution (the full width at half maximum for elastic scattering neutron peak is 1.6 ns). Therefore, the threshold of the detector can be extended to 1 MeV or a little bit lower without difficulty for 10 MeV incident neutrons. For the incident neutron energy of 13 MeV, the bias of the neutron detector should be higher (about 2.2 MeV).

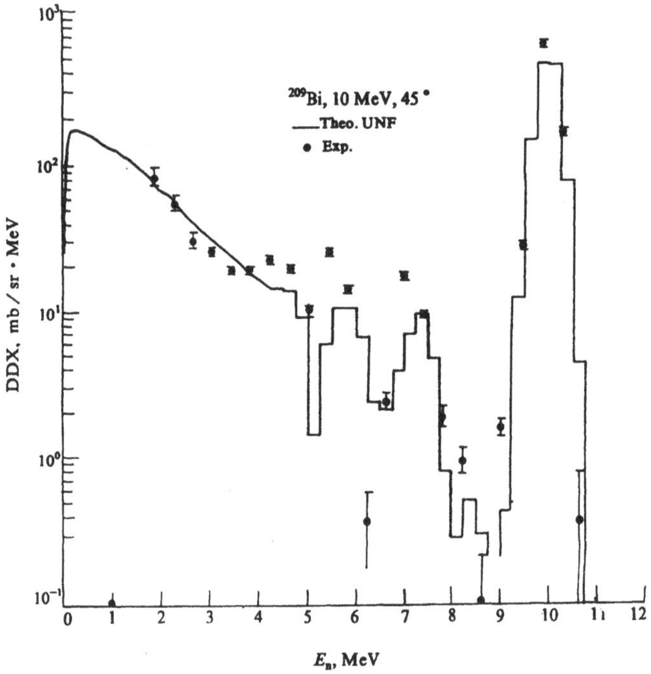

Fig. 12. Double differential cross sections of ^{209}Bi at 10 MeV, θ_1=45°
∘ present work
– calculation

The energy resolution is about 2.4% for 10 MeV neutrons measured by the normal TOF spectrometer and about 12% for 4 MeV neutrons measured by the abnormal TOF spectrometer.

In conclusion, the preliminary results show that our method can be used for secondary neutron double differential cross section measurement in the incident neutron energy range 9 to 13 MeV.

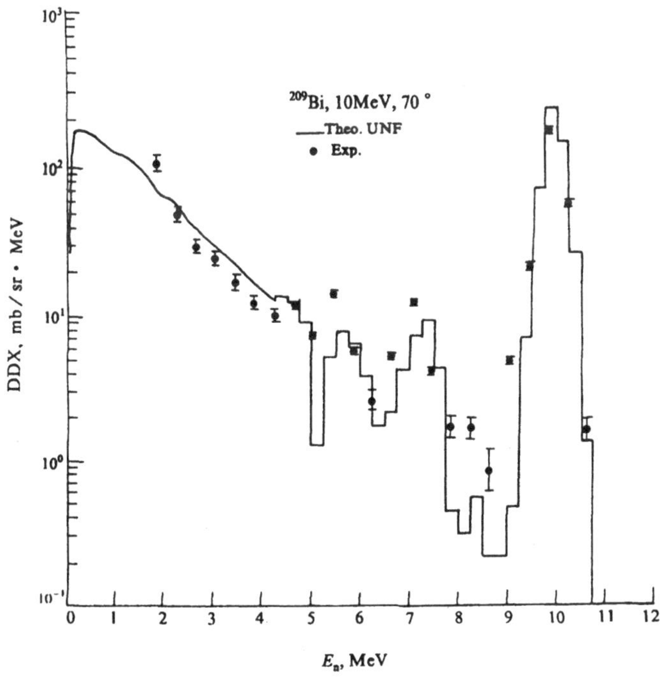

Fig. 13. Double differential cross sections of ^{209}Bi at 10 MeV, $\theta_1 = 70°$

∘ present work

– calculation

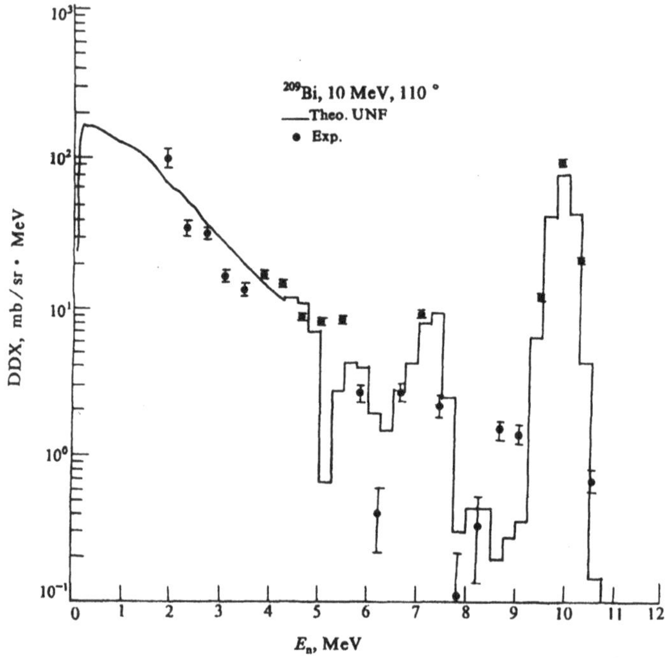

Fig. 14. Double differential cross sections of ^{209}Bi at 10 MeV, $\theta_l=110°$

∘ present work

– calculation

Table 1 Flight time of secondary neutrons induced by monoenergetic and break-up neutrons of the d+D reaction under the following arrangement:
The distance between neutron source and sample (d1) is 220 cm.
The distance between neutron detector and sample (d2) is 70 cm

primary neutron		secondary neutron		total
E_n (MeV)	d1=220cm	E_n' (MeV)	d2=70cm	
10	50.3ns	10	16.0ns	66.3ns
10	50.3ns	1.5	41.3ns	91.6ns
10	50.3ns	1.0	50.6ns	100.9ns
3.5	85.0ns	3.5	27.1ns	112.1ns
3.5	85.0ns	1.5	41.3ns	126.3ns
3.5	85.0ns	1.0	50.6ns	135.6ns

References

1. A. Takahashi, *the present proceedings*
   A. Takahashi et al., *J. Nucl. Sci. Technol.* **26** (1989) 15
2. M. Baba et al., *Proc. of Inter. Conf. on Nucl. Data for Sci. and Tech.* (Mito, Japan, 1988) P. 209
3. H. A. Obiesi et al., *Proc. of Inter. Conf. on Nucl. Data for Sci. and Tech.* (Jülich, Germany, may,1991)
4. D. M. Drake et al., *Nucl. Sci. and Eng.* **63** (1977) 401
5. M. Drosg et al., *Rad. Effect* **92** (1986) 145; P. W. Lisowski et al., LANL, *LA8342* (1980)
6. E. Dekempeneer et al., *Nucl. Sci. and Eng.* **97** (1987) 353
7. R. C. Byrd et al., *Annual Report-TUNL* **XXI** (1981-1982) p. 18
8. Wang Xiaozhong et al., Neutron Physics Lab, CIAE, *Laboratory Report* (1989)
9. Qi Bujia, Tang Hongqing, Zhou Zuying et al., *Proc. of Inter. Conf. on Nucl. Data for Science and Tech.* (Jülich, Germany, May, 1991)
10. S. Chiba et al., *the present proceedings*
11. Sa Jun, Tang Hongqing, Zhou Zuying et al., "A muti-detector fast neutron TOF spectrometer at the HI-13 tandem accelerator", *Beijing National Tandem Accelerator Laboratory Progress Report* (1988 - 1989)P. 119
12. Zhang Jingshang, *the present proceedings*

SECONDARY NEUTRON DOUBLE DIFFERENTIAL CROSS SECTIONS FROM ^{209}Bi AT 14.2 MeV

SHEN GUANREN, XIA HAIHONG, TANG HONGQING,
ZHOU ZUYING, SUI QINGCHANG, QI BUJIA,
KE ZUNJIAN AND ZHOU CHENWEI
China Institute of Atomic Energy
P. O. Box 275-46, Beijing 102413, P. R. China

ABSTRACT

The secondary neutron double differential cross sections from ^{209}Bi at 14.2 MeV .have been measured at 6 angles between 25° - 150° using an associaed particle TOF spectrometer. Flight path was 2.7m. The neutron detector was biased at 1.3 MeV. The time resolution was about 1.2 ns. The data were compared with existing data and theoretical calculated results. Good agreement is achieved.

1. Introduction

^{209}Bi is a typical spherical odd-even nucleus. Accurate secondary neutron double differential cross sections (SNDDX) are interesting for examining nuclear models. SNDDX from ^{209}Bi induced by 14.2 MeV neutron have been measured using an associated particle fast neutron TOF spectrometer at 600 KV Cockcroft Walton accelerator in CIAE.

2. Experimental Procedure

The schematic diagram of the spectrometer is shown in Fig. 1. It was previously described in details[1, 2]. Here the experimental condition is given briefly. 14.2 MeV neutrons were produced by the T(d,n)^{4}He reaction. The ^{209}Bi sample is a cylinder of bismuth metal with 3 cm in diameter and 3 cm in length. The flight path was 2.7 m. The neutron detector was biased at 1.3 MeV neutron energy. The background produced by γ-rays was rejected by means of a pulse shape discriminator CANBERRA MODEL 2160A. The overall time resolution was about 1.2 ns for primary neutrons. The energy resolution was about 5% for 14.2 MeV neutrons. The neutron detector efficiency was measured by the n-p scattering method.

The typical secondary neutron TOF spectrum from ^{209}Bi at 90° is shown in Fig. 2. The measuring time of this spectrum was about 80 hours.

To get the absolute values of SNDDX, the n-p scattering differential cross section at 25° was used for normalization. The n-p scattering neutron counts $N_H(25°)$ were accurately measured with the statistical uncertainty <2%.

3. Data Reduction and Results

The data reduction method was described in details else where[2] . Here only the expression is presented. The channel number N_i corresponding to energy E_i is determined by the formula:

$$N_i = N_{el} - 72.298[(1+0.798\times E_i)/E_i - (1+0.798\times 10^{-3}\times E_{el})E_{el}]L/W, \qquad (1)$$

where E_{el} is the energy of the elastic scattering neutron(MeV), E_i is the secondary neutron energy, W is the time width per channel (ns), N_{el} is the channel number of the elastic scattering neutron peak, and L is the secondary neutron flight path(m).

The net secondary neutron TOF spectrum was obtained by substracting accident background from the TOF spectrum of ^{209}Bi. Formula (1) is used to determine the position of the ith ΔE interval in the TOF spectrum. The secondary neutron counts $N^{Bi}(E_i, \theta)$ from ^{209}Bi in the ith ΔE_i at angle θ are obtained.

The double differential cross sections $\sigma(E_i, \theta)$ from ^{209}Bi is obtained by the following expression:

$$\sigma(E_i, \theta) = \frac{N_{Bi}(E_i,\theta)}{N_H(25°)} \times \frac{N_H}{N_{Bi}} \times \frac{\varepsilon_H(E)}{\varepsilon_{Bi}(E_i)} \times \sigma_H(25°), \qquad (2)$$

where θ is the measuring angle in lab. system, N_H and N_{Bi} are the numbers of hydrogen and ^{209}Bi nuclei in polyethylene and ^{209}Bi samples, respectively. $\varepsilon_H(E)$ and $\varepsilon_{Bi}(Ei)$ are the neutron detector efficiencies at neutron energies E and E_i. $N_H(25°)$ is the total counts in the n-p scattering peak at 25°. $\sigma_H(25°)$ is the n-p scattering differential cross section at 25° and is equal to 192.1 mb/Sr.

The net TOF spectrum is converted into secondary neutron energy spectrum using a code based on Eqs.(1) and (2) on IBM pc/xt personal computer. The corrections for multiple scattering, incident neutron flux attenuation and finite geometry were made using a Monte Carlo code named DCM. The correction error is about 1.5-3% for ^{209}Bi sample and 1% for n-p scattering at 25°.

The SNDDX from ^{209}Bi are obtained at 6 angles between 25° - 150°. The error sources of the data mainly come from counting statistics (It's about 10 to 20% for most of energy bins), the efficiency of the neutron detector (It's about 3%) and the correction of Monte Carlo code.

4. Comparison and Analysis

The SNDDX of the present work are compared with the data of A.Takahashi et al. (En=14.1 MeV) and they are in agreement within the error for most of experimental points. The data are also analyzed by the semi-classical theory of multi-step nuclear reaction processes considering angular momentum and parity conservation. The theoretical energy spectrum formula is as follows:

$$\frac{d\sigma}{d\varepsilon}=\sum_n \sum_{J\pi} P^{J\pi}(n)\, \sigma_a^{J\pi} \frac{T_b^{J\pi}(n, E, \varepsilon)}{T_t^{J\pi}(n, E)}, \tag{3}$$

where E is the excitation energy, ε is the energy of emitted particle, $\sigma_a^{J\pi}$ stands for the absorption cross section in $J\pi$ channel, $T_b^{J\pi}(n, E, \varepsilon)$ is the T factor of emitted particle b with excitation energy E at n exciton state in $J\pi$ channel, $T_t^{J\pi}(n, E)$ is the total T factor. $P^{J\pi}(n)$ stands for the $J\pi$ channel occupation probability of n exciton state.

The UNF code based on the above theory was used to calculate the SNDDX. It was described in details in Refs.[3, 4]. The optical parameters were adjusted by a code based on spherical optical model. The contributions of direct reaction was calculated using DWUCK-4 code.

The measured and calculated SNDDX are shown in Figs. 3 - 8. The angular integrated spectra are shown in Fig. 9. From Fig. 9, the (n, 2n) and total reaction cross sections induced by 14.2 MeV neutrons were obtained by integration of measured SNDDX from 0 to 6.5 MeV and 0 to 15 MeV (from 0 to 1.5 MeV, the calculated results are used), respectively. They are 5.30±0.25 b and 2.027±0.16 b. The comparion of the present work, previous experimental data, calculation results, ENDF/B-6 and JENDL-3 is listed in Table 1.

Table 1 The comparison of reaction cross sections

RCS (mb)	σ(TOT)	σ(N, 2N)
PRESENT WORK	5300.0±0.25	2027.0±0.16
CAL. RESULTS	5340.0	2126
ENDF/B-6	5347.07	2140
JENDL-3	5370	2201
L.Rosen		2300±0.30
Flerov		2420±0.20
A.Takahashi		1940

5. Conclusion

The SNDDX of ^{209}Bi measured in present work agree with previous data and are reproduced well by calculation of the semi-classical theory of multi-step nuclear reaction processes. It shows that the semi-classical theory model meets with success. The calculated total and (n, 2n) reaction cross sections agree with ENDF/B-6, JENDL-3 and the results of A. Takahashi et al.[7] within error, and are a little bit lower than the data of L. Rosen and Flerov[5, 6].

Acknowledgements

The authors would like to thank Prof. Zhang Jingshang for kindly providing UNF code, the Cockcroft Walton accelerator group for providing good beam and Prof. Ling Yude for the Monte Carlo correction. We also would like to thank Miss Yuan Yuan, Miss Yu Ruixiang and Mr. Jia Guoxiang for their help during the experiment.

References

1. Shen Guanren et al., *Chin. J. Nucl. Phys.* **6** (1984) 193.
2. Shen Guanren et al., *Chin. J. Nucl. Phys.* **8** (1986) 298; *Chinese Physics* **8** (1988) 1000.
3. Zhang Jingshang *Proceeding of "Beijing International symposium on fast neutron physics"* (9-13 sep. 1991. Beijing, China).
4. Xia Haihong et al., to be published.
5. L. Rosen et al., *Phys. Rev.* **107** (1957) 824.
6. Flerov *AE* **5** (1958) 657.
7. A. Takahashi et al., *JAERI-M88-102* (1988).

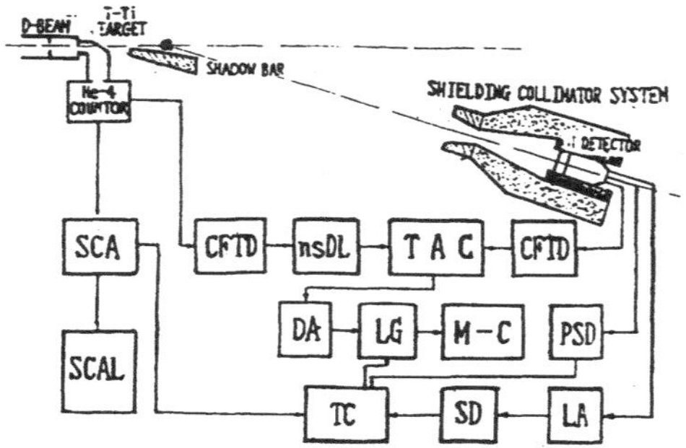

Fig.1 THE SCHEMATIC DIAGRAM OF APTOF SPECTROMETER.

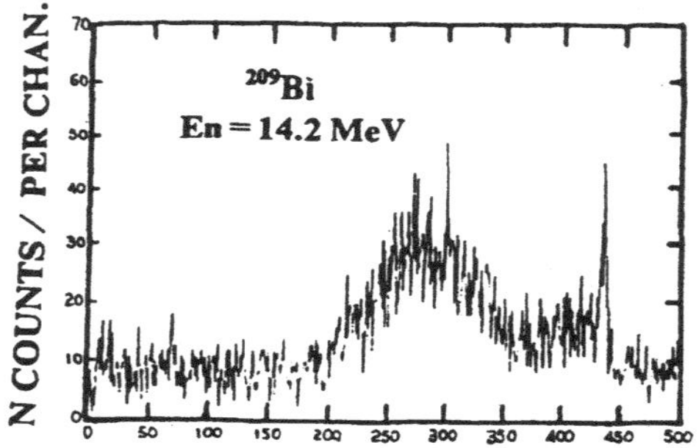

Fig.2 THE TOF SPECTRUM AT 90 DEG.

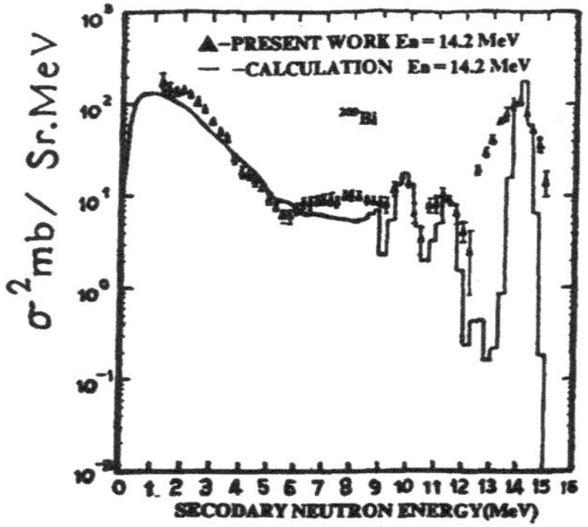

Fig.3 THE SNDDX AT 25 DEG.

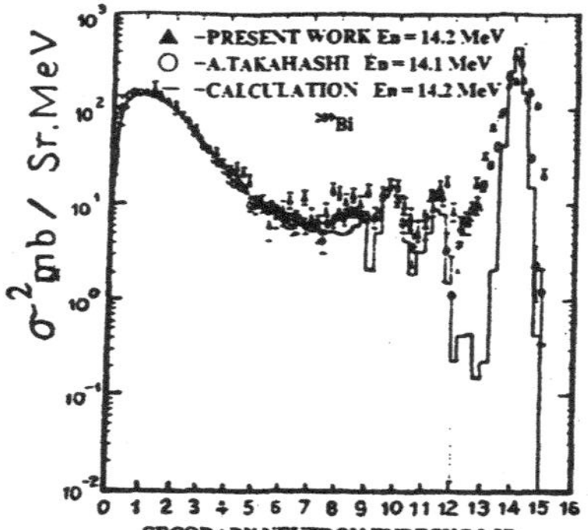

Fig.4 THE SNDDX AT 30 DEG.

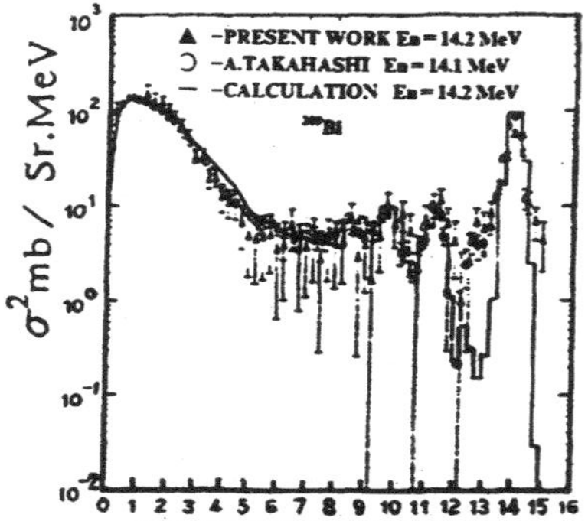

Fig.5 THE SNDDX AT 60 DEG.

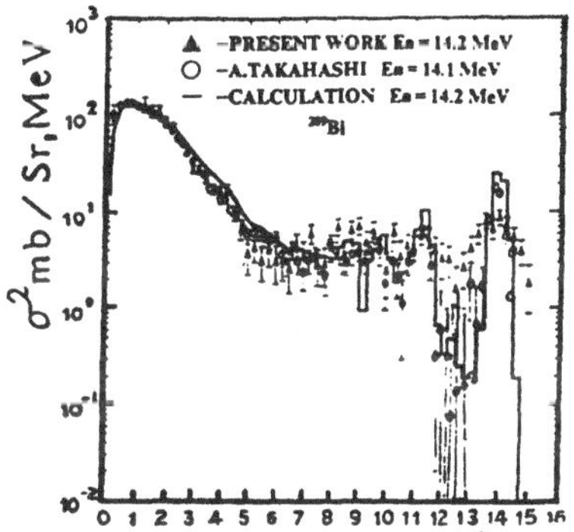

Fig.6 THE SNDDX AT 90 DEG.

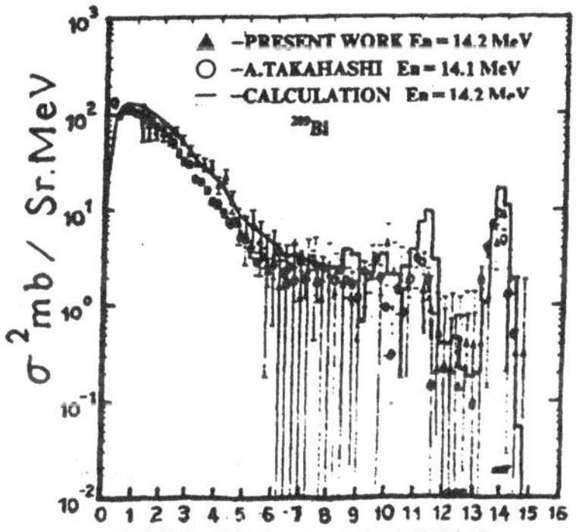

Fig.7 THE SNDDX AT 120 DEG.

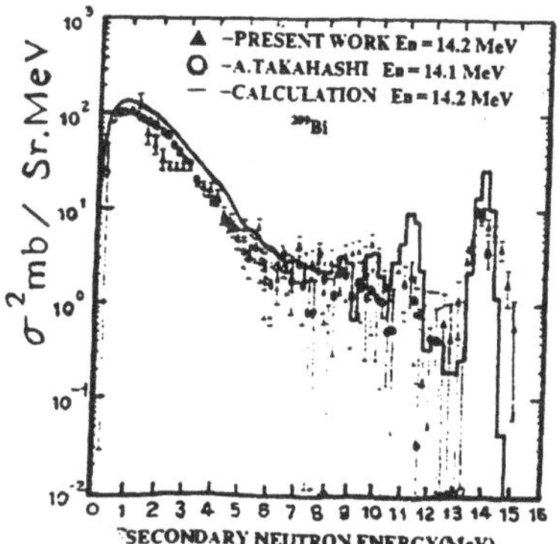

Fig.8 THE SNDDX AT 150 DEG.

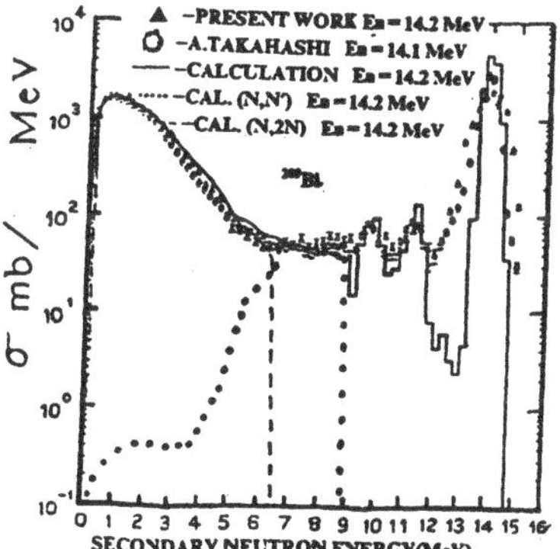

Fig.9 THE ANG. INTEG. SPECTRUM.

NUCLEAR FISSION

NEUTRON EMISSION IN NUCLEAR FISSION

M. V. Blinov
V. G. Khlopin Radium Institute,
Shvernic Ave. 28, 194021 Leningrad, USSR

ABSTRACT

Some new results on the study of the neutron emission in fission obtained in the recent years are presented. The integral and differential data on the neutron emission and the results of the analysis for these data are given. The "open" questions of the mechanism of the fission neutron emission are discussed in this report too.

1. Introduction

Understanding the property of fission neutron emission is very important for design of various installations based on the chain nuclear reaction. The characteristics of fission neutrons are high priority nuclear data for practical use. From theoretical point of view the study of the fission neutron emission is of interest because a lot of information of fission process can be derived from the measurement of the fission neutrons. By using knowledge of neutron emission characteristics the excitation and deformation energies of fissioning system, the viscosity of the nucleus and its fragments etc. can be determined on various stages of its evolution.

A significant process to understand the character and the mechanism of fission neutron emission was going on in the recent years as a result of the studies from a number of the experimental and theoretical groups in USA, USSR, Germany, Belgium, France, China, India and Japan.

In this report I shall very briefly present some new results of integral data. Then I would like to attract your attention to some problems connected with the differential data and that information which can be obtained for understanding fission process and de-excitation of high-excited fragments.

2. Integral Energy Spectra and Multiplicities of Fission Neutrons

It is very important to have a good knowledge of the multiplicity and energy spectrum of ^{252}Cf spontaneous fission neutron, because they were recommended by IAEA as an international standard. Mannhart made evaluation of this spectrum[1] in the region 0.05 - 20 MeV on the basis of a careful analysis of the experimental results of several scientific groups. High precision data from this evaluation were obtained in the energy range 0.5 - 12 MeV. In the region of low (below 0.5 MeV) and high energies (more than 12 MeV) it is necessary to carry out new experimental studies. There is a good agreement of the theoretical calculations[2, 3, 4] with the evaluation[1] in the energy range. But in the region of low energies there is an essential difference (more than 10%), which may be connected with influence of some non-equilibrium mechanism of fission neutron emission.

New measurements of ^{252}Cf spontaneous fission neutron multiplicity (intern. standard) were not carried out because the accuracy of existing evaluated data is rather high.

The experimental dependence of the multiplicity on the excitation energy of the fissioning nuclei E^* was considered in detail in Boldeman's review published recently[5]. So I shall not discuss this problem. The calculations of the dependence $\nu(E^*)$ were made by Madland et al.[6] and Märten et al.[7]. It should be mentioned that as a whole there is an essential success in the theoretical description of these dependences $\nu(E^*)$. In Fig. 1 experimental data and results of the calculations[7] are shown for fast neutron induced fission of ^{235}U.

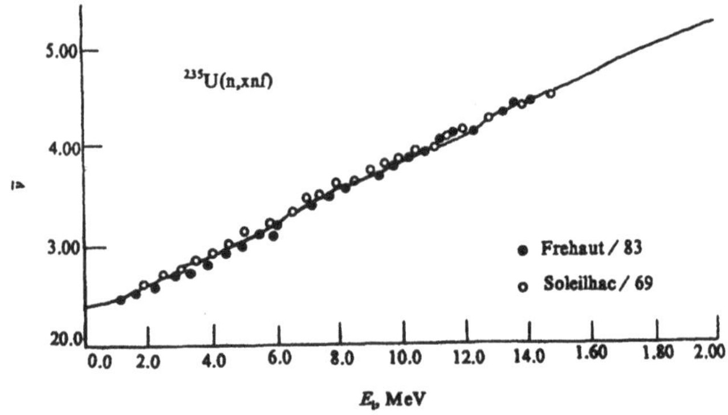

Fig. 1. Calculated multiplicites for ^{235}U[7] in comparison with experimental data[13]

Concerning neutron energy spectra the situation is much more complicated because the accuracy of spectrum measurements is not sufficiently high. The use of the standard spectrum for the efficiency determination allow to improve the accuracy of the measurements rather considerably. The calculations of the spectra are connected with many difficulties. I shall briefly mention the latest experimental results. In the study of Boykov et al.[8] the measurements of neutron spectra in fission of ^{235}U, ^{238}U and ^{232}Th induced by 2.9 MeV and 14.7 MeV neutrons (below and above the fission threshold respectively) were carried out. The contribution of the non-equilibrium neutrons at 14.7 MeV was determined and the excess component of neutrons in low-energy portion in comparison with statistical calculations was found. Lovchikova et al.[9, 10] measured fission neutron spectra of ^{232}Th, ^{235}U and ^{239}Pu. Li Anli et al.[11] measured the spectrum of ^{238}U induced by 12 MeV neutrons. Baba et al.[12] measured the spectrum of ^{238}U and ^{232}Th induced by 2 MeV neutrons. In those studies the parameters of Maxwellian and Watt distributions were determined. In Ref. [8] and [12] it was obtained that the neutron spectrum in the energy region above E_n = 6 MeV goes below Maxwellian distribution for 2 MeV neutron induced fission. This conclusion agrees with the data for the ^{252}Cf spontaneous fission. The very high accuracy measurement of the ^{248}Cm spontaneous fission neutron spectrum was carried out in the cooperation work of the Radium Institute (RI) and Peking University[25].

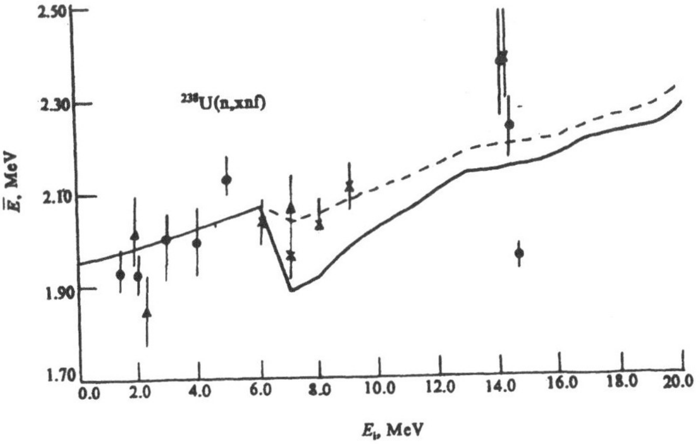

Fig. 2. Experimental data[8, 10] on average neutron energy $\bar{E}$ in comparison with the calculated data[7] for ^{238}U. The figure is taken from ref. [7]

The calculations of the energy dependence neutron spectra for various nuclides were carried out in Refs. [6] and [7]. It should be stressed that the modification of the models, taking into account the specific features of the individual nuclides and fragments allows essential improvement to the fit to the experimental data. Fig. 2 and Fig. 3 show the results of the theoretical calculations[7] and experimental data which are the dependence of average energy

of neutrons $\bar{E}$ on neutron multiplicity for ^{232}Th and ^{238}U. It can be seen that there are too few experimental points and their accuracy of the average energy determination is rather low.

If one considers the problem as a whole, it should be stressed that the lack of experimental data at high accuracy on neutron spectra is obvious. So it is necessary to carry out the new measurements both for spontaneous and neutron induced fissions for various nuclides, especially for neptunium, americium and curium isotopes.

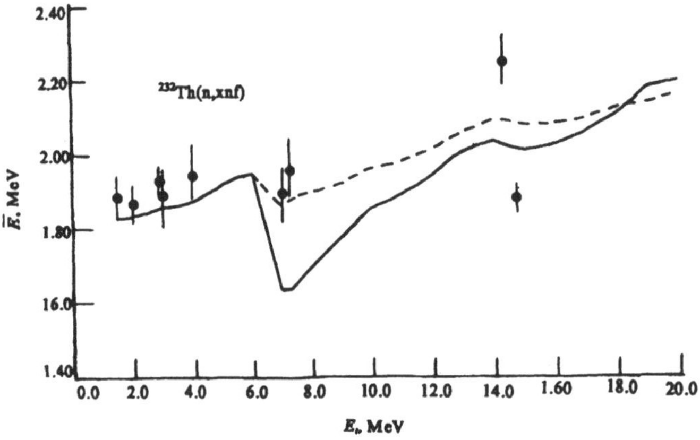

Fig. 3. Experimental data[8, 10] on average neutron energy $\bar{E}$ in comparison with the calculated data[7] for ^{232}Th

3. Differential Characteristics of Neutron Emission and Results of Their Analysis

The mechanism of the fission neutron emission will be discussed in this section according to the evaluation of the fission process from the moment of particle interaction with nucleus up to the final stage that is concerning with the formation and the de-excitation of the equilibrium fragments.

At initial stage of the interaction of the incident particle with nucleus at energies more than 10 MeV/nucleon non-equilibrium neutrons with temperature about 5 MeV (described with Marxwellian distributions for heavy ion-induced reactions) are emitted. For example, in the collaborative work of RI and JINR(USSR)[14] it was found that in the fission of ^{238}U induced by ^{12}C (energy about 105 MeV) the multiplicity of such neutrons is equal to 0.3 $\pm$ 0.1 n/fission (the total multiplicity is 10.3 $\pm$ 0.6 n/fission). The emission mechanism of non-equilibrium neutrons in the reactions was studied in many papers, but the full understanding for the complicated problem has not yet been reached up to now.

Usually it is assumed that the motion to the saddle point begins after the formation of the thermal equilibrium in a compound nucleus. It is very interesting to have the information about the transition time. From the large number of experimental results got by different scientific groups with various ions at various excitation energies it became clear that the multiplicity of neutrons emitted before scission point, ν_{pre}, is much larger than that predicted by statistical theory. Even the descent time after saddle point has been taken into account, the transition time calculated from pre-scission neutron multiplicity is rather large (10^{-20}s) (descent time was determined from calculation assuming the viscosity of the definite type). The calculated fission time on the basis of the experimental multiplicity for ^{205}Fr at excitation energy 77 MeV is shown in Fig. 4[15]. From measurements of the same type for the compound nucleus ^{250}Cf at excitation energy 75 MeV it was obtained that the fission time is equal to 6×10^{-20}s[14], which agrees with time for light fissioning nucleus ^{205}Fr. At the high initial excitation energies (about 500 MeV) the fission time is also large and equals 10^{-20}s.

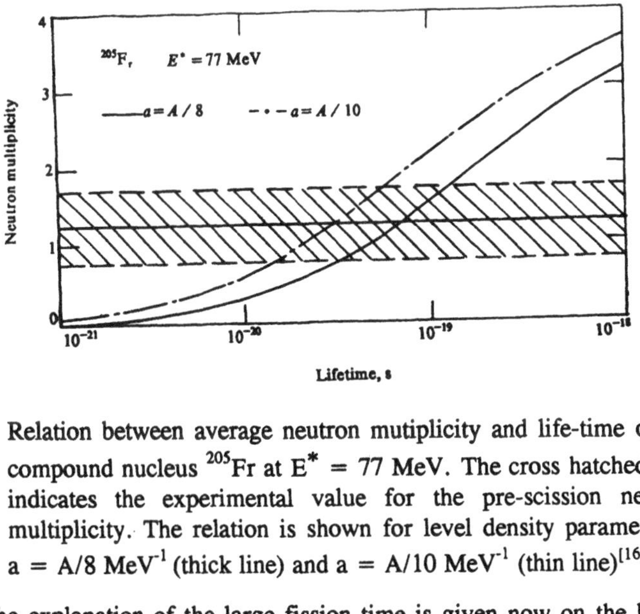

Fig. 4. Relation between average neutron mutiplicity and life-time of the compound nucleus ^{205}Fr at $E^* = 77$ MeV. The cross hatched area indicates the experimental value for the pre-scission neutron multiplicity. The relation is shown for level density parameter of $a = A/8$ MeV^{-1} (thick line) and $a = A/10$ MeV^{-1} (thin line)[16]

The explanation of the large fission time is given now on the basis of the influence of the nuclear viscosity on fission process. The viscosity is a fundamental nuclear characteristic and it is determined by the combination of the collective and one-particle degrees of freedom. But there are not many existing methods to determine the viscosity. And the existing methods are often connected with very limited region of the influence of the viscosity. In this case the inner viscosity of the whole nucleus takes place and this is very important. As a result of the analysis it was found that the fission is the process of high viscosity. It should be stressed that the study of the dissipation in fission now is still in the first stage. In order to understand the fission process, it is necessary to take into account such important effects very carefully as neutron emission in the fragment acceleration and the influence of the non-equilibrium processes. From the analysis of the neutron angular distribution, it is obtained that the excitation energy at the scission point is not essentially changed with excitation energy of the initial nucleus above $E^* = 50$ MeV[16]. Therefore the nucleus undergone fission is at rather low excitation energies. But in this aspect the new investigations are necessary because neutron emission during the fragment acceleration strongly changes the neutron angular distribution and influences the ratio ν_{pre}/ν_{post} and influences the conclusion about the fission process character at high and intermediate excitation energies.

The determination of neutron emission on the stage between the saddle and scission point (ν_{ss}) is a difficult task. But it is necessary to know the excitation energy at the saddle point where the formation of mass and energy distributions of fragments is begun. The direct experimental information on ν_{ss} is absent practically. For the spontaneous fission of ^{252}Cf in Ref. [17] the upper limit $\nu_{ss} <$ 0.1 n/f was obtained. From evaluations of the probabilities of the equilibrium and non-equilibrium emission the diescent time limits were found $5 \times 10^{-21}s < t_{ss} < 5 \times 10^{-20}s$[17]. But it is necessary to improve the accuracy of ν_{ss} to determine the definition of viscosity (one-body or two-body).

The interesting stage of fission is the rupture of the nucleus. At the moment nuclear potential changes rapidly and the formed fragments have essentially non-equilibrium shape. In ref. [17] it was found that in spontaneous fission of ^{252}Cf there is a neutron component which is emitted at the time close to scission moment. This low-energy component has a preferential yield in direction of fragment motion (preliminary data of Ref. [25]). If these data by careful measurements of neutron angular distribution are approved, the states of the neutrons in the strong distort nuclear field will arise.

In a number of papers the neutron emission during fragment acceleration was considered. Refs. [18-22] are theoretical calculations. The results of the calculations are contradictory in some papers. The great influence of this kind of neutron emission was found in some papers, and rather low influence also was found in other papers. Ref. [23] is the first experiment that investigated the neutron emission mechanisms. This experiment is a measurement of differential angular and energy distribution of neutrons. The analysis of the experimental result showed the large deviation from the hypothesis that all neutrons are evaporated from fully accelerated fragments. But in the neutron energy range of 1-6 MeV, the neutrons of spontaneous fission of ^{252}Cf can be completely explained by the hypothesis in the acceleration process. The calculated probability of the emission during fragment acceleration is shown in Fig. 5. The experimental result derived from the model with hypothesis that all neutrons are emitted from fully accelerated fragments and the comparison with model calculations is shown in Fig. 6. As can be understood from Fig. 6 the emission of three neutrons from one fragment gives essential change of the neutron angular distribution. These data are very important for interpretation of high energy fission.

A new aspect of the analysis of differential energy spectra of fission neutrons is to study how fast the deformation energy of fragments transforms into the excitation energy(in comparison with neutron emission time), This dissipative process is directly connected with the nature and value of the nuclear viscosity. The theoretical calculations resulted in rather large scatters of the values of the dissipation time. Ref. [17] for the first time compared the fission neutron spectra and spectra of neutron got from various nuclear reactions of (n, xn), (p, xn) and (α, xn) with the comparative masses and excitation energies. From the comparison it was found that for the masses of the fragments far from closed shells there is rather good agreement of average energies of neutrons for equilibrium component for these two cases (up to excitation energy 30 MeV) (Fig. 7)[24]. Near closed shells Z = 50 and N = 82 the essential difference in average energies was found (Fig. 8)[25]. It is difficult to understand now the origin of this phenomenon. Probably it is connected with a peculiarity of the excitation mechanism of the fragments. In fission the fragment excitation is of the collective character, but in the reactions of (n, xn) and (p, xn) single particle process is of importance. To investigate the origin of this interesting phenomenon which may be

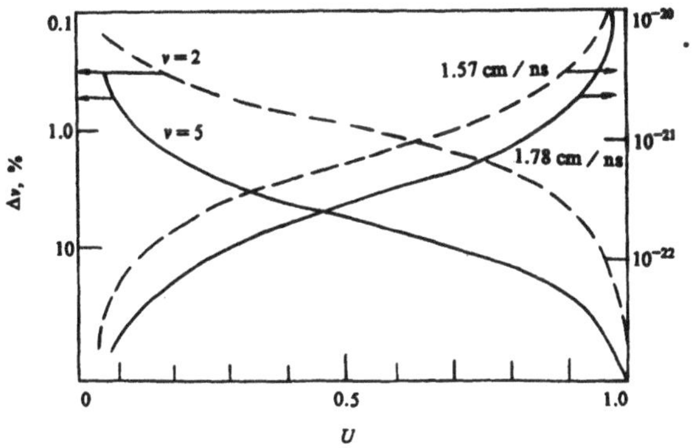

Fig. 5. The dependence of fragment velocity $u = v/v_\infty$ on the acceleration time t

- the share of neutrons emitted during the time t[23]

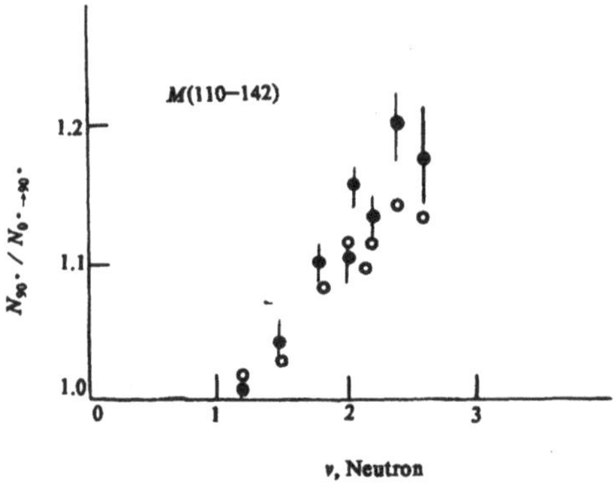

Fig. 6. Ratio of the numbers of neutrons in the energy range 0.7 - 10 MeV in the LS for M = 110 (142) amu[23]

- ∘ measured at the angle 90° to that calculated from the data at ϕ = 0° without account of neutron emission during fragment acceleration
- • calculated from the data at ϕ = 0° with account of neutrons emitted during acceleration to the one calculated from ϕ = 0° without account of neutron emission during fragment acceleration

connected with dissipative effects. It is necessary to do further studies for both neutrons, the fission neutrons and the neutrons emitted from nuclear reactions.

On the final stage neutrons are emitted from equilibrium fragments. But information obtained in such late stage gives essential data about the earlier stage, the stage of nucleus scission. The data $\nu_{(M)}$ reflects the stage of the scission moment. "Saw-tooth" shape of $\nu_{(M)}$ was obtained again for fission of a number of nuclei induced by thermal neutrons in ref. [29]. The explanation of the "saw-tooth" dependence was given in the framework of the influence of the fragment shell structure on their shape in the scission moment. The fragments in the region of closed shells Z = 50, N = 82 (A = 132) have a spherical shape and emit few neutrons. But in magic region near A = 82 where though few neutrons are emitted, the rise of $\nu_{(M)}$ in the direction to the smaller masses had not been observed earlier. Recently in ref. [27] it was reported that the authors registrated such rise (Fig. 9) and so the model of the influence of the shell effects on $\nu_{(M)}$ was further approved. The course of the curve $\nu_{(M)}$ agrees qualitatively with Brosa's calculations based on the two fission models of multimode and the neck random rupture model[28].

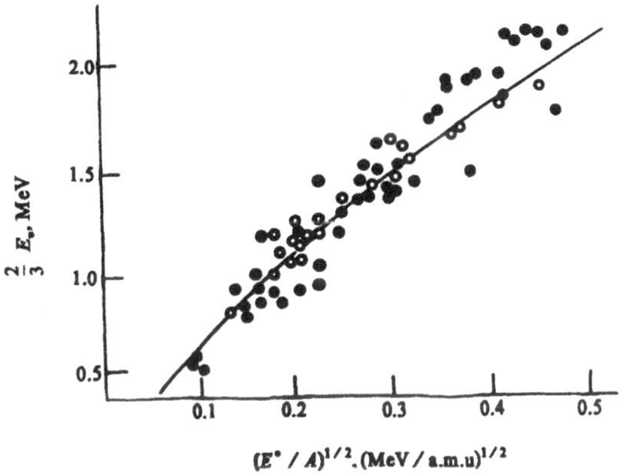

Fig. 7. Average neutron energy in dependence on the ratio of the average thermal energy of the excitation E^* of the emitting nuclei to the mass number A[24]

- • the data on reaction (p, xn), (n, xn) and (α, xn)
- ∘ the data for fission neutrons (the data for the region of the closed shell A = 125 - 135 are excluded from comparison)

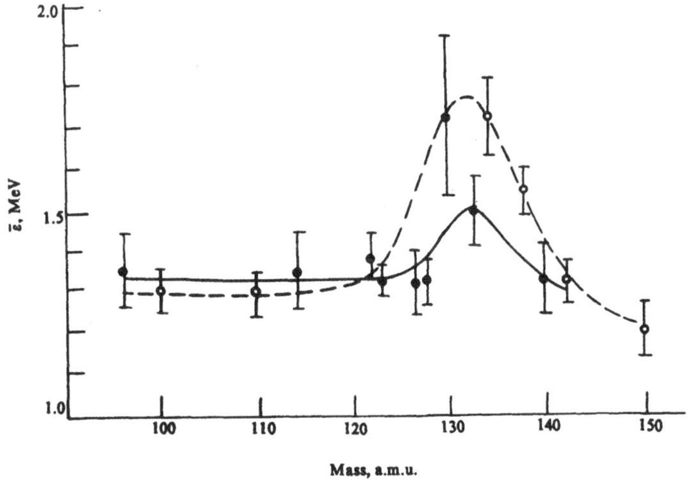

Fig. 8. Average neutron energies as a function of the nucleus (fragment) mass[25]

o- the data for fission neutrons

•-the data for neutrons emitted from interaction of 14 MeV neutrons with nuclei

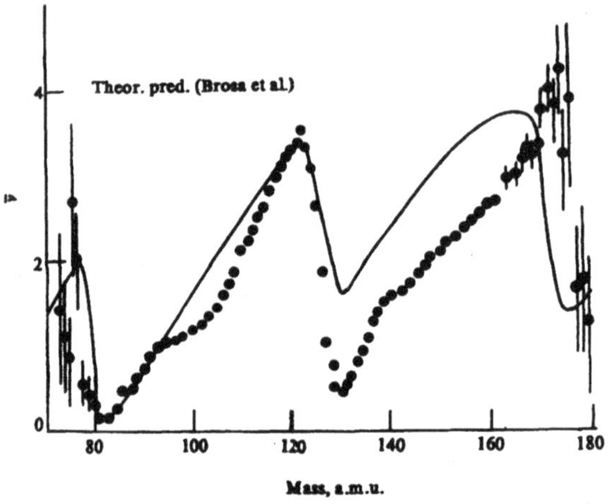

Fig. 9. Average neutron multiplicity in dependence on the fragment mass (spontaneous fission of ^{252}Cf)
points:experimental data of Ref. [27]
Curve:calculated data of Brosa[28]

When the excitation energy becomes larger, the minimum in the region A = 132 disappears gradually and the curve becomes smooth. Its slope has to be determined by the condition of the equal temperatures of both fragments at the scission point. Therefore it is reasonable that the multiplicity would be proportional linearly to fragment masses. But a number of studies show that it is not right. The slope is much larger than that from this linear dependence on A. One of the examples is the fission of ^{238}U induced by protons with 150 MeV energy[21] and the other, induced by carbon with energy 105 MeV[14]. In both cases the slopes are much larger than the slope calculated from the equation $T_1 = T_2$. In these literatures the various hypothesises were proposed and the reasons of this effect were given. To my opinion this effect is connected with the great contribution of the deformation energy to the total excitation energy of fragments. This changes essentially the slope of the curve.

As the fission fragment excitation energy reaches 40 - 50 MeV, the equilibrium process of the nuclei becomes unique, and the study of pure statistical properties of highly excited fragments becomes possible. Fig. 10 shows that even the fragment excitation energy E^* is only 20 MeV in fission neutron spectrum, there is no hard non-equilibrium neutron component[17].

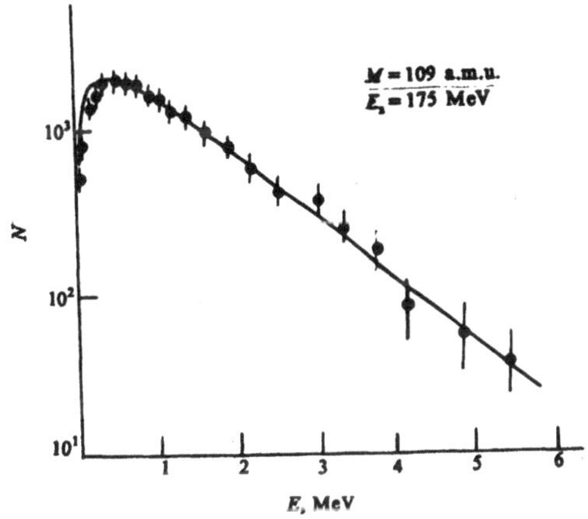

Fig. 10. Fission neutron spectrum in CMS for fragment mass M= 109 amu and the excitation energy E^{*} = 20 MeV[17] (spontaeous fission of ^{252}Cf)

The behaviour of the level density of fission fragments in dependence on their mass number was studied in refs. [17], [27], [29], and on their excitation energy in ref. [17]. In ref. [17] the essential low value of the parameter "a" in the region of magic nuclei with A = 132 and the rise of the parameter "a" with the decrease of the total kinetic energy E_k were obtained. These data showed the essential rise of excitation energy of fragments in the scission point with the decrease of E_k.

It is necessary to be stressed that fission neutrons give the unique possibility to study the behaviour of neutron inverse cross-section (the cross-section of the capture of neutron by the excited nucleus) based on the energy of the incident neutrons. The sensitivity of such determination in low energy neutrons is sufficiently high. In Radium Institute the first results of the determination of the behaviour $\sigma_{inv}(E_n, E^{*})$ for spontaneous fission fragments of ^{252}Cf were obtained just recently.

In the end of my report I would like to attract your attention to the problem of the determination of the lifetime of highly excited nuclei. It is known that the methods of the determination of the lifetime of the equilibrium excited nuclei for excitation energies 10 - 50 MeV are absent now. The study of angular and energy distributions of fission neutrons gives the possibility to determine the lifetime of the excited fragments up to 40 - 50 MeV excitation energy. Such type of the study is carried out now in the Radium Institute.

4. Conclusion

Thus it can be said that in recent years the essential progress was achieved on the problem of fission neutron emission study. The recent achievements in the experimental techniques give a new possibility to study to fission process and neutron emission from excited nuclei.

For the "open" questions of the fission neutron emission mechanism, it is desirable to solve the problems of fission dynamics in the nearest future, and the neutron emission in non-equilibrium processes near scission point, the influence of neutron emission during fragment acceleration, physical basis for the behaviour of the excitation energy in scission point (the slow dependence on the initial excitation energy of nucleus), the partition of the excitation energy between fragments, the statistical properties of fission fragments and the lifetimes of highly excited fragments are possible to get progress in near future. Of course, this is only a incomplete list of the important questions to be investigated.

For practical goals, it is necessary to have a precision information on multiplicities and spectra of neutrons for fission of Pu, Am, Np and Cm isotopes. That is important for the problem of the transmutation of heavy nuclides accumulated in the reactors.

References

1. W. Mannhart, *Physics of neutron emission in fission* (Proc. of the IAEA meeting, Mito, Japan. Vienna IAEA 1989) p. 305
2. H. Märten et al., *ibid* p. 245
3. D. Madland et al., *ibid* p. 259
4. B.G. Gerasimenko, V. A. Rubchenya, *ibid* p. 283
5. J. W. Boldeman, *ibid* p. 47
6. D. Madland, *Nuclear data for neutron emission in the fission process* (Proc. of the IAEA meeting, Vienna, Austria, 1990, to be published)
7. H. Märten, *ibid*
8. G. S. Boykov et al., *Nuclar Physics* (in Rus.) **53** (1991) 628
9. G. V. Lovchikova et al., *Nuclear constants* (in Rus.) **4** (1990) 17
10. G. V. Lovchikova et al., *Proc. of the Intern. Conf. on 50 Anniversary of nuclear fission* (Leningrad, USSR, 1989, to be published)
11. Li Anli et al., *Nuclear data for neutron emission in the fission process* (Proc. of the IAEA meeting, Vienna 1990, to be published)
12. M. Baba et al., *Physics of neutron emission in fission* (Proc. of the IAEA meeting, Mito, Japan. Vienna IAEA 1989) p. 149
13. J. Frehaut et al., *Proc. of the Intern. Conf. on Nuclear data for science and technology*(Antwerp, Belgium, 1982. D. Reidel Publ. Comp. Endhoven, 1983) p. 78

14. M. V. Blinov et al., Preprint *JINR* (1989) p-7-89-227
15. H. Rossnor et al., *Phys. Rev. C*, **V.40** (1989) 2620
16. D. Hilscher et al., *Proc. of the Intern. school seminar on heavy ion physics* (Dubna, USSR, 1989) *JINR* 1990
17. O. I. Batenkov et al., *Atomic Energy* (in Rus.) **4** (1988) 429
18. V. P. Eismont, *Atomic Energy* (in Rus.) **19** (1965) 113
19. K. Skarsvag, *Phys. Scripta* 7 (1973) 160
20. G. A. Pik-Pichak, *Nuclear Physics* (in Rus) **10** (1969) 321
21. E. Cheifetz et al., *Phys. Rev.* **2** (1970) 256
22. H. Märten and D. Seeliger, *J. Phys. G. Nucl. Phys.* **V.14** (1988) 21
23. O. I. Batenkov et al., *Proc. of the Intern. Conf. on 50 Anniversary of nuclear fission* (Leningrad, USSR, 1989, to be published)
24. O. I. Batenkov et al., *Nuclear data for science and technology* (Proc. of the Intern. Conf., Mito, Japan, 1988) *JAERI* (1989) P. 79
25. O. I. Batenkov et al., *Nuclear data for science and technology* (Proc. of the Intern. Conf., Julich, Germany, 1991, to be published)
26. Y. Nakagone et al., *Physics of neutron emission in fission* (Proc. of the IAEA meeting, Mito, Japan, 1988, Vienna, 1989), p. 65
27. C. Budtz-Jorgensen and H. H. Knitter *ibid* p. 181
28. U. Brosa et al., *Z. Naturforschung*, **V. 41a** (1986) 1141
29. M. Samant et al., *Nuclear data for neutron emission in the fission process* (Proc. of the IAEA meeting, Vienna, Austria, 1991, to be published)

MEASUREMENT OF THE PROMPT NEUTRON SPECTRUM OF ^{238}U FISSION

LI ANLI, BAI XIXIANG, WANG XIAOZHONG,
MENG JIANGCHEN, LIU WEIPING, YUAN YUAN,
HUANG SHENGNIAN and CHIH TAHAI
China Institute of Atomic Energy
P. O. Box 275-46, Beijing 102413, China

ABSTRACT

The prompt neutron spectrum of ^{238}U fission induced by 10 MeV neutrons was measured using a multi-segment fission chamber and a two-arm TOF spectrometer. The flight time of incident neutrons was event by event recorded. Use of coincidence between the fission fragments and neutrons enabled the fission neutrons to be distinguished from other secondary ones. The events induced by 10 MeV neutrons were software-selected by gating the 10 MeV peak of incident neutron TOF spectrum.

The measured spectrum can be roughly described by the Maxwellian distribution with T equal to 1.38 ± 0.04 MeV.

1. Introduction

The knowledge of the neutron spectrum of ^{238}U fission and its dependence on incident neutron energy is of great importance for designing fast neutron devices such as fast breeder reactor and fusion-fission hybrid reactor. It is also interesting for the study of fission mechanism. There are no experimental data in the incident energy range 8 to 13 MeV because of lack of monoenergetic neutron source. In the case of incident energy greater than 6 MeV, second chance fission (n, n'f) might occur which perhaps has an observable influence on fission neutron spectrum. The present measurement at E_n = 10 MeV should be helpful to verify the theoretical models.

2. Experimental Arrangement

The measurement has been carried out at HI-13 Tandem Van de Graaff Accelerator Laboratory of CIAE. The experimental arrangement is shown in Fig. 1. The primary neutrons of 10 MeV with breakup neutrons were produced at 0° direction via the D(d, n) reaction. A pulsed deuteron beam of 7.4 MeV, with a pulse width of 1.0 ns, a repetition rate of 4 MHz and an average beam current of 1.3 μA impinged on a deuterium gas cell at pressure of 4.3 atm.

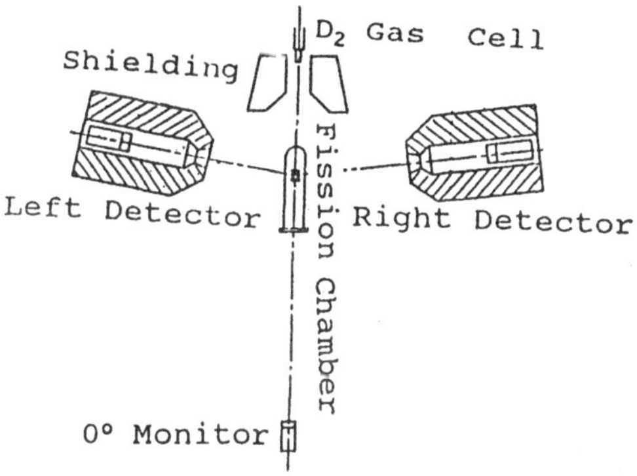

Fig. 1. The experimental arrangement

The natural uranium has been both-side electro-deposited on the 0.2 mm thick plates of stainless steel with deposition thickness of about 0.5 mg/cm^2. 103 plates in all were loaded into a parallel-plate ionization chamber in spacing of 1.5 mm. The total amount of the deposited natural uranium was about 5 g. The chamber operated in multi-segment mode and filled with pure methane at pressure of 4.3 atm. The distance between centres of the gas cell and the chamber was 64 cm.

The fission neutrons were detected by two well shielded liquid scintillators of 25 cm in diameter by 5 cm thick. The detectors were placed at $\pm$ 100° to the incident neutron beam. Time of flight technique was used with flight path of 2.3 m for both detectors. A small liquid scintillator located at 0° and 3.1 m downstream of the gas cell served as a monitor of incident neutron TOF spectrum.

Use of coincidence between the signals from fission fragments and those from neutron detectors enabled the fission neutrons to be distinguished from other secondary ones. The timing output of fission chamber and pick-off signal of pulsed beam were fed into a TAC to measure the incident neutron TOF spectrum. Events induced by 10 MeV neutrons were finally selected by using coincidence between software-gated peak of the 10 MeV neutrons and the TOF spectrum of secondary neutrons detected with neutron detectors, while events from breakup neutrons were rejected.

As mentioned above, the background was suppressed in a great measure, a small amount of it, however, still leaked in via random coincidence. Such a random coincidence TOF spectrum was measured by the following means. The timing output of fission chamber was delayed for some integer cycles of pulsed beam and then fed into the same TAC used to measure fission neutron TOF spectrum. The TOF spectrum obtained in this operation reproduced fairly the random coincidence in the same cycle. Fig. 2(A) shows the measured TOF spectra, the real plus random and the random alone. Actually both of them were measured simultaneously and routed into different areas of computer memory. As can be seen from the figure, a peak of elastic scattering neutrons leaked in and distorted the

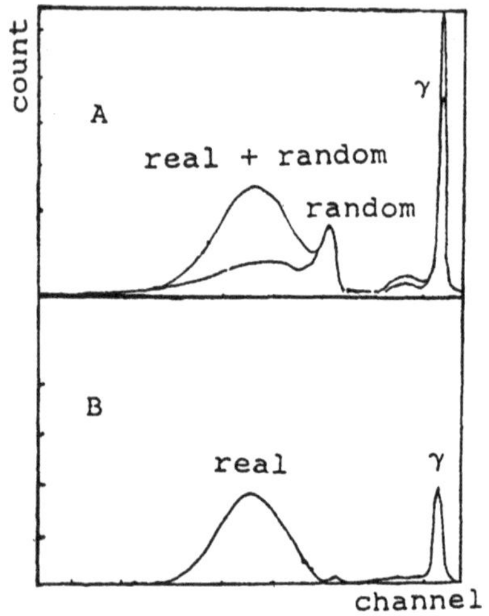

Fig. 2. Fission neutron TOF spectrum at low bias
(A) real + random and random alone
(B) real prompt fission neutron TOF spectrum after subtracting the random coincidence background

fission neutron spectrum. The peak disappeared after subtracting the random coincidence, as shown in Fig. 2(B).

The schematic diagram of electronics is shown in Fig. 3. The 103 plates were grouped into eight electro-separated chambers in order to reduce the pile-up of interplate capacitances and make rise time of fission fragment pulses fast. Each chamber was connected to individual preamplifier, fast filter amplifier and constant fraction discriminator. The eight timing outputs were mixed together for enabling eight singles of TOF spectrum to be measured in a single ADC, while these outputs were fed into an input register of computer interface to distinguish which chamber the

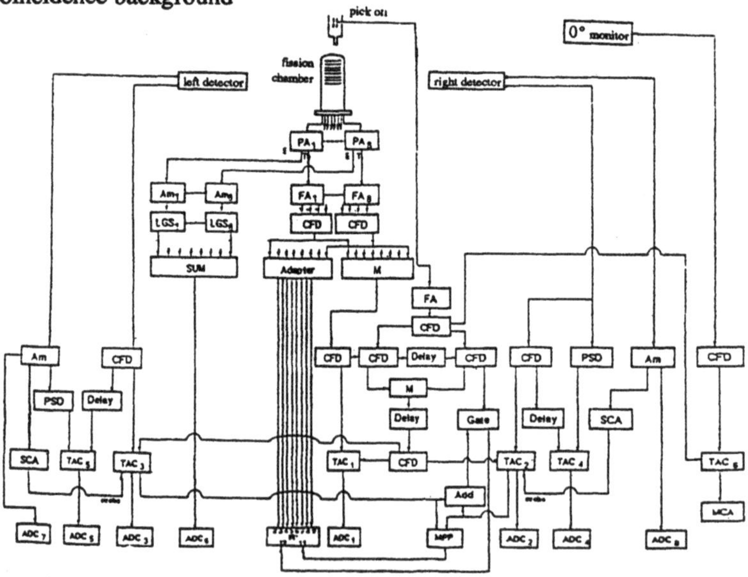

Fig. 3. The block diagram of electronics

fission event belongs to.

Two biases at Ep = 0.65 and 1.65 MeV were set for each neutron detector by electronics and software respectively. The higher was aimed at upgrading the ratio of effect to background in higher energy region of spectra. Moreover, the pulse shape discrimination of neutrons from gamma rays was used for eliminating γ ray background. The efficiency response of neutron detectors was determined by combining Monte-Carlo calculation in standard code NEFF4 with experimental calibration using a low-mass ^{252}Cf ionization chamber. The measured spectra at two biases connected with each other spontaneously. The available energy range in the present experiment is 0.75 to 11.5 MeV.

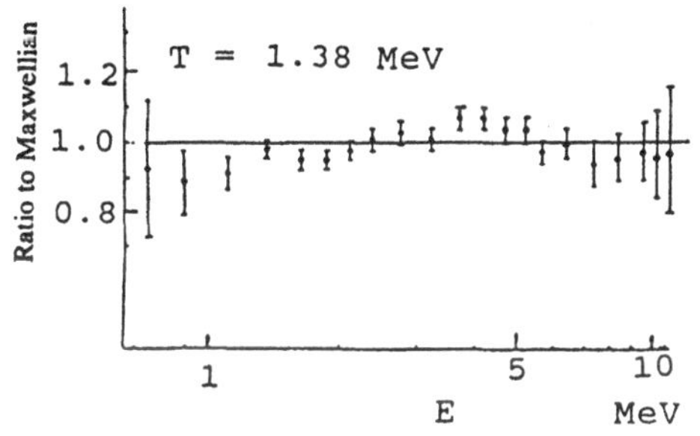

Fig. 4. The fitting result. Points are experimental data

3. Result and Discussion

The measured spectrum of fission neutrons can be roughly described by the Maxwellian distribution. The optimum fitting yields

$$T = 1.38 \pm 0.04 \text{ MeV} \quad \text{or}$$

$$\bar{E} = \frac{3}{2}T = 2.07 \pm 0.06 \text{ MeV}$$

as shown in Fig. 4. The error refers to statistic uncertainty and that from determination of detection efficiency. The theoretical calculation predicts that the average energy increases slowly as incident neutron energy goes up. The present measurement in comparison to previous data[1] at different incident energies shows such a tendency, as can be seen from Fig. 5.

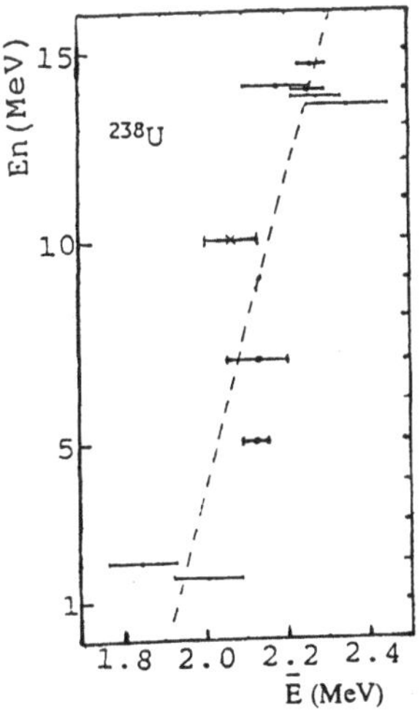

Fig. 5. The comparison of present result with previous data[1]
◆ previous data
✱ present result

In principle, there might appear two Maxwellian components in the present measurement corresponding to (n, f) and (n, n'f) reactions respectively because the post-fission neutrons were emitted from fragments having different excitation energies. Unfortunately the two components can not be clearly separated, since accuracy of our data is not good enough. The fitting Maxwellian temperature here is thus a weighted average of two components. The theoretical calculation of spectrum based on standard nuclear evaporation model will be presented later on.

Reference

1. André BERTIN, Robert BIOS, and Joël FREHAUT, *Rapport-R-4913*, **Juin** (1978)

STUDY OF PROMPT NEUTRON SPECTRUM IN LOW ENERGY PORTION FOR ^{252}Cf AND ^{248}Cm SPONTANEOUS FISSION

BAO SHANGLIAN
Institute of Heavy Ion Physics,
Peking University, Beijing 100871, P. R. China

ABSTRACT

Measurements of integral prompt neutron spectrum of ^{252}Cf spontaneous fission (SF) and a relative prompt neutron spectrum of ^{248}Cm to ^{252}Cf SF in low energy portion were carried out in two separated experiments. The results of the first experiment shows the neutron yield above 150 KeVwas consistent with Maxwellian distribution with T = 1.42 MeV. The result of the second experiment shows that the neutron component below 500 KeV is systematically higher than Maxwellian distribution.

1. Introduction

The fission neutron spectrum of ^{252}Cf spontaneous fission (SF) is a standard. But the spectrum in low energy portion is still not precision enough[1-4] and the existing experimental data are systematically higher than calculated data by various models[5]. The knowledge of SF neutron spectra for other heavy elements is very poor and is interesting both for studying the process of heavy nucleus fission and for practical requirement concerning the accumulation of heavy elements in nuclear reactors.

Up to now, the mechanism of neutron emission in nuclear fission is not fully understood. Existing results are contradictory[5]. Therefore the information on energy and angular distribution of fission neutrons correlated with fragment characteristics is very useful to understand the fission and neutron emission mechanism.

2. Description of the Experiments

2.1. Measurement of ^{252}Cf

The detailed description of the experiment can be found elsewhere[7,8]. The experiment was performed in a new neutron hall, with a space of 20 × 12 × 8 m^3. The neutron detector is an NE912 lithium glass detector (45 mm in diameter and 9.55 mm in thickness) matched with RCA8850 phototube and a same size ^{7}Li glass detector was used to estimate the gamma-ray background. The fission fragment detector is a mini-ion chamber (total weight is 1.6 g), which was hung up in the middle of the hall. The ^{252}Cf source was made by means of transfusion in vacuum on a highly polished stainless steel disc and had intensity of 3.8×10^3 f/s.

The neutron detector efficiency was calculated using a Monte Carlo code[9]. The neutron multiple scattering effect was considered through the prompt neutron detector efficiency. The result was compared with A. Lajtai's relatively measured data[1,10] for the detector of same size. The result showed that except the three points in the resonance peak of the $^{6}Li(n, \alpha)$ reaction and some systematic deviation in very low energy portion the consistence was good.

Some corrections in the measurement were considered: (i) the scattering on preamplifier was measured: 1%; (ii) the scattering on air column between the two detectors was estimated by a Monte Carlo calculation: 0.5%; (iii) the scattering on the source backing was measured: <1%; (iv) the scattering in the environment was measured by a shadow bar; (v) the correction for delayed gamma-rays was calculated in the time range 30 -100 ns.

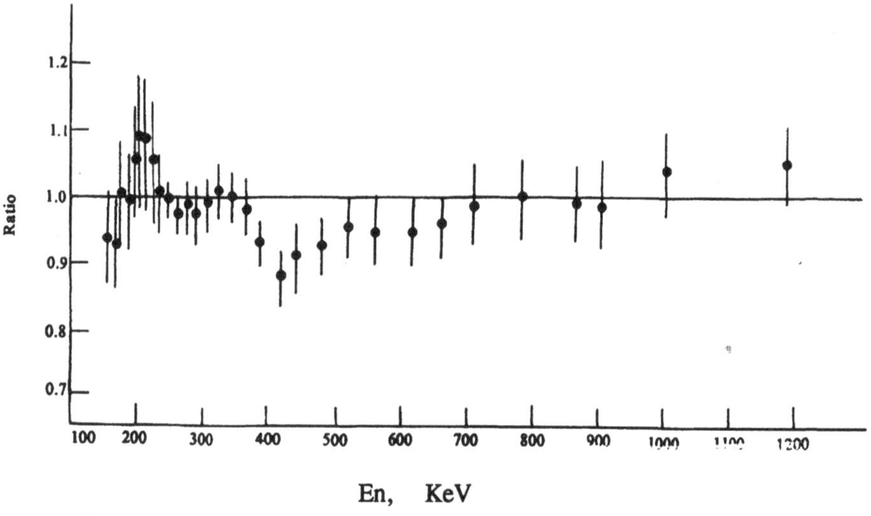

Fig. 1. The Ratio of the experimental data to the Maxwellian distribution with T = 1.42 MeV

After all the corrections the result below 1 MeV is shown in fig. 1. The uncertainties included are: (i) the uncertainty of statistics: 4% (ii) the uncertainty of detector efficiency calculation:6%; (iii) the uncertainty caused by transformation from time to energy scale: 0.66%; (iv) the uncertainty caused by the delayed gamma-ray correction: 2%.

2.2. *The Relative SF Neutron Spectrum Measurement of ^{248}Cm to ^{252}Cf*

The measurement was carried out in V. G. Khlopin Radium Institute as a cooperation program[6], which was done with a multiple parameter system. The fission layers of ^{252}Cf on one side and ^{248}Cm on other side with a 0.1mm platinum disk as the backing were positioned in the center of a thin stainless steel spherical shell(0.3mm). The intensities were 10^5 f/min for ^{252}Cf and 2×10^4 f/min for ^{248}Cm. Both sides of the source were covered by Al_2O_3 films in thickness of 40 $\mu g/cm^2$. The micro-channel plates were used as the starting and stopping signals for fragments from ^{248}Cm and ^{252}Cf, respectively. The fission fragment detectors were the surface barrier detectors (SBD), which located 10 cm far from the source. The SBD determine the energy and emission angle of the fission fragments. The two stilbene neutron detectors were symmetrically arranged and located 15, 30, and 60 cm away from the source with various angles relative to the source plane. The n-gamma discrimination was used. The time resolution of TOF spectrometer was 1 ns. The prompt SF neutron spectrum integrated from all fragments at 0° relative to the normal of the source plane is shown in Fig. 2.

The data of ^{248}Cm were treated relative to that of ^{252}Cf. Therefore the systematic uncertainty was greatly reduced. In order to check the results, the source was turned by 180° and the results were well consistent (2-8%). From Fig. 2 one sees that the neutron yield below 500 KeV is systematically higher than the Maxwellian distribution.

3.Discussion and Conclusion

^{252}Cf SF neutron spectrum is the most important SF neutron spectrum of heavy elements and has been studied in detail. But the integral spectrum below 0.5 MeV is still not precision enough to determine the shape so well (uncertainties about 10%)[1-4]. The discrepancy between the experiments and various model calculations below 0.5 MeV under the assumption that all neutrons emitted from full accelerated fission fragments[12,13,15] was large (about 10%).

For the discrepancy there are various explanations. One of them can be found in reference [17], which was a differential neutron spectrum measurement of ^{252}Cf. The authors assumed that the most part of the excess neutrons below 0.5 MeV could be the scission neutrons. But some arguments exist. The most arguments are caused by the uncertainties in experiments and theoretical

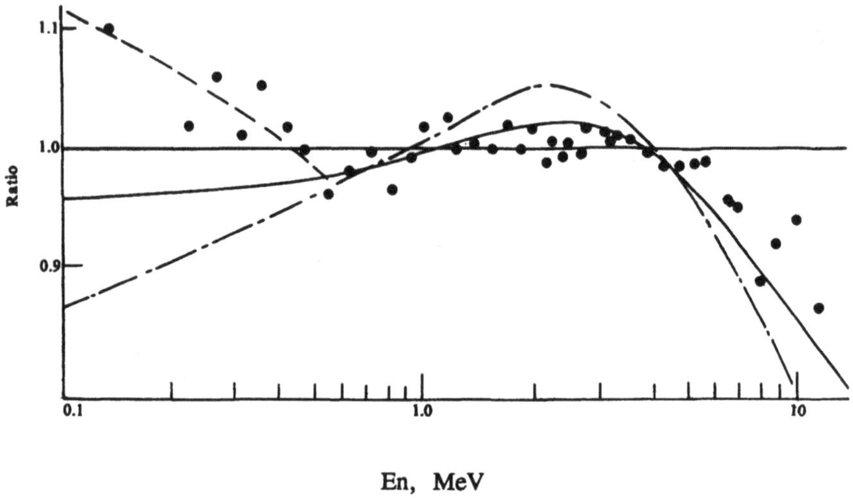

Fig. 2. The deviation of the integral neutron spectrum for ^{252}Cf and ^{248}Cm from Maxwellian distributions

- • The data from the experiment of ^{248}Cm (T=1.38MeV)
- – The evaluation data[11] for ^{252}Cf
- –•– The results of theoretical calculation for ^{252}Cf[15,18]

calculations. Some experiments lead to the conclusion that the scission neutrons are very rare[20]. Some others estimated the scission neutrons which were no more than 5%[5]. Therefore the question of scission neutrons is not firmly determined up to now. From the fact the conclusion is clear: More careful experimental and theoretical study should be performed to make sure whether the discrepancy in measured low energy spectrum and theoretical models based on neutron emissions from fully accelerated fission fragment is firmly established or not.

The result from relative measurement of ^{248}Cm to ^{252}Cf showed the much higher component below 0.5 MeV, which is a support for fiction scission neutrons. A checking measurement and more careful data analysis are needed. From the experiment one knows that the detailed and precision correlation measurement between various quantities correlated with fission fragment and neutron emission will be very helpful to elucidate the neutron emission mechanism, which is one of the important process after saddle point. And the differential measurement of SF neutron spectra for other heavy elements is absolutely necessary.

Up to now, measurement and direct theoretical calculation are very difficult for so called scission neutrons, which may be considered as followings:

(1) Neutrons emitted after saddle point but before scission point. Calculations for the neutrons are very difficult.

(2) From ternary fission (neutrons emitted during fission), only the neutrons from decay of ^{5}He can be determined, which has very low intensity[18]. The direct ternary fission by neutron emission has not been studied.

(3) Neutrons emitted during acceleration of the fission fragments. These neutrons are difficult to be separately measured and also no satisfactory theory exists. Some theoretical estimation[19] showed that dynamically changing fragment forms by one body dissipation exhibit large uncertainty, which will easily smear out the fiction scission neutrons since the input data have very large uncertainties.

References

1. A. Lajtai, P. P. Dyachenke, L. S. Kuzayva, V. N. Kononov, P. A. Androsenko and A. A. Androsenko, *Report INDC(NDS)-46/L* (1983) 177.
2. M. V. Blinov, G. S. Boykov and V. A. Vitenko, *Proc. Int. Conf. on Nuclear Data for Science and Technology*, (Antwerp, Belgium, September, 1982).
3. W. P. Ponitz and T. Tamura: *ibid* (1983) p. 465.
4. J. W. Boldeman, D. E. Clancy and D. Culley, *Nucl. Sci.* Eng. **93** (1986) 131.
5. H. Marten, D. Richter and D. Seeliger, *Proc. IAEA Consultants Meeting on Physics of Neutron Emission in Fission,* (Mito, Japan, May, 1988.)
6. I. O. Batenkov, M. V. Blinov and Bao Shanglian, Proc. Int. Conf. Nuclear Data for Science and Technology, Julich, Germany, May, 1991.
7. G. Y. Tang, S. L. Bao, J. Wang, W. G. Zhong, Y. L. Li, Z .M. Shi, F. Z. Huang J. X. Chen, J. C. Meng, A. L. Li, Z. Y. Bao and S. N. Huang, Proc. Int. Conf. on Nuclear Data for Science and Technology, Mito, Japan, May, 1988, p. 755.
8. S. L. Bao, G. Y. Tang, J. H. Ran and W. T. Cao, *50 years with Nuclear Fission*, (Edited by James W. Behrens and Allan D. Carlson, Published by the American Nuclear Society, Inc. La Grange Park, Illinois, USA), P. 951.
9. G. Y. Tang, S. L. Bao, Y. L. Li and W. G. Zhong, Institute of Heavy Ion Physics, Peking University, *Report CNIC-00316, PU-0003*(CNIC, Beijing, 1988).

10. A. Lajtai, Monte Carlo calculation for Thick Glass Detector, Private Communication, May, 1987.
11. W. Mannhart, IAEA-TECDOC-410, June, 1986, p. 158.
12. D. G. Madland and J. R. Nix Nucl. Sci. Eng. 81(1982)213.
13. H. Marten and D. Seeliger, Proc. Int. Conf. on Nuclear Data for Basic and Applied Science, Santa Fe, May, 1985.
14. H. Marten and D. Seeliger, *INDC(DGR)-30/L*, 1984.
15. B. F. Gerasimenko and V. A. Rubetchenya, *Neutron Physics*, Vol. 1,(Moscow, 1984) p. 349.
16. U. Brosa, S. Grossmann and A. Muller, *PEPLCM* **197** (1990) 167-262.
17. O. I. Batenkov, A. B. Blinov, M. V. Blinov and S. N. Smirnov, IAEA Consultants Meeting on Physics of Neutron Eimission in Fission, Mito, Japan 1988, p.25.
18. E. Chefetz, B. Eylon, E. Fraenkel and A. Gavron, *Phys. Rev. Lett.* **29** (1972) 805
19. H. Marten and D. Seeliger, *J. Phys.* **G10** (1984) 349.
20. C. Budts-Jorgensen and H.-H. Knitter, *J. Nuclear Physics* 1989.

BEHAVIOUR OF PROMPT GAMMA EMISSION IN COLD FRAGMENTATIONS OF BINARY FISSION OF ^{252}Cf

HAN HONGYIN, LI ANLI, CHEN JINHUA
AND HUANG SHENGNIAN
China Institute of Atomic Energy,
P. O. Box 275-46 Beijing 102413, P. R. China

ABSTRACT

The property of the gamma-ray emission from the cold fragmentations of ^{252}Cf spontaneous fission was studied in a four parameter correlation experiment.The kinetic energies and the emission angles of fragments were measured using a twin gridded ion chamber, while a NaI(Tl) detector was employed to measure the gamma-ray energy spectra. This paper presents the first observed results about the gamma-ray emission from the fragment pairs with excitation energy less or higher than 6 MeV.

1. Introduction

So called cold fission is a kind of rare fission events in which the total kinetic energy of fragment pairs (TKE) is close to the fission reaction Q-values. Therefore the difference between the cold fission and the normal one is there is no neutron emission in the cold fission. The investigation on the cold fission will provide us useful knowledge including:

(1) pairing, shell and liquid drop effects in fragments;
(2) ground state deformations of nuclei in a large mass range covered by the fragments;
(3) mechanism of gamma-ray emission and neutron emission onset in fission process as excitation energy increases from nearly zero to higher values.

2. Experiment

A twin gridded ion chamber was employed to measure the kinetic energies of fragment pairs and the fragment emission angles with respect to the axis of the cylindrical chamber. A NaI(Tl) detector in size ϕ10cm × 10cm was accepted to measure γ-ray spectra. Angles subtended by the NaI(Tl) detector from the fission source was 18 degree. Fission source was prepared using the transfer method on the carbon foil with thickness of 20 μg/cm^2. The strength of source was 65 fiss./s.

Data aquisition was completed on VAX-11/780 computer. The signals from the two anodes and the one grid of the twin ion chamber as well as the signal from the NaI(Tl) detector were recorded event-by-event on the tapes.

In order to reduce the influence of pulse pile-up for investigation of cold fission process with TKE values close to the fission reaction Q-values, a pulse pile-up rejector was accepted. calibration and the response function of γ-detector were measured using the lines of a series of the γ-ray sources.

3. Analysis of data

1) Neutron background was corrected for measured γ-ray spectra for normal binary fission because NaI(Tl) detector is sensitive to neutrons emitted.

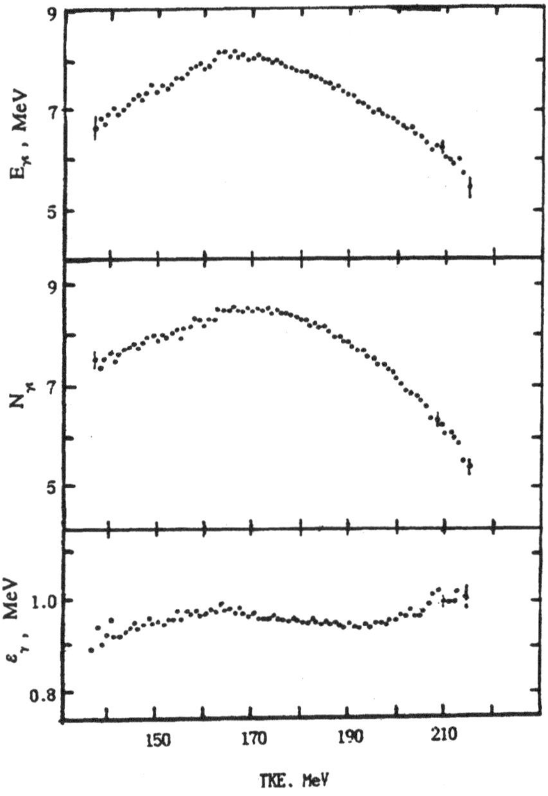

Fig. 1. Average total energy $E_{\gamma t}$, average total number $N_{\gamma t}$ of γ - rays per fission and average energy per photon ε_γ as functions of TKE

(2) Doppler shift in energy of γ-rays emitted from moving fragments was taken into account.

(3) Iterative procedure was used to find the pre-neutron emission total kinetic energy of fragments on the basis of corrections of neutron emission, hight defect of fragment pulses and fragment energy losses in the backing.

(4) `Weighting method' proposed by Meier-Leiblitz[1] was employed to determine the total energy $E_{\gamma t}$ and the total number $N_{\gamma t}$ of any γ-spectrum. Weighting factors came from the measured respone function. In the way, the average energy per photon ε_γ can be expressed dirctly,

$$\varepsilon_\gamma = E_{\gamma t}/N_{\gamma t}$$

(5) Jϕrgensen's experimental results[2] show that no neutrons can be emitted from light fragments or heavy ones if TKE value of a fission event is higher than TKELmax- or TKEHmax-value of the corresponding mass split. From the criteria, the cold fission events or the events with neutron emissions from one fragment only were selected.

(6) Qmax(A_H)-values were calculated using the mass tables given by Möller and Nix's paper[3].

4. Experimental results and discussion

(1)*Normal binary fission events*

(A)Average energy per photon ε_γ, average total number $N_{\gamma t}$ and average total energy $E_{\gamma t}$ of γ-rays per fission as functions of TKE are given in fig. 1. In the case of TKE-value higher than 160 MeV, variation of $E_{\gamma t}$ with TKE is in agreement with Nifenecker's results[4], the latter was obtained using large liquid tank. It has been found that $E_{\gamma t}$ decreases as TKE goes down from 160 MeV to 140 MeV. It must be pointed that in Jϕrgensen's experiment of ^{252}Cf spontaneous fission[2], the variation of neutron multiplicity with TKE deviates downwards from the fitting straight line and the deviation becomes greater and greater when TKE goes down from 160 MeV to 140 MeV. These experimental facts imply that the variation of measured $E_{\gamma t}$ with TKE

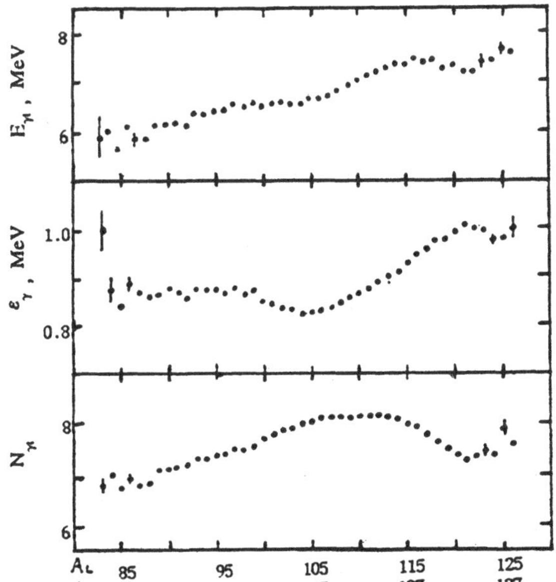

Fig. 2. Average energy per photon ε_γ, average total number $N_{\gamma t}$ and average total energy $E_{\gamma t}$ of γ-rays per fission as functions of mass number of light fragment or heavy one

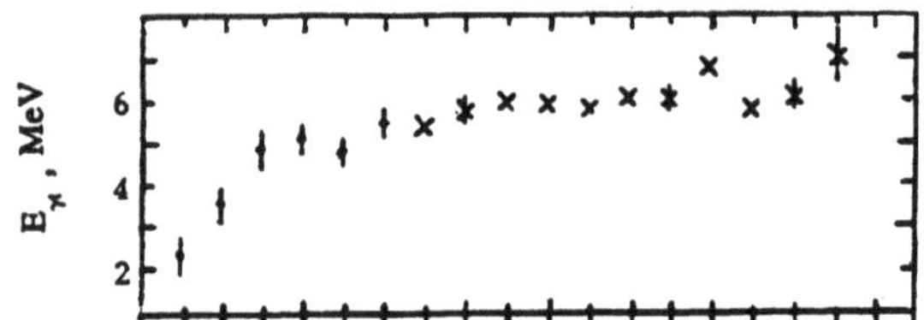

Fig. 3. Average total number $N_{\gamma t}$ and average total energy $E_{\gamma t}$ of γ-rays per fission and average energy per photon ε_γ as functions of excitation energy E_x = Qmax-TKE

┼┼┼ E_x less than 7 MeV, event without neutron emission.

✱✱✱ E_x greater or equal to 7 MeV, event only with neutron emission from heavy fragments.

may come from the unknown reason.

(B) A pronounced peak occurs in the average energy per photon ε_γ at heavy fragment mass number A_H = 132, and the minima occur in both $N_{\gamma t}$ and $E_{\gamma t}$ at the same place A_H = 132, see fig. 2. These facts are explained on the double closed shell in the heavy fragments A_H = 132.

(2) *Cold fission events*

(A) It was found that both $N_{\gamma t}$ and ε_γ increase as excitation energy E_x = Qmax-TKE goes up from zero to 6 MeV, while $E_{\gamma t}$ increase with excitation energy because $E_{\gamma t}$ is product of $N_{\gamma t}$ and ε_γ, see fig. 3. The increase of $E_{\gamma t}$-value with excitation energy in the case of E_x less than 6 MeV is expected becuase no neutrons emit and all excitation energy will release in the form of γ-rays.

Variation of $E_{\gamma t}$ with excitation energy E_x is determined by increments of both $N_{\gamma t}$ and ε_γ and is in discrepancy with that observed in neutron induced fission reactions. Supposing that the gamma-ray spectra in fission reactions are the gaussion distributions and using neutron multiplicity data, J. Frehaut[5] deduced that increment of $E_{\gamma t}$ with increasing E_x is caused mainly by variation of ε_γ with

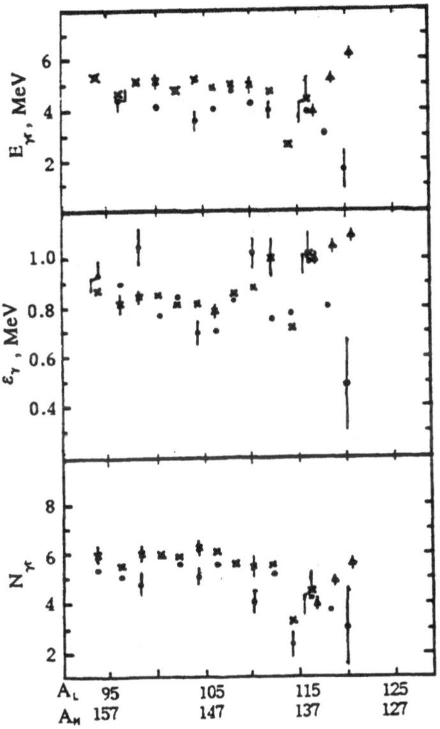

Fig. 4. Average total number $N_{\gamma t}$ and average total energy $E_{\gamma t}$ of γ-rays per fission and average energy ε_γ per photon as functions of mass number of light fragments or heavy ones at fixed excitation energy Ex=Qmax-TKE.

⸸⸸⸸ Ex less than 7 MeV, cold events without neutron emission.

⁑⁑⁑ Ex greater than 7 MeV and less than 16 MeV, events only with neutron emission from the heavy fragments.

⸸⸸⸸ Ex greater than 7 MeV and less than 16 MeV, events only with neutron emission from the light fragments.

excitation energy E_x.

In order to give a tentative explanation for the facts that both $N_{\gamma t}$ and ε_γ increase with increment of excitation energy E_x, we suppose the fragments are good rotators and the radiation is the pure E2 transition. The following relationships hold.

$$E_{\gamma t} \propto J^2 \text{ and}$$

$$N_{\gamma t} \propto J; \ \varepsilon_\gamma = \frac{E_{\gamma t}}{N_{\gamma t}} \propto J$$

where J is the angular momentum corresponding to the E_x value.

The above explanation is only tentative. A good theory is really required. when excitation energy E_x increases from 6 MeV to 16 MeV, the above three physical quntities increase slowly and aprouch the corresponding values in the normal binary fission because neutron emission is possible from the heavy fragments. This is expected on the basis of fission statistic model.

(B) The average total number $N_{\gamma t}$ and the average total energy $E_{\gamma t}$ of γ-rays per fission and the average energy per photon ε_γ as functions of mass number of heavy fragments or light ones are shown in fig. 4. It is found that as excitation energy E_x goes up from zero to 16 MeV, namely fission mode changes from the cold fission into the mode in which neutron emission is possible from heavy fragments, $N_{\gamma t}$ and $E_{\gamma t}$ increase with increasing excitation energy E_x in fragment pairs for the measured mass splits,but variation of ε_γ with E_x is complex.These facts also is useful for understanding of γ-ray emission mechanism in fission process.

References

1. E. Pleasonton et al., *Phys. Rev.* **C6** (1972) 1023
2. C. Budtz- Jφrgensen and H. -H. Knitter, *Nucl. Phys.* **A490** (1988) 307
3. P. Möller and J. R. Nix, *Atomic Data and Nucl. Data Tables* **26** (1981) p. 165
4. H. Niferecker et al., *Nucl. Phys.* **A189** (1972) 285
5. J. Frehaut, *proceedings of IAEA consultants meeting on physics of neutron emission in fission* (organized by IAEA, Mito, Japan, 1988)

RECENT PROGRESS IN FISSION CROSS-SECTION MEASUREMENTS

A. J. Deruytter
Central Bureau for Nuclear Measurements
Joint Research Centre
B-2440 Geel, Belgium

ABSTRACT

Recent progress in fission cross-section measurements is discussed. Special attention is given to the problems associated with the determination of the number of fissile nuclei in the sample and the determination of the efficiency of detection of a fission event. Improvements for precise fission cross-section measurements follow from the detailed biparametric study of fissile layers in the 2π gridded ionisation chambers used for the σ_f measurements. Refined methods for σ_f measurements of nuclides with large λ_α-value are also discussed.

Measurements of thermal fission cross-sections below the fission barrier and the measurement of the low-energy fission cross-section of the isomeric state of ^{235}U are presented.

1. Introduction

The methods for neutron induced fission cross-section measurements depend on the neutron energy, the α-activity of the fissile sample and the nature of the measurement, in particular an absolute measurement at a spot-point energy (e.g. 2200 m/s and 14 MeV) is much more demanding than a relative measurement with respect to a reference cross-section.

In order to measure σ_f at a specified neutron energy the fission rate in a sample with well-known number of nuclei exposed to a well-known neutron flux has to be determined. This requires (a) the production of neutrons with well-know energy or energy-distribution; (b) the determination of the neutron flux at the sample position, (c) the determination of the number of nuclei in the sample, (d) the detection of the fission events and consequently the reaction rate (e) the determination of the efficiency of detection, i.e. the ratio of the number of detected fission events to the total number of fissions induced in the sample.

A detailed discussion of the neutron sources and the determination of the neutron flux is not within the scope of this paper. Recent measurements [1] at very low neutron energies (μeV) at the Institut Laue-Langevin, Grenoble and at very high energies (400 MeV) at the spallation neutron source of Los Alamos [2] indicate the present limits of the accessible energy range. We will limit the discussion to some remarks related to recent fission cross-section measurements.

The detection of fission events is done by detection of the fission fragments, the neutrons or the gamma-rays emitted in the fission process. The detection of prompt neutrons and gamma-rays is used for thick samples. The paper will deal mainly with measurements using thin samples and detection of the fission fragments.

Recent developments are discussed with respect to the points (c), (d) and (e) and they follow often from a refined use of gridded ionisation chambers for the detection of fission fragments. Indeed we discuss the detailed study of thin fissile layers for precise cross-section measurements in gridded ionisation chambers with a 2π-geometry and measurement techniques to obtain cross-sections of nuclides with a large radioactive constant for α-emission.

To conclude two examples of the stringent requirements on neutron beam purity and (or) sample preparation are presented : (1) very small thermal sub-barrier fission cross-sections and (2) the fission cross-section of an isomeric state in ^{235}U.

2. Measurement Techniques

Fission cross-sections are measured by three different methods : by using mono-energetic neutron sources, by spectrum average techniques and by the production of neutrons with a wide energy range (white sources) which are sorted into energy bins by the time-of-flight technique. It is beyond the scope of this paper to discuss the different techniques in detail. Earlier publications deal with the subject, e.g. refs [3], [4] and [5].

Recent developments specially aimed to increase the potential for neutron fission cross-section measurements were :

(1) the installation of a liquid nitrogen (77°K) cooled methane moderator at GELINA (Geel Electron Linear Accelerator) which yields about five times more neutrons below 20 meV compared with the usual water-beryllium moderator at room temperature, enabling measurements of fission cross-sections of the common fissile nuclides down to 2 meV, resulting in more reliable values for the Westcott g_f-factors, and their temperature dependency [6]. The improvement of the fission cross-section below 50 meV is striking, e.g. for ^{241}Pu in ref [7]..

(2) The vertical Very Cold Neutron source at ILL, originating from a 25°K liquid D_2 moderator and covering neutron wavelengths between 20 and 600 Å (with a peak around 80 Å) was used for σ_f measurements in the μeV range. Via a set of vertical and curved neutron guides with excellent reflectivity the neutrons are brought to the experimental area and energy selection is performed with a slow chopper enabling for the first time measurements of σ_f in the neutron energy range from 6-60 μeV [1]. This facility allows to look for the possible solid state effects in this very low energy region.

(3) The Rensselaer Intense Neutron Spectrometer (RINS) consisting of a pure lead cube, 1.8 m on a side which is pulsed by the RPI electron linac provides a very intense neutron field from 1 eV to 80 keV which is used for measuring the fission cross-section of microgram quantities of highly active actinides as well as gram quantities of nuclei with very weak cross-sections [8]. The most difficult measurements were for ^{242}Cm and ^{248}Cm; e.g. ^{242}Cm has a half-life of only 163 days and consequently a very high α-activity ($\sim 10^8$ α decays s^{-1} μg^{-1}) and spontaneous fission background (~ 9 SF $s^{-1} \mu g^{-1}$). A sample of only 1.1 μg of ^{242} Cm was available. The sensitivity of the RINS system expressed as the product of the minimum required sample mass times the sample cross-section was estimated to be 1 microgram-barn. For many nuclides this is the only technique available to measure fission cross-sections

over this energy-range and it also extends the neutron energy range below the 25 eV cut off of the nuclear explosion experiments.

(4) Neutron induced fission cross-section ratios for several nuclides relative to ^{235}U have been measured from 1 to 400 MeV [9] and a measurement of the fission cross-section of ^{235}U in the range from 3 to 30 MeV has been performed [10] at the high-intensity proton spallation neutron source of Los Alamos Scientific Laboratory [11]. This facility allows a very important extension of the available neutron-energy range for σ_f measurements. The ^{235}U (n,f) cross-section was obtained relative to the hydrogen scattering cross-section and the shape data have been normalized to the accurately known value near 14 MeV.

The neutron flux determination is an essential part of the process of fission cross-section determination and the methods available are the associated-particle method, the associated activity method, flat-response counters, the manganese-bath method, and the use of a standard reference cross-section. The Time Correlated Associated-Particle method (TCAPM) is considered as the most precise method of flux determination for fast neutrons and was used in recent absolute measurements of neutron induced fission cross-sections of ^{235}U, ^{237}Np and ^{239}Pu at 5 neutron energies in the energy-range 2 to 20 MeV [12]. These results for ^{235}U are very important for bringing the shape measurements on an absolute basis in this energy range.

3. Detailed study of thin fissile layers in gridded ionisation chambers for the measurement of σ_f

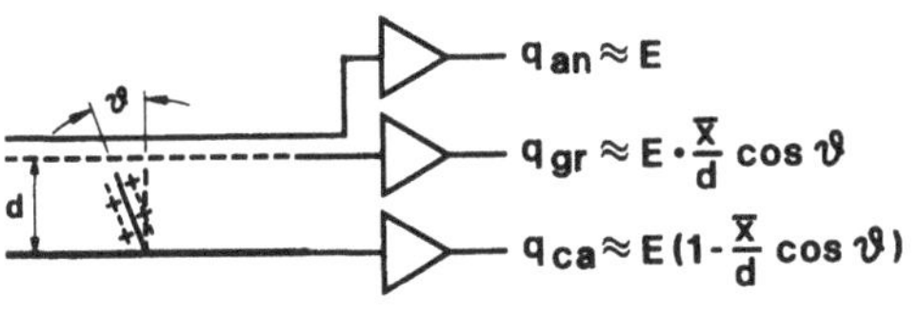

Fig. 1 (a) - Schematic drawing of gridded ionization chamber and signal taking.

The fission cross-section for ^{235}U is requested with an accuracy of 1% for technological and metrological applications in a broad energy range. The techniques for σ_f measurements of ^{235}U based on the absolute determination of masses and detection efficiencies have to be preferred with respect to the shape measurements of σ_f and normalisation at spot point energies.

The gridded ionisation chambers constructed at CBNM allow a precise measurement of the α-activity of a fissile layer and of the efficiency for the detection of fission fragments in a 2π-geometry in the chamber used for the fission cross-section measurements. The methods are described in ref. [13].

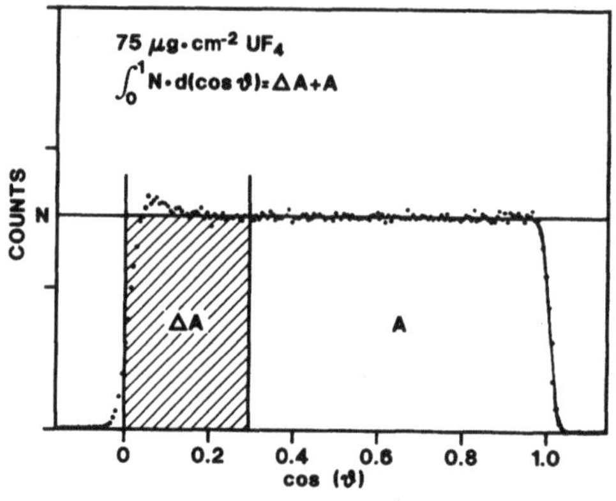

Fig. 1 (b) - The cos Θ distribution measured for a 75 µg. cm^{-2} UF_4 sample.

In a fast parallel-plate ion chamber with a grid inserted between the anode and the cathode, the induced signal on the cathode by an ionising particle depends on cos Θ, the induced signal on the anode only on the energy of the particle. From the two signals one deduces the value of cos Θ, where Θ is the angle of the particle with respect to the normal on the electrodes. Consequently from the values of cos Θ for each particle, the distribution in cos Θ of all particles emitted in the 2π geometry can be deduced. Such a distribution in shown in Fig. 1(b) for a sample of 75 µg cm^{-2} of UF_4, prepared by evaporation. It has a rectangular shape as expected for an isotropic angular distribution. Nevertheless a small peak is visible close to cos Θ=0. This is very likely due to backscattered alpha-particles, which leave the sample with angles close to Θ = 90°. A lower limit for cos Θ for which the effects of backscattering and self-absorption are negligible can be selected. So for cos Θ >(cos Θ)$_{min}$ = 0.3, the height of the plateau, N, has the same value as for the ideal case where no correction for diffusion or self-absorption has to be applied. The α-activity of the layer Φ may then be determined as:

$$\Phi = 2 \int_0^1 N.d(\cos\Theta) = 2.(A + \Delta A)$$

where A is the number of counts above (cos Θ)$_{min}$, and ΔA=N. (cos Θ)$_{min}$is the integrated area from (cos Θ)$_{min}$ till cos Θ = 0. A total estimated systematic error of about 0.2% can be reached, comparable to α-measurements in a low-geometry counter. In the present method the systematic error originates from the detector-associated electronics, and not from the geometrical uncertainties. But it has to be underlined that the detection efficiency for the present method is about one order of magnitude larger than for usual low-geometry counting and is not dependent of the sample shape and thickness inhomogeneities as long as they remain within

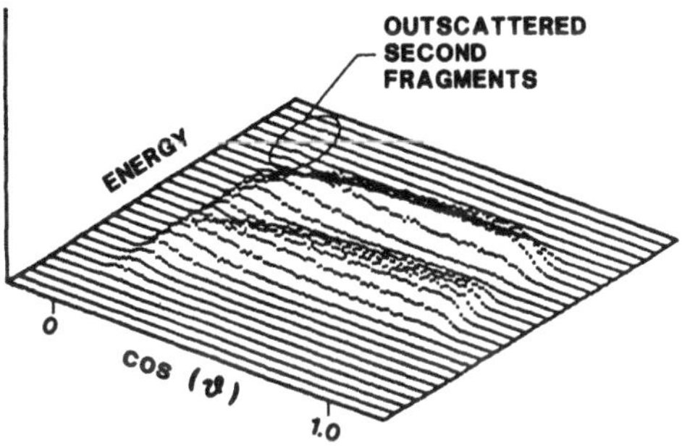

Fig. 2 - The number of fission fragments versus the fragment energy and versus cos Θ with high resolution in the cos Θ parameter, for the 75 μg.cm^{-2} UF_4 sample.

reasonable limits. With the knowledge of the specific α-activity of the fissile material, the mass of the layer can be determined.

This method can also be applied to absolute fission fragment counting. At CBNM [13] measurements were done with UF_4 samples, but also with a sample prepared by electrospraying of a U_3O_8 suspension. The UF_4 samples ranged in thickness from 15 to 500 μg·cm^{-2}. The simultaneous biparametric measurement of energy and angle with respect to the normal on the sample plane (Θ), allows a clear picture of the trajectories of the fission fragments leaving the sample in the ionisation chamber gas. It is illustrated in Fig. 2, where a biparametric distribution is presented of energy versus cos Θ for the same sample of 75 μg·cm^{-2} UF_4 : it is seen that energy degradation of the fragments occurs for cos Θ approaching zero. However in this case there is also a high-energy tail (barely visible on the figure) distributed close to cos Θ = 0. This tail stems from fission events where the fragment initially emitted downward is scattered back into the detector volume where its ionisation charge is added to the ionisation of the unscattered fragment. As in the case of alpha particles a cos Θ value, $(\cos\Theta)_{min}$, can be found such that the fission spectra have no tailing for $\cos\Theta > (\cos\Theta)_{min}$. However, a similar improvement was not found for a 100 μg·cm^{-2} suspension-sprayed U_3O_8 sample. The difference between the two types of samples is illustrated in Fig. 3, which shows the fraction of events falling in the low-energy tail between 10 and 35 MeV as a function of cos Θ. Whereas the tail content for the evaporated UF_4 sample decreases rapidly with cos Θ, becoming negligible for $\cos\Theta > 0.3$, the tail content for the sprayed sample changes weakly with cos Θ, and is not negligible at all even for large values of cos Θ. This means that even fission fragments emitted perpendicular to the sample can loose appreciable energy and can even be completely stopped in the sample. Although the nominal thickness, only 100 μg·cm^{-2}, is small compared to the fission fragment ranges, the sample must contain grains, which individually can have sizes larger than the fission fragment ranges.

The cos Θ distributions for evaporated UF_4 samples can be obtained in a similar way as for α-particles and are shown in Fig. 4. However, due to the shorter range of the fission fragments, it was necessary to increase $(\cos\Theta)_{min}$

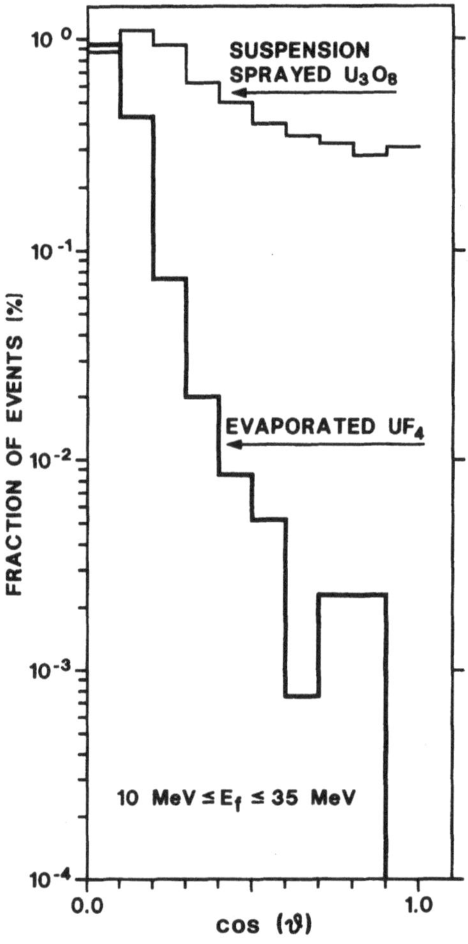

Fig. 3 - The fraction of fission events in percent of the total in cos Θ intervals of 0.1 in the range between 10 and 35 MeV versus cos Θ for the 75 µg. cm^{-2} UF_4 sample and for the 100 µg.cm^{-2} U_3O_8 sample.

up to 0.6 for the thickest UF_4 sample to avoid loss of fission events below the bias level. Budtz-Jørgensen and Knitter [13] then also determine the ratios $N_{2\pi}/N_{cos}$ between 2π counting and intensities found from the cos Θ distributions. They are represented in Fig. 5. Indeed the inefficiency 1 - $N_{2\pi}/N_{cos}$ of the 2π-counting for the UF_4 samples could be deduced. Backscattering of fission fragments does not increase the counting in a 2π detector, since it results in one sum pulse for the two fragments. It should be remembered that the 2π counting is affected both by self-absorption and by the extrapolation of the fragment spectrum down to zero pulse height. The calculated absorption shows somewhat weaker thickness dependence than experimentally observed. But the most important observation resulted from the $N_{2\pi}/N_{cos}$ ratio obtained from the sprayed U_3O_8 sample. The biparametric distribution of cos Θ and E showed that there did not exist a cos Θ value above which there was no selfabsorption. Therefore the cos Θ distribution for this sample can only be used to determine a lower limit for the loss of fission fragments in the sample. However this limit is 3.7 ± 0.3 %, a factor of ~ 3 larger than for an evaporated UF_4 layer of the same thickness. This makes it extremely difficult to correct for self-absorption in the sample, since standard correction formulae usually applied are no longer valid. This is more serious since suspension-sprayed or painted samples often

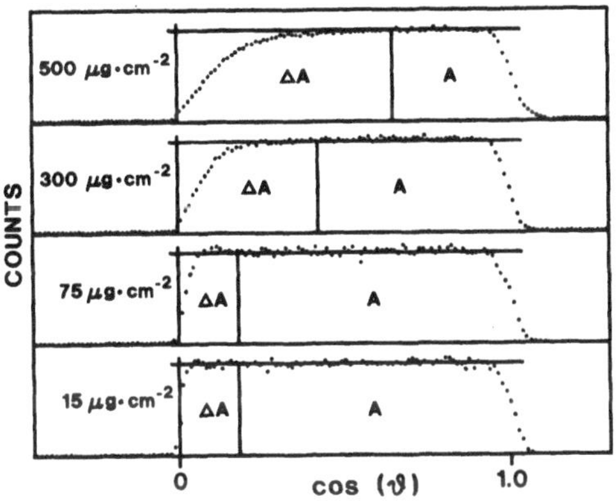

Fig. 4 - The cos Θ distributions of the fission fragments of the four UF_4 vacuum-evaporated samples.

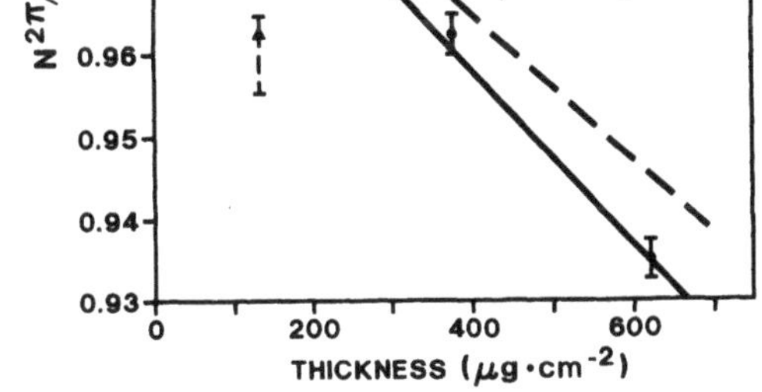

Fig. 5 - The ratio of the 2π count-rate and the total number of fission events determined from the cos Θ distribution plotted versus the sample thickness.

have been used in σ_f-measurements, especially with more exotic samples, where the material losses in connection with vacuum evaporation cannot be tolerated. For fast neutron induced fission cross-section measurements [10] [12], corrections for fission fragments that do not escape the target are often made using the expressions derived by Carlson [14], which are however deduced for ideally flat layers. They only give valuable expressions taking into account both angular distribution and momentum effects under ideal circumstances. Under conditions such as for a 100 μg·cm^{-2} electrosprayed U_3O_8 layer, application of incorrect absorption corrections can lead to wrong determination of σ_f and even of the cross-section shape, due to the energy dependence of the fission fragment angular distributions. It confirms that 2π measurements, with conventional corrections for such type of targets are not reliable.

To conclude in the biparametric spectra it can clearly be seen in which parameter range they are undisturbed. This information then permits, in the case of α-particles a determination of the source strength and in the case of fission fragments, a determination of the absolute number of fission events, with an accuracy better than 0.3 % and 0.5 % respectively. Error sources of earlier methods like backscattering, absorption processes, source shapes, and source inhomogeneities may be circumvented. An important strength of the method of Budtz-Jørgensen and Knitter is the large solid angle for the particle detection.

4. Measurements of fission cross-sections for nuclides with very large value of λ_α

At CBNM two methods were developed : (1) intrinsic suppression of the α-particle background [15] and (2) use of the characteristics of ionisation chambers [16].

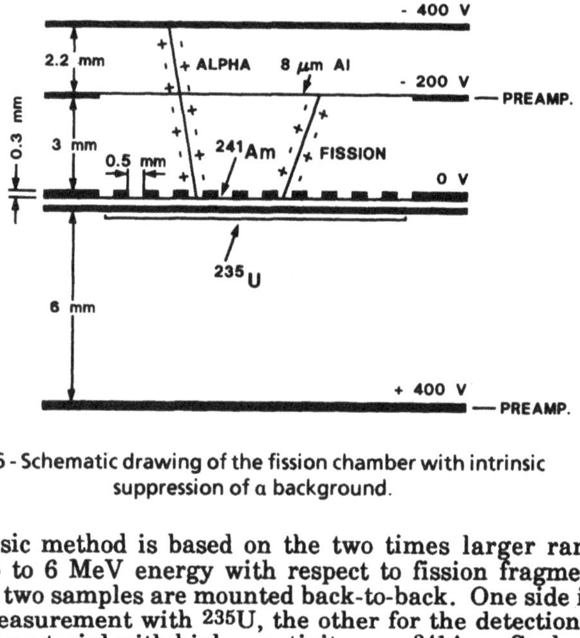

Fig. 6 - Schematic drawing of the fission chamber with intrinsic suppression of α background.

The intrinsic method is based on the two times larger range of α-particles with 5 to 6 MeV energy with respect to fission fragments. In a double chamber two samples are mounted back-to-back. One side is used for the reference measurement with ^{235}U, the other for the detection of fission fragments of the material with high α-activity, e.g. ^{241}Am. Such a chamber is schematically represented in Fig. 6. The ^{241} Am side is divided in two parts by an aluminium foil of 8 μm thickness which is kept at an intermediate potential, and from which a signal can be taken. The ^{241} Am sample is covered by a collimator in such a way that all detected α-particles pass through the foil. The detection geometry is chosen in a way that the α-particles produce about the same ionisation on both sides of the foil, so that there is no net current in the intermediate electrode. On the other hand fission fragments cannot traverse the foil and this side behaves as a normal planparallel ionisation chamber for the fission fragments. The detection limit realised for this chamber was 1 fission fragment in the presence of 5×10^{11} alphas. A disadvantage of the method is the reduced transmission of the collimator (e.g. 13%).

The other method exploits the properties of an ionisation chamber where the interelectrode distance is larger than the fission fragment range in the gas. This 2π chamber has a good discrimination against α-particles, signals proportional to the energy and a resolution time of nanoseconds.

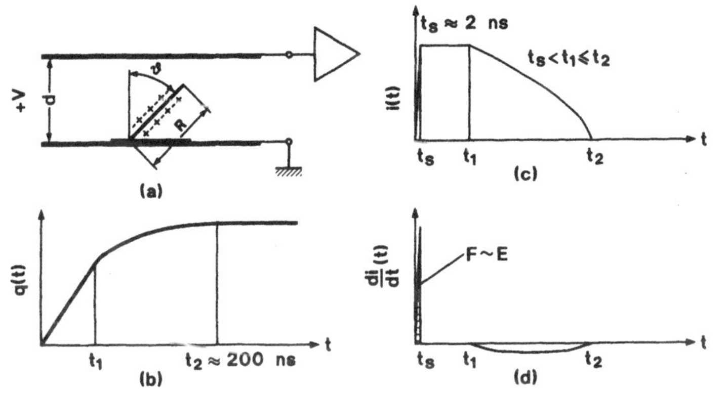

Fig. 7 - The parallel plate chamber with an ion trace is shown in (a), where R is the range of a fission fragment. Figures (b), (c) and (d) show the charge, the current, and the time differentiated current as a function of the time.

In Fig. 7 the operation principles of the chamber are illustrated. The time variation of the collected charge q (t) of the electrons, formed in the ionisation track, is fast and linear up to t_1, when the first electrons arrive at the collector. Then the rise is slower until all electrons are collected ($t_2 \approx 200$ ns). The current impulse i(t) has a height proportional to the number of electrons in the ionisation track and is hence proportional to the kinetic energy of the fission fragments. The rapid rise only depends on the stopping time of the fission fragments in the gas. This time t_s is only 2 ns. As a result the differentiation of the current pulse is very short and the area remains proportional to the fission fragment energy. The ratio of pulse height from fission to α-particles is > 12, much larger than in a chamber with electrode distances smaller than the fission fragment range. The short width and the energy proportionality ensures that the chamber, operating in a high alpha background is able to discriminate between signals from α pile-up and fission fragments. The limiting factor for that type of ionisation chamber are the commercially available preamplifiers whose rise time of ~ 10ns and electronic noise set a limit on how fast an electronic differentiation can be performed. In practice up to 5×10^7 alpha particles s^{-1} can be separated from fission fragments, while keeping the detection efficiency of the latter larger than 96 %, and the time resolving power better than 1.5 ns.

This method has recently been applied for the fission cross-section measurements of ^{243}Am [17]. An example of the fission fragment spectrum of ^{243}Am is shown in Fig 8.

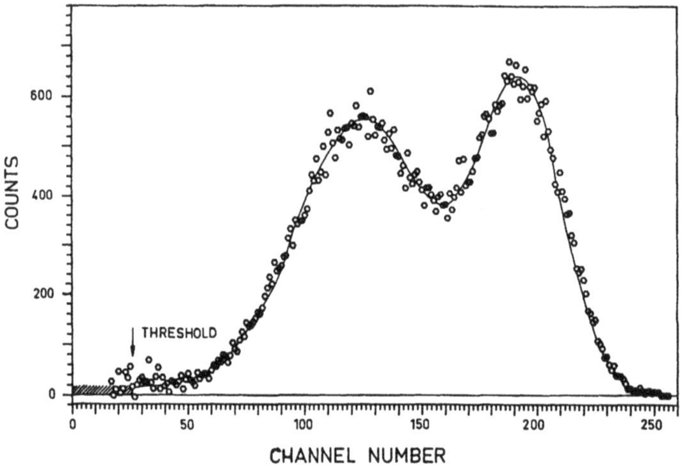

Fig. 8 - Fission fragment pulse-height spectrum from ^{243}Am (n,f). The counting threshold and the part to be interpolated to zero pulse height are indicated.

5. Measurements of thermal cross-sections below the fission barrier.

Typical examples of such very small fission cross-sections are the 2 200m/s values for ^{232}Th equal to 3 $\pm$ 1μb [18] and for ^{238}U equal to 11 $\pm$ 2 μb [19]. These measurements of such small cross-sections pose stringent conditions on the chemical and isotopical purity of the samples and on the intensity of extremely pure thermal neutron beams. For the ^{238}U measurements the material only contained 12 ppb of ^{235}U, for 1 ppm of ^{235}U would lead to a background fission of 587 μbarn, fifty times larger than the net signal from ^{238}U. On the other hand a contamination of 1 % of fast neutrons in the neutron beam would lead to a fast fission contribution in ^{238}U of several mbarn, 1000 times larger than the thermal signal. Nevertheless thanks to the availability of very pure beams of thermal neutrons at the curved neutron guides at ILL, Grenoble (F) and freshly separated ^{238}U at the IPN, Lyon (F) the experimental conditions could be realised.

The same weak cross-sections have been measured with samples of the order of grams, consistent with the RINS lead slowing-down spectrometer sensitivity of the order of 1 microgram-barn. The thermal values are obtained by extrapolation from the measured cross-section in the 1 eV to 80 keV range [8]. The extrapolated values, in particular the value for ^{238}U of 54 $\pm$ 6 μb, are in contradiction with the results obtained at ILL Grenoble. The source of the discrepancy is still unknown.

A more recent example was the measurement of σ_f° for ^{243}Am [20]. In this particular case, ^{239}Pu with its very large thermal fission cross-section is formed by decay of ^{243}Am following the process :

$$^{243}Am \xrightarrow[7370yr]{\alpha} {}^{239}Np \xrightarrow[2.355d]{\beta-} {}^{239}Pu$$

Using target material, only one year after the (Am, Pu) separation the ^{239}Pu (n_{th},f) reaction will contribute for 70 mb in the fission rate. The σ_f° value of ^{243}Am is only (74±4)mb. The measurements were done at the curved guide (87m) H22 at the ILL High Flux reactor, with a ratio of thermal neutrons to epi-thermal and fast neutrons of 10^6 and a neutron intensity of $5x10^8$ $n.cm^{-2}.s^{-1}$. In this particular case 1 % of fast neutrons would lead to a fast fission contribution in ^{243}Am of about 15mb. Again the stringent experimental conditions were reached.

6. <u>Measurement of the fission cross-section of the isomeric state of ^{235}U</u>

^{235}U has an isomeric state 1/2+ with an excitation energy of only 76.8 eV above the ground-state, which has spin and parity $7/2^-$. This metastable state ^{235m}U disintegrates with a half-life of $T_{1/2}=26$ minutes to the ground-state by internal conversion of the E3 transition. The neutron captures in ^{235}U and ^{235m}U lead respectively to states (3^-,4^-) and (0^+,1^+) for energies close to the binding-energy in ^{236}U. For this reason the study of the fission of ^{235m}U nuclei is interesting to compare the fission characteristics of states with very different spins and different parities, but about equal excitation energies.

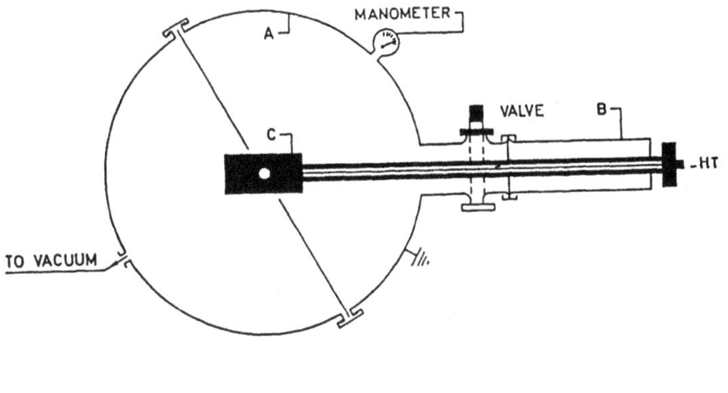

Fig. 9 - Target preparation set-up. A: aluminium sphere (Ø = 40 cm) with inner coating of $^{239}PuF_3$; B: lock chamber; C: sliding target holder.

The major difficulty of the study of ^{235m}U (n,f) is the preparation of the target. The production of nuclei ^{235m}U and the preparation of samples of this nuclide are based on the α-disintegration of ^{239}Pu that feeds exclusively the isomeric level ^{235m}U. The recoil products ^{235m}U have an energy of about 90 keV and are positively charged. They can then be collected on a target. Evidently the ^{239}Pu mother source has to be thin, homogeneous and it should cover a large surface area in order to allow a sufficient number of recoil nuclei ^{235m}U to be collected. Pauwels et al [21] conceived and realised at CBNM, Geel, a mother source in the form of a sphere in aluminium with a 40 cm diameter, coated on the inside with a layer of 10μg·cm^{-2} of $^{239}PuF_3$ deposited in vacuum. In Fig. 9 the experimental equipment is shown for the collection of the ^{235m}U nuclei. The sphere A is evacuated to 1 Pa. The chamber B is connected and contains a sample-carrier C electrically isolated and mobile. An electric voltage of up to -5kV can be applied to the target-holder to guide the ^{235m}U recoils to the target. The lock chamber allows exchanging of targets. In Fig. 10 the number of nuclei ^{235m}U and the number of nuclei ^{235g}U are shown as a function of the collection time. After 30 minutes 8x10^9 nuclei of ^{235m}U were collected, which allowed D'Eer et al [22]

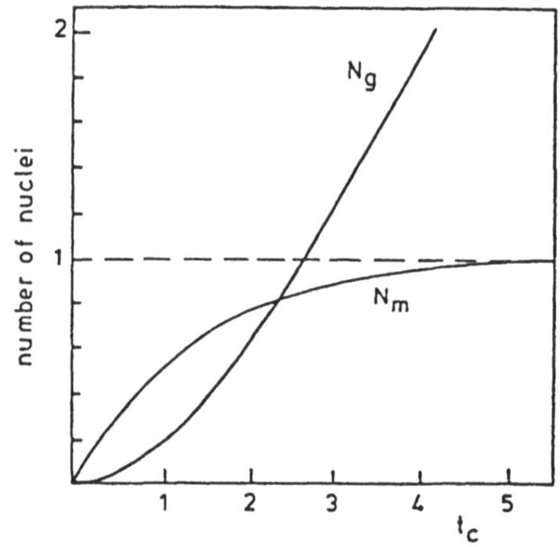

Fig. 10 - Buildup of isomeric (N_m) and ground state (N_g) ^{235}U nuclei on the target in units of 7 x 10^{10} nuclei as a function of the collection time t_c (in units of $T_{\frac{1}{2}}$ = 26 min).

to measure the ratio of the fission cross-sections of ^{235m}U and ^{235g}U at two neutron energies 5 meV and 28 meV, using again the neutron beams at the ILL, Grenoble (F). The results 1.61±0.44 and 2.47±0.45 for σ_m/σ_g were obtained for the cold and the thermal neutrons, respectively. The results suggest the possible presence of a low-energy resonance in the $^{235m}U(n,f)$ cross-section. The result for the thermal ratio disagrees with a recent determination by Talbert et al [23] using a quite different target preparation technique. The reason for the discrepancy is not understood, but all measurements confirm $\sigma_m/\sigma_g > 1$.

7. Conclusion

Recently several developments at neutron production facilities led to the widening of the neutron energy region where fission cross-sections can be measured. In particular the extension towards the high energy region at the spallation neutron source of the Los Alamos facility is noteworthy. The measurement techniques have been adapted to measure fission rate and neutron flux in the full energy range.

For the standard fission cross-section of ^{235}U recent new information is available from the μeV region [1], over the thermal and resonance region [6], [24] [25], several spot point measurements between 2 and 20 MeV [12], and up to 400 MeV at the Los Alamos spallation source [9] [10].

The sensitivity of the lead slowing-down neutron spectrometer, RINS, at Rensselaer was estimated at 1 microgram-barn. It is the only available facility for σ_f measurements of certain special nuclides in its energy range.

Important improvements in the precision measurements of fission cross-sections are based on the use of biparametric (energy and emission angle) information obtained from gridded ionisation chambers : for the α-particles in view of the determination of the source strength for mass-determination of the fissile layer; for the fission fragments in view of the determination of the absolute number of fission events. The knowledge of the characteristics of the fission foils is extremely important for the validity of the application of standard formulae for the correction of self-absorption effects in the layers, especially for absolute cross-section measurements. In the fast neutron region, where anisotropy effects occur in the fragment emission this remark is also valid for shape measurements.

Techniques were also developed to measure cross-sections of very α-active nuclides, by an intrinsic suppression of the α-particle background through compensation and by the use of ionisation chambers where the inter-electrode distance is larger than the fission fragment range. Quite high α-activity can be tolerated in these cases.

Very small thermal sub-barrier fission cross-sections and the fission cross-section of the isomeric state of ^{235}U have been measured thanks to the availability of ultra-pure neutron beams and (or) improved techniques for sample preparation.

References

[1] C. Wagemans, A.J. Deruytter, R. Barthélémy, W. Mampe, P. Ageron and A. Michaudon, Proceedings Nuclear Data for Science and Technology, MITO, Japan, 131-132 (1988).

[2] P.W. Lisowski, J. L. Ullman, S.J. Balestrini, A.D. Carlson, O. Wasson, N.W. Hill, Proceedings Nuclear Data for Science and Technology, MITO, Japan, 97-99 (1988).

[3] G.D. James and G. de Saussure, pp 89-156, in "Nuclear Fission and Neutron Induced Fission Cross-Sections", Ed. A. Michaudon, Pergamon Press, 1981.
[4] S. Cierjacks and A. Smith, "Neutron sources: present practice and future potential", Proceedings Nuclear Data for Science and Technology, MITO, Japan, 325-336 (1988).
[5] K.H. Böckhoff, A.D. Carlson, O. Wasson, J.A. Harvey and D.C. Larson, Nuclear Science and Engineering 106, 192-207 (1990).
[6] C. Wagemans, P. Schillebeeckx, A.J. Deruytter and R. Barthélémy, Proceedings Nuclear Data for Science and Technology, MITO, Japan, 91-95 (1988).
[7] C. Wagemans, P. Schillebeeckx, A.J. Deruytter and R. Barthélémy, Proceedings Nuclear Data for Science and Technology, Jülich, Germany, May 1991, to be published.
[8] R.C. Block, R.E. Slovacek, Y. Nakagome, R.W. Hoff, Proceedings of the Conference 50 Years with Nuclear Fission, Gaythersburg, USA, April 1989, (ANS, La Grange Park, Ill, 1989), Vol. I, p. 354.
[9] P.W. Lisowski, J.L. Ullmann, S.J. Balestrini, N.W. Hill, A.D. Carlson, O.A. Wasson, Proceedings of the Conference 50 Years with Nuclear Fission, Gaythersburg, USA, April 1989, (ANS, La Grange Park, Ill, 1989), Vol. I, p. 443.
[10] A.D. Carlson, O.A. Wasson, P.W. Lisowski, J.L. Ullmann, N.W. Hill, Proceedings of Nuclear Data for Science and Technology, Jülich, Germany, May 1991, to be published.
[11] S.A. Wender and P.W. Lisowski, Nuclear Instruments & Methods B24/25, 897 (1987).
[12] K. Merla, P. Hausch, C.M. Herbach, G. Musiol, G. Pausch, U. Todt, L.V. Drapchinsky, V.A. Kalinin and V.I. Shpakov, Proceedings of Nuclear Data for Science and Technology, Jülich, Germany, May 1991, to be published.
[13] C. Budtz-Jørgensen and H.H. Knitter, Nuclear Science and Engineering, 86, 10-21 (1984).
[14] Gary W. Carlson, Nuclear Instruments and Methods, 119, 97-100 (1974).
[15] C. Budtz-Jørgensen and H.H. Knitter, Nuclear Instruments and Methods, 154, 121 (1978).
[16] C. Budtz-Jørgensen and H.H. Knitter, Nuclear Science and Engineering, 79, 380-392 (1981).
[17] H.H. Knitter and C. Budtz-Jørgensen, Nuclear Science and Engineering, 99, 1-12 (1988).
[18] P. D'Hondt, C. Wagemans, G. Barreau and A.J. Deruytter, Annals of Nuclear Energy, 7, 367-370 (1980).
[19] P. D'Hondt, C. Wagemans, A. Emsallem and R. Brisot, Annals of Nuclear Energy, 11, 485-488 (1984).
[20] C. Wagemans, P. Schillebeecks and J.P. Bocquet, Nuclear Science and Engineering, 101, 293-297 (1989).
[21] J. Pauwels, R. Eykens, A. D'Eer, M. Nève de Mévergnies and C. Wagemans, Nuclear Instruments and Methods, A 257, 21 (1987).
[22] A. D'Eer, C. Wagemans, M. Nève deMévergnies, F. Gönnenwein, P. Geltenbort, M.S. Moore, J. Pauwels, Phys. Rev. C, 38, 1270-1276 (1988).

[23] M. Talbert, J. Starner, S. Balestrini, M. Attrep, D. Efurd and F. Roensch, Phys. Review C, 36, 1896 (1987).
[24] L.W. Weston and J.H. Todd, Nuclear Science and Engineering, 88, 567 (1984).
[25] R.A. Schrack, Proceedings Nuclear Data for Science and Technology, MITO, Japan, 101-104 (1988).

MASS DISTRIBUTION IN 11.3 MeV NEUTRON-INDUCED FISSION OF ^{238}U

LI ZE, WANG XIUZI, JING KEXING, CUI ANZHI,
CHIH TAHAI, LI DAMING, LIU YONGHUI, LI XUELIANG,
LIU CONGGUI, SU SHUXIN, TANG PEIJIA, LIU DAMING,
ZHANG SULAN, ZHANG SHENGDONG and GUO JINGRU
China Institute of Atomic Energy, P. O. Box 275-46
Beijing 102413, People's Republic of China

ABSTRACT

Fission product yields for 40 mass chains were determined for the fission of ^{238}U induced by 11.3 MeV neutron for the first time. Absolute fission rate was monitored with a double-fission chamber. Fission product activities were measured by HPGe or Ge(Li) γ-ray spectrometry of irradiatted ^{238}U foils and by chemical separation of the fission product elements followed by β- counting and/or γ- ray spectrometry. Time of flight technique was used to measure the neutron spectrum in order to estimate the fission events induced by break-up neutrons and scattering neutrons. A complete mass distribution curve has been obtained and dependence of fission yield with neutron energy is discussed.

1. Introduction

The fission product mass distribution of ^{238}U has been extensively investigated, but only a few of those investigations is for fission induced by the monoenergetic neutrons. S. Nagy et al. studied the mass distribution of ^{238}U with essentially monoenergetic neutrons of 1.5, 2.0, 3.9, 5.5, 6.9 and 7.7 MeV[1]. T. C. Champman et al. determined the mass distribution for the fission of ^{238}U induced by the neutrons with energy of 6.0, 7.1, 8.1 and 9.1 MeV[2] . Li Ze et al. measured the mass distribution for the fission of ^{238}U with 8.3 MeV neutron[3]. Zhang Chunhua et al. investigated the mass distribution with 3.0 MeV neutron[4,5] and Wang Xiuzi et al. measured the yields of 9 fission products with 5.0 MeV neutron[6]. Several other measurements of the mass distribution are for 14 MeV neutron[7,8], since 14 MeV neutron is readily available. The fission product yield of ^{238}U induced by neutron of energy between 9.1-14 MeV is, however, still absent. A complete set of the fission product yield data of ^{238}U with monoenergetic neutron at a set of energies up to 14 MeV is of important application in determining the burn-up factor ,decay heat etc.of nuclear devices, therefore, to complete the data set, a measurement with neutron of about 11 MeV becomes essential. Also, the second chance fission takes place in this energy region, thus the mass distribution to be measured is helpful to understanding the fission process.In the present work the product yield of ^{238}U fission induced by 11.3 MeV neutron was measured by means of the gamma-ray spectrometry and the chemical separation of the fission product elements followed by β or γ-counting. 11.3 MeV neutron were produced by D(d, n)^{3}He reaction.

2. Expirement

The experiment was carried out at HI-13 Tandem in CIAE. The deuterium gas chamber, which was used to produce neutron by the bombardment of deuteron beam, was of the size $\phi 22 \times 30$ mm and made of stainless steel. In order to reduce the background neutron, the bottom of the chamber was covered with Pt foil of 0.2 mm thickness, and the cylindrical wall with Ta foil of 0.2 mm thickness. The entrance window of the chamber was made of Mo with thickness of 15 or 25 μm.The purified deuterium gas of 5 atomesphere pressure was filled in the chamber. The energy of deuteron was 9.0 MeV. The neutron spectrum was measured with TOF technique in order to estimate the fission events induced by the background neutrons of energy between 2-6 MeV from D(d, np)D reaction channel and from scattering. The ratio of background fission events to 11.3 MeV neutron fission events was estimated to be 0.264.The samples used in the irradiation were $\phi 16$ mm disks of natural uranium metal with weight of about 2g. The uranium disks were sealed in pure Al foil of 0.2 mm thickness. The sample, which was sandwiched between standardized thin samples when irradiating to monitor the fission rate absolutely, was mounted in a double fission chamber. The standardized thin samples were made of the same natural uranium as the thick samples. The double fission chamber was covered with Cd of 1 mm thickness in order to shield from the thermal neutron from the environment. 26 samples were irradiated for the period varying from 0.5 to 30 hrs at the distance of about 5 cm from the neutron source in the direction of zero degree.

After the irradiation, the samples were taken either to undergo the chemical separation or to collect their gamma-ray spectra directly by Ge(Li) or HPGe gamma spectrometer. In the former case the chemical separation was followed by gamma or beta counting to determine the fission products quantitatively. In the latter case the gamma spectra were collected successively over the period varying from several hours up to two months to encompass the wide range of half-lives of fission products involved, and to eliminate the cross interference of the product nuclides emanating gamma rays of almost the same energy even Ge(Li) or HPGe can not resolve.

3. Result and discuss

The yields of 40 product mass chains have been obtained absolutely, among which 31 were from the gamma spectra directly collected and 14 from the chemical separation, and 5 of them were obtained by both methods. The corrections were made for the self-absorption of gamma rays, the cascade coincidence losses, the pile-up losses in counting system, and the influence of the background neutron. The complete mass distribution was thus given from the mass yields we determined in Fig. 1. The sum of the yields directly measured was 59.42% for the light mass group, and 67.44% for the heavy mass group. Including the mass yields not measured but obtained by interpolating or extrapolating, the sum was 100.76% for the light mass group and 99.51% for the heavy mass group, both of which are in agreement with 100% excellently. This implies that we monitored the absolute fission rate and detected fission products to very good accuracy. The mean mass numbers are 97.69u and 137.73u for the light and heavy mass group respectively, which gives the mean fission neutron number of each fission $\bar{\nu}$=3.58. The number agrees with the evaluated experimental data of the mean fission neutrons of ^{238}U 3.9646 fairly well[8].

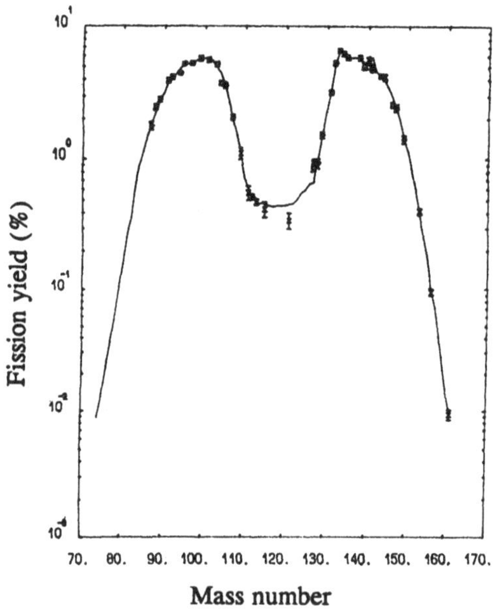

Fig. 1. Mass distribution for 11.3 MeV neutron-induced fission of ^{238}U.
∘ GS, × RC

To compare our data with other data published in literature, some product yields are showed in Fig. 2 and Fig. 3. The data points from the present

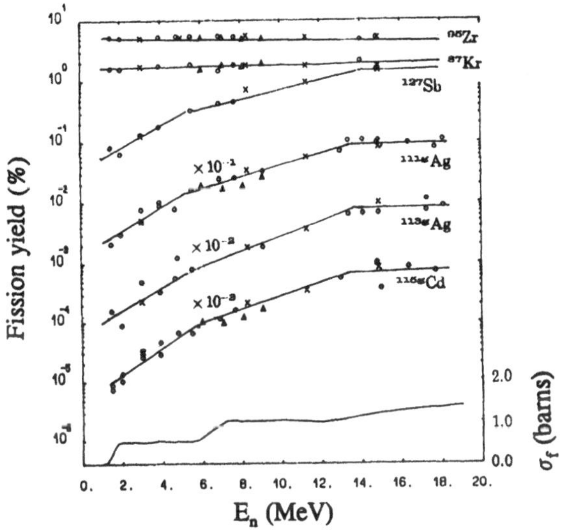

Fig. 2. Fission yields and cross section σ_f for fission of ^{238}U by monoenergetic neutrons as a function of neutron energy
∘ S. Nagy, Δ C. T. Champman, + D. R. Nethaway,
× present lab.

experiment fall in the line determined by other data very well. In semi-logarithm coordinate represent the dependence of fission yield in different mass range of mass distribution on neutron energy is following:

(1) Fission yield decreases with increasing neutron energy for the products in the peak region of mass distribution.

(2) Fission yield increases with increasing neutron energy for the products in the valley region and the slope of line changes at the energies corresponding to second and third chance fission.

(3) The tend of fission yield change is different for different mass number of product in right wing of heavy mass group. The slope of line increases with increasing mass number of product, e.g. the slope of line for ^{147}Nd, ^{153}Sm and ^{156}Eu is negative, almost zero and positive, respectively.

Acknowledgement

The authors are indebted to Tang Hongqing, Zhou Zuying and Sa Jun for the measurement of the neutron spectrum.

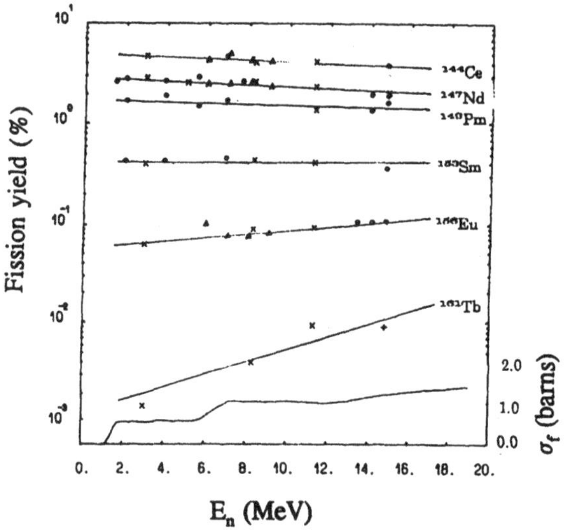

Fig. 3. Same as Fig. 2, but for the some prdudts of right wing of heavy peak of mass distribution curve

References

1. S. Nagy et al., *Phys. Rev.* **C17** (1978) 163.
2. T. C. Champman, et al., *Phys. Rev.* **C17** (1978) 1089.
3. Li Ze et al., *Chinese J. of Nucl. Phys.* **7** (1985) 97.
4. Zhang Chunhua et al., *J. of Nuclear and Radiochemistry* **7** (1985) 1.
5. Su Shuxin et al., (to be published).
6. Wang Xiuzi et al., *J. of Nuclear and Radiochemistry* **6** (1984) 183.
7. Liu Conggui et al., *Chinese J. of Nucl. Phys.* **7** (1985) 235.
8. D. R. Nethaway et al., *Phys. Rev.* **182** (1969) 1251.
9. F. Manero et al., *Energ. Rev.* **10** (1972) 637.

γ–RAY SPECTROSCOPY AND
(n, γ) REACTION MECHANISM

MECHANISM OF (n,γ) REACTION AT LOW NEUTRON ENERGIES

S. Raman
Oak Ridge National Laboratory, Oak Ridge, Tennessee 37831

and

J. E. Lynn
Los Alamos National Laboratory, Los Alamos, New Mexico 87545

Abstract

We discuss the interplay between direct capture, valence capture, and compound-nuclear capture in attempting to explain the vast amount of capture data for light-mass nuclei.

1. Introduction

Information that has been accumulated in a vast range of nuclides on the radiative transitions that follow slow-neutron capture has been one of the main sources of knowledge on the detailed nuclear level structure of these nuclides. Most of this knowledge comes from γ-ray energies and level placements, the relative yields, branching ratios of secondary transitions between low-lying states, and angular correlations among successive secondary transitions. Spectroscopic information may also be contained in the *cross sections of the primary transitions* originating from the capturing state, and this fact is demonstrated by the success of "direct" theories of neutron capture for many nuclides, especially those of light and near closed-shell character.

2. $E1$ transitions

2.A. Direct capture

Although low-energy neutron capture can be regarded as the classical compound-nuclear process (the original experimental evidence for large neutron capture cross sections motivated Bohr in the creation of the compound-nucleus concept), it has also been known for a long time [1] that far away from resonances a *direct form of radiative capture* not involving the compound nucleus can exist. In this process, the neutron-target interaction is represented in zero order by a potential well, and the neutron that is initially in an s orbit simply falls into a p-wave orbit in the final nucleus resulting in the emission of a primary $E1$ transition. The theoretical analysis of this process requires a knowledge of the coherent scattering length and (d,p) spectroscopic factors.

This form of neutron capture is amenable to calculation because the major contribution to the radial factor of the $E1$ matrix element arises from a region outside the potential well where the wave functions are known. These wave functions depend only on the scattering length (for the initial state) and the binding energy plus single-particle spectroscopic factor (for the final state). Indeed, many partial $E1$ cross sections can be reproduced by a formalism that treats the potential with a sharp cutoff at the well radius and ignores all internal contributions to the matrix element. This version is known as *channel capture* (see Fig. 3 of Ref. [2]).

2.B. Channel capture

The channel-capture formula [1,3] is simple:

$$\sigma^{\rm CH}_{\gamma(i\to f)} = \frac{0.0614}{R\sqrt{E_{\rm lab}}}\left(\frac{Z}{A}\right)^2\left(\frac{M+m}{M}\right)^3 \sum_J g_J \frac{1}{|1-iP_0\mathcal{R}_J|^2}$$

$$\times \left| y^2\left(\frac{y+3}{y+1}\right)^2 + 2\,\mathrm{Re}\mathcal{R}_J \frac{y^3(y+1)(y+2)}{(y+1)^2} + |\mathcal{R}_J| y^2\left(\frac{y+2}{y+1}\right)^2 \right| \mathcal{W}_{J_f jJ}\,\theta_f^2,$$

where

$$y = R\sqrt{\frac{M}{M+m}}\sqrt{\frac{2mE_f}{\hbar^2}},\quad \mathrm{Re}\mathcal{R}_J = 1-(a_{sJ}/R),\quad R=\left(1.16A^{1/3}+0.6\right)\mathrm{fm},$$

M – target mass, m – neutron mass, g_J – statistical weight factor, P_0 – penetration factor for s-wave neutrons, $\mathcal{R}_J$ – reduced $\mathcal{R}$ function, $\mathcal{W}$ – spin coupling factor, θ_f^2 – spectroscopic factor, E_f – binding energy of the final state, and a_{sJ} – scattering length. This simple analytical formula has been found to be very successful in estimating the cross sections of primary $E1$ transitions in many nuclides [3].

2.C. Potential capture

In the more detailed version of capture theory [4-8], called *potential capture*, the wave functions are computed numerically from a realistic optical model (such as Woods-Saxon) and the internal contributions are included. The optical-model parameters (well depth, diffuseness, and imaginary potential amplitude) are varied within physically reasonable limits to reproduce the measured scattering length (which may be affected by local compound-nuclear states). The depths of the real potential are then varied to reproduce the binding energies of final bound states of largely single-particle character. This is the *specialized optical-model* [*S*] procedure. This model works well even when (as in ^{9}Be) the scattering length (7.0 fm) is large compared to the potential radius (3.0 fm), resulting in large cancellations in the $E1$ matrix elements [5].

It can be argued that forced adjustment of the optical-model parameters is not the best way to proceed to explain thermal-neutron cross sections. Also, occasionally, certain optical-model parameters get pushed to values that are at the limits of physical acceptability. To circumvent this problem, the direct-capture amplitude can be written as a sum of two terms—a potential-capture amplitude (based on a global [*G*] potential such as Moldauer's) and a valence [*V*] correction due to the wave function of local resonance levels projected on the single-particle motion. The valence correction can be assessed from the difference between the calculated (from the optical model) potential scattering length and the measured scattering length of the target nucleus. We call this the *global optical model + valence correction* [*G+V*] procedure.

From the global optical-model calculations, a complex capture amplitude is obtained. The real part is the potential capture and the imaginary part gives a cross section that is the average of a resonance-resonance interference term (not a cross section averaged over resonances in the Hauser-Feshbach sense). From this, the valence radiation strength function can be obtained and hence the valence correction. The direct-capture cross section is given by

$$\sigma^{\mathrm{dir}}_{\gamma(i\to f)} = \sum_J g_J \left| \sqrt{4\pi} R \mathcal{R}^{\mathrm{loc}}_J \left[\frac{\overline{\Gamma}_{\lambda\gamma(i\to f)(\mathrm{val})}}{\Gamma_{\lambda n}} \right]^{1/2} + \sigma^{1/2}_{\mathrm{pot},\gamma} \right|^2 ,$$

where $\mathcal{R}^{\mathrm{loc}}_J$ is the contribution from local levels to the reduced $\mathcal{R}$ function

$$R\mathcal{R}^{\mathrm{loc}}_J = \left(a_{\mathrm{pot}} - a_J\right),$$

$\overline{\Gamma}_{\lambda\gamma(i\to f)(\mathrm{val})}$ is the average valency radiation width, and $\Gamma_{\lambda n}$ is the neutron width derived from the neutron strength function calculated from the optical-model scattering. In general, we find that the [*G+V*] and [*S*] methods give very similar results [6].

2.D. <u>Compound-nuclear contributions</u>

Differences between experimental and calculated cross sections are attributed to a *compound-nuclear radiative amplitude* due to the tails of nearby resonances. Compound nucleus (CN) is a generic term for mechanisms involving more general features of the wave functions than the simple picture of neutron motion in the field of the unexcited core of the target. It is uncorrelated with the single-particle content of the initial or final state. The experimental cross section $\sigma_\gamma(X)$ is

$$\sigma_\gamma(X) = \left(\sigma^{1/2}_{\mathrm{dir},\gamma} + \sigma^{1/2}_{\mathrm{CN},\gamma}\right)^2 .$$

The extracted CN cross section can be converted to a radiation width if it is assumed that the discrepancy between the potential and measured scattering lengths is due to a single capturing resonance level. Thus

$$\frac{k\left\langle \sigma_{\mathrm{CN},\gamma} / E^3_\gamma \right\rangle}{2\pi R\mathcal{R}^{\mathrm{loc}}} \approx \frac{\left\langle \Gamma_{\lambda\gamma,\mathrm{CN}} / E^3_\gamma \right\rangle}{E_\lambda} .$$

This radiation width can be compared with the *Cameron estimate*

$$\Gamma_{J,\gamma,\mathrm{CN}}(\text{in eV}) \approx 0.33\times10^{-9}\,E_\gamma^3(\text{in MeV})A^{2/3}D_J(\text{in eV}) ,$$

which usually underestimates the average strength of high-energy (~ 5 MeV) transitions. Alternatively, the compound-nuclear radiation width can be derived from the *Brink's model* of the damped giant resonance built on each final state, but it has been established that this model overestimates the widths for transitions in the 2 to 3 MeV range.

In general, the $\left\langle \Gamma_{\lambda\gamma,\mathrm{CN}} / E_\gamma^3 \right\rangle / E_\lambda$ deduced from experiment agrees (within an order of magnitude) with Cameron's $\left\langle \Gamma_{\lambda\gamma,\mathrm{CN}} / E_\gamma^3 \right\rangle / D_J$. By comparing them, we can suggest which bound or unbound level might be responsible for the CN contributions [9].

2.E. Generalized valence correction

For some nuclides the discrepancies between theory and experiment are not only greater than would be expected from the Cameron estimate, but also appear to be systematic. The softness of ^{44}Ca allows virtual excitation of quadrupole collective vibrations with sufficient amplitude to interfere with the direct-capture amplitude. The potential field that such a nucleus presents to an incoming neutron contains higher multipole terms that couple the initial configuration (target nucleus ⊗ single-particle motion within the spherical potential) to other configurations (target excited collective state ⊗ single-particle motion). The higher multipole terms have a similar mixing effect on the final states. This *generalized valence capture* model [9] leads to a systematic reduction (by a factor of 2) in the capture cross sections of transitions to the group of low-lying states in ^{45}Ca from the values computed in the direct-capture theory, thereby improving agreement with experiment. [There will also be a systematic enhancement for transitions to some higher-lying states but these have not been measured.]

2.F. Valence capture

The total (n,γ) capture cross sections discussed above are small (< 1 b) indicating that the influence of resonance wings is small. In some cases, nearby resonances clearly have a major influence and more complex capture mechanisms, such as the *valence mechanism*, have to be considered. We base our calculations on the $\mathcal{R}$-matrix theory of nuclear reactions and the assumption of a real potential. The fractionation of the modeled single-particle state into the actual resonance is known from the experimentally determined neutron width of the resonance. It is therefore unnecessary, in this approach, to model the expected average fractionation by the formulation of the imaginary part of the optical model. The valence-capture amplitude, while being closely related to the potential-capture amplitude, is now more sensitive to the components of the wave function in the internal region because the initial-state wave function has an antinode rather than a node close to the potential radius. This also allows significant valence capture by p-wave resonances (see Fig. 5 of Ref. [2]).

The valence contribution is proportional to $\Gamma^{1/2}_{\lambda(n)}\Gamma^{1/2}_{\lambda(\gamma,\text{val})}/(E_\lambda - E)$. The valence γ width for resonance λ has the important property of $\Gamma^{1/2}_{\lambda(\gamma,\text{val})} \propto \Gamma^{1/2}_{\lambda(n)}$. The valence term in the capture amplitude is constructive or destructive with respect to the potential-capture amplitude depending on whether the resonance level E_λ is above or below the thermal neutron energy E, respectively. In the case of ^{43}Ca (whose thermal capture cross section of 6.2 b is an order of magnitude greater than the direct capture cross section), the resonance influence is clearly dominant but not overwhelming. (Our analysis is approximate because the total scattering cross section and the parameters of the 1.48-keV resonance are poorly known.) Even though the resonance valence radiation width is small (~0.13 eV out of 1.9 eV), constructive interference yields an estimate of 3.7 b (a substantial fraction of the 6.2 b) for the direct component, indicating that even in this quasicompound nuclear situation, direct capture plays an important rolc.

At shown in the following table, direct-capture theory, at first sight, fails badly in the case of the ^{14}N(n,γ) reaction with thermal neutrons [10]. The deduced CN component is of similar magnitude to the direct component.

Final state E_f (keV)	Primary E_γ (keV)	$\sigma_{\gamma,\text{dir}}$ (theory) (mb)	σ_γ(exp.) (mb)	σ_γ (CN) (mb)
0	10829	330	10	220 or 450
6324	4509	1	14	5 or 25
9152	1681	0.1	1	1

But a resonance at –24 keV is known from the ^{14}C(p,γ) reaction and we use its γ-ray spectrum and our calculated valence widths to deduce the compound-nuclear contributions. We then combine everything to estimate

Final state E_f (keV)	Primary E_γ (keV)	σ_γ (theory) (mb)	σ_γ(exp.) (mb)
0	10829	8	10
6324	4509	16	14
9152	1681	3	1

The agreement between theory and experiment is now satisfactory. The concept implicit in valence capture, namely that the neutron valence radiative width is of similar magnitude to the compound-nuclear radiative width, is shown explicitly here for the first time in a resonance formed by a non-neutron reaction.

We have applied the valence-capture theory to calculate (see table below) the expected total radiation width of the 153-keV resonance in ^{13}C [11]. The resulting width of ~ 0.2 eV disagrees strongly with the previous measured value of 2.4 ± 0.9 eV.

if $1p_{3/2}$	$\Gamma_{\gamma\to 6.1}$	$= 0.16$ eV	$\Gamma_{\gamma\to 6.7}$	$= 0.06$ eV
if $2p_{3/2}$		$= 0.14$ eV		$= 0.03$ eV

The partial radiation widths of the 6.1- and 6.7-MeV transitions have been remeasured at the Tokyo Institute of Technology [11].

$\Gamma_{\gamma\to 6.1} = 0.151^{+0.076}_{-0.033}$	$\Gamma_{\gamma\to 6.7} = 0.030^{+0.030}_{-0.013}$

These results offer a striking confirmation of the valence-capture theory. (Scattered-neutron sensitivity was probably underestimated previously.) Of the two possible scenarios for nucleosynthesis—red-giant He burning and nonstandard Big Bang—the new result has little significance for the former, but affects the latter in terms of a significant reduction in the production of $A \geq 14$ isotopes [11].

3. $M1$ transitions

Some tendency for the strong $M1$ transitions to be associated with high spectroscopic factors suggests the operation of a direct mechanism quite analogous to that for $E1$ transitions. In this case, the chances of arriving at a successful theory are slimmer for two reasons: (1) In the $E1$ case the matrix elements involve the electrostatic dipole moment, whereas in the $M1$ case, the operator depends on currents that may be described much more poorly by the wave functions of a simple direct-capture model. (2) The radial component of the $E1$ matrix element is more strongly weighted to the channel region—where the wave functions are well established by the energies of the initial and final states—than is the radial $M1$ element. In fact, in the $M1$ case, it is often stated that there can be no direct capture analogous to $E1$ because the radial wave functions are necessarily orthogonal in a simple potential well. However, because of the complexity of the nucleus, we usually find it necessary to generate the radial wave functions of the initial and final states with rather different mutual potentials (just as we usually do for $E1$ capture) and $M1$ transitions become possible. The $M1$ operator has the form [12]

$$H'_{M1,\mathfrak{M}} = \frac{e\hbar}{2mc}\left[\frac{3}{4\pi}\right]^{1/2}\left[\frac{\mu_l I_{\mathfrak{M}}}{I} + \frac{\mu_n \sigma_{n,\mathfrak{M}}}{\sigma}\right].$$

From this expression the reduced matrix element for the spin factor can be computed, while the radial matrix element is simply the integral of the product of the projection of the radial wave function of the initial state onto the entrance channel and the radial wave function of the single-particle component of the final state. Because there is no orbital contribution in the above $M1$ operator, the single-particle components of the initial and final states have zero orbital angular momentum in slow-neutron direct $M1$ capture.

In very light nuclei, direct $M1$ capture (with the $M1$ operator substituted in place of the $E1$ operator) appears to play a qualitatively important role.

Target	E_f (keV)	E_γ (keV)	σ_γ(theory) (mb)	σ_γ(exp.) (mb)
^{9}Be	6265	547	$37 \times \theta_f^2$	14
^{12}C	3089	1857	$25 \times \theta_f^2$	6
^{13}C	6094	2080	$24 \times \theta_f^2$	34
^{13}C	6900	1274	$68 \times \theta_f^2$	67

In ^{20}F, the direct-capture $M1$ model underestimates the cross sections to the two strongest $s_{1/2}$ states (at 3.488 and 3.526 MeV) by a factor of ~ 4 and to the state at 1.057 MeV by three orders of magnitude. In ^{21}Ne, the calculated cross section of the primary $M1$ transition to the 2.794-MeV state is an order of magnitude greater than experiment. In ^{23}Ne, the calculated and measured cross sections to the state at 1.02 MeV nearly agree (0.6 mb and 0,9 mb, respectively). In this region, the simple direct process has a substantial but not predominant role to play.

We have recently analyzed the ^{7}Li(n,γ) data in the 0 – 400 keV region [12]. We have (1) deduced the direct-capture cross section for thermal-neutron $E1$ capture as 39 mb, (2) analyzed the total cross section and angular distributions of scattered neutrons to deduce $\mathcal{R}$-matrix parameters, (3) showed that the potential-capture cross section behaves as $1/v$, and (4) calculated the valence $M1$ radiation width for the 255-keV resonance as 0.094 eV. The measured thermal-neutron capture cross section is 45.4 ± 3.0 mb. The capture data obtained at Lockheed (Palo Alto) are consistent with $1/v$, whereas those obtained at Karlsruhe are not. If we normalize the Karlsruhe data to thermal (and $1/v$), we infer a radiation width of 0.09 ± 0.02 eV (average of Lockheed and renormalized Karlsruhe data) which is now in agreement with theory. The Karlsruhe conclusion that the ^{7}Li(n,γ) reaction is not the predominant link to the primordial r process is not supported.

4. Conclusion

The nuclides that we have investigated till now are listed in the following table [13]. After more than half a century of study, we still do not have a fully quantitative or universally applicable theory of neutron radiative capture. What we have is a number of models that are applicable in different mass or energy regions and a certain amount of basic theory that explains, in principle, why some of these models work in certain situations. Fortunately, high-quality data continue to be produced in this area and this is leading to continued quantitative development and application of capture models.

Mechanism of (n,γ) reaction at low neutron energies

Mechanism	Explains[a]
*E*1 transitions	
Potential + valence correction + compound	^{7}Li, ^{9}Be, ^{12}C, ^{13}C, ^{19}F, ^{24}Mg, ^{25}Mg, ^{26}Mg, ^{28}Si, ^{29}Si, ^{30}Si, ^{32}S, ^{33}S, ^{34}S, ^{40}Ca, ^{42}Ca, ^{46}Ca, ^{48}Ca
Potential + generalized valence correction + compound	^{44}Ca
Valence + potential + compound	^{43}Ca
Compound + valence + potential	^{14}N
Valence	^{13}C (153 keV)
*M*1 transitions	
Potential	^{9}Be, ^{12}C, ^{13}C
Compound	^{14}N
Valence	^{7}Li (255 keV)

[a]Thermal-neutron capture unless stated otherwise.

5. Acknowledgments

This work was sponsored by the Department of Energy under Contract DE-AC05-84OR21400 with Martin Marietta Energy Systems, Inc. (Oak Ridge) and W-7405-eng-36 with the University of California (Los Alamos). We thank our colleagues who are listed in the references and especially Ed Jurney and Bill Starner of Los Alamos.

[1] A. M. Lane and J. E. Lynn, Nucl. Phys. **17**(1960) 563; **17**(1960)586; See also J. E. Lynn, *Theory of Neutron Resonance Reactions*, Clarendon, Oxford (1968).

[2] S. Raman, in *Neutron Capture Gamma Ray Spectroscopy and Related Topics*, Proceedings of the Fourth International Conference at Grenoble, France, 1981 (Edited by T. von Egidy, F. Gönnenwein, and B. Maier), p. 357, Institute of Physics, Bristol (1982).

[3] S. Raman and J. E. Lynn, in *Neutron Induced Reactions*, Proceedings of the Fourth International Symposium at Smolenice, Czechoslovakia, 1985 (Edited by J. Kristiak and E. Betak), p. 253, VEDA, Bratislava (1986).

[4] S. Raman, R. F. Carlton, J. C. Wells, E. T. Jurney, and J. E. Lynn, Phys. Rev. C **32**(1985)18.

[5] J. E. Lynn, S. Kahane, and S. Raman, Phys. Rev. C **35**(1987)26.

[6] S. Kahane, J. E. Lynn, and S. Raman, Phys. Rev. C **36**(1987)533.

[7] S. Raman, S. Kahane, R. M. Moon, J. A. Fernandez-Baca, J. E. Lynn, and J. W. Richardson, Phys. Rev. C **39**(1989)1297.

[8] S. Raman, J. A. Fernandez-Baca, R. M. Moon, and J. E. Lynn, Phys. Rev. C **44**(1991)518.

[9] S. Raman, S. Kahane, and J. E. Lynn, in *Proceedings of the International Conference on Nuclear Data for Science and Technology, Mito, Japan, 1988* (Edited by S. Igarasi), p. 645, Saikon, Tokyo (1988).

[10] E. T. Jurney, J. E. Lynn, J. W. Starner, and S. Raman, to be published.

[11] S. Raman, M. Igashira, Y. Dozono, H. Kitazawa, M. Mizumoto, and J. E. Lynn, Phys. Rev. C **41**(1990)458.

[12] J. E. Lynn, E. T. Jurney, and S. Raman, Phys. Rev. C **44**(1991)764.

[13] J. E. Lynn and S. Raman, in *Capture Gamma-Ray Spectroscopy*, Proceedings of the Seventh International Conference at Asilomar, California, 1990 (Edited by R. W. Hoff), p. 355, AIP, New York (1991).

THE TEST OF NUCLEAR MODELS USING E2 TRANSITIONS FOR RARE EARTH DEFORMED NUCLEI

Z. R. Shi
China Institute of Atomic Energy,
P. O. Box 275 (67), Beijing 102413, China

ABSTRACT

The collective model of Bohr and Mottelson and interacting boson model of Arima and Iachello are tested through comparing the changes of B(E2) values with angular momenta for rare earth deformed even-even nuclei. Some conclusions are drawn such as β surface vibrations are absent or not important and the γ surface vibrations or deformations are important for most of rare earth deformed nuclei. The B(E2) values for transitions within the ground state band can not be fitted by the models.

1. Introduction

A. Bohr and B. R. Mottelson collective model (CM)[1, 2] and A. Arima and F. Iachello interacting boson model (IBM)[3, 4, 5] are two main models in the nuclear structure. IBM is a algebraic approach that is simple and easy to use, in particular for the description of transition nuclei. However, the CM has the explicit physical picture and four symmetries: spherical surface vibration (SSV), axial rotator, asymmetric rotator and γ-unstable vibration. The IBM also has a set of symmetries: U(5), SU(3), SU(3)* and O(6) corresponding to the above. Both CM and IBM have achived some successes in some contexts.

Previously the test of models is in terms of comparing the theoretical level scheme and the B(E2) with experimental ones. As well known, Hamiltonian operators are not highly sensitive to the nuclear structure because they contain many multipole terms. To test the models more sensitive operators are needed. The transition operators, such as E2 and M1 are more sensitive because they have only one or two terms, in particular the later is sensitive to single particle motions. But yet there are no enough experimental data of M1 transition intensities to do systematic study. In this work the dependence of B(E2) values on angular momenta for intraband and interband transitions of rare earth deformed nuclei (REDN) are used to test the CM and IBM. The calculated B(E2) values are obtained by rotational coupling scheme (I-coup.), axial rotator (AR),rotation and vibration model (RVM), asymmetric rotator model (ARM) and ARM with β surface vibration (ARM_β) in the CM and IBM_2 in the IBM.

The present objective is to find out which collective motions are important and how are the E2 transitions described by the CM and IBM.

Some conclusion are obtained, especially the β surface vibration is absent or not important for most of REDN. As Iachello in 1982[6] and Arima in 1986[7] metioned that the nature of the lowest 0^+ excited states is not clear.

2. Theoretical background

Under the adiabatic condition the collective model Hamiltonian can be written as

$$H = H_c + H_i, \tag{1}$$

where H_c and H_i are the collective and the intrinsic motion part, respectively. H_c can be divied into three parts: collective rotation due to nuclear deformations, surface vibrations in the β and the γ directions for quadrupole deformed nuclei. In the body fixed - frame the H_c is written as

$$H_c = T_r + T_\beta + T_\gamma + V(\beta, \gamma), \tag{2}$$

$$T_r = \Sigma_k(\hbar^2/2J_k)I_k^2 ,$$

$$J_k = 4B\beta^2\sin^2(\gamma - 2\pi k/3),$$

$$T\beta = -(\hbar^2/2B\beta^4)(\delta/\delta\beta)\beta(\delta/\delta\beta),$$

$$T_\gamma = -(\hbar^2/B\beta^2\sin3\gamma)(\delta/\delta\gamma)\sin3\gamma(\delta/\delta\gamma).$$

Usually the vibrations in the β and the γ directions are considered as intrinsic motions of rotational bands. The four symmetries, RVM and ARM_β can be deduced from eq. (2). The dependences of models on deformations and surface vibrations are listed in Table 1. The important collective motions will be found through comparing the B(E2) values.

An expression of the B(E2) values of intraband and interband transitions for AR, ARM, RVM and ARM_β is obtained and shown as

$$B(E2:I_iK_i\text{-}I_fK_f) = <I_iK_i2K_i\text{-}K_f \mid I_fK_f>^2M_0[1+Z_\beta F_\beta(I_iI_f)+Z_\gamma F_\gamma(I_iI_f) + Z_{\beta\gamma}F_{\beta\gamma}(I_iI_f)]^2, \tag{3}$$

where M_0 is the amplitude of E2 transitions for AR and the Z_β, Z_γ and $Z_{\beta\gamma}$ are the relative E2 amplitudes for the β vibrations, γ vibrations or deformations and the β-γ coupling, respectively. These amplitudes can be calculated through the excitation energies of headbands. The $F_j(I_iI_f)$ are the analytic functions of angular momenta of initial and final states.

Table 1. The dependences of models on deformations and vibrations

model		deformation		vibration	
CM	IBM	β	γ	β	γ
SSV	U(5)			×	
AR	SU(3)	×			
ARM	SU(3)*	×	×		
γ-unstab.	O(6)	×	unstable		
ARM_β		×	×	×	
RVM		×		×	×

The coupling between the collective and the intrinsic motions due to Coriolis and centrifugal forces is called as the rotation coupling scheme, in brevity I-coup.. The B(E2) values for transitions within a ground state band (g) can be expresses as

$$B(E2: I_f + 2 - I_f) = M_0[1 + 2\alpha I_f(I_f + 3)]^2, \quad (4)$$
$$M_0 = (5/16\pi)e^2Q^2.$$

That of interband E2 transitions can be written as

$$B(E2: I_iK_i - I_fK_f) = k<I_iK_i2K_i - K_f \mid I_fK_f>^2\{M_1 + M_2[I_f(I_f + 1) - I_i(I_i + 1)]\}^2, \quad (5)$$

where k = 1 for the β-g band and k = 2 for the γ - g band E2 transitions. As well known, the eq. (5) is called as Mikhailov rule. The M_1 and M_2 are called as Mikhailov parameters.

The Hamiltonian and E2 transition operator of IBM_2 can be written in the forms

$$H = \varepsilon_d n_d + k_{\pi\nu}Q_\pi Q_\nu + k_\pi Q_\pi Q_\pi + k_\nu Q_\nu Q_\nu + bLL + 2aM, \quad (6)$$
$$Q_\rho = Q_\rho = (s^+\tilde{d} + d^+\tilde{s})^{(2)}_\rho + \chi_\rho(d^+\tilde{d})^{(2)}_\rho, \; k_{\pi\nu} = 2k_\pi = 2k_\nu$$

$$T(E2) = e_\pi Q'_\pi + e_\nu Q'_\nu$$
$$Q'_\rho = (s^+\tilde{d} + d^+\tilde{s})^{(2)}_\rho + \chi_\rho(d^+\tilde{d})^{(2)}_\rho$$

$$\chi_\rho \neq \chi'_\rho$$
$$\rho = \pi \text{ or } \nu$$

The $Q_\rho Q_\rho$ terms are remained in the eq. (6) for SU(3) symmetry. The parameters b and a are taken as constant values obtained from those of SU(3) limit. Sets of six excitation energies and six B(E2) values measured precisely are taken as reference values for ^{154}Sm, ^{156,158}Gd and ^{158}Dy four nuclei to find out the optimal Hamiltonian parameters. The code NPBOS written by Otsuka et al in 1985[8] is run at VAX-780 computer. The detail for above four nuclei is given in ref. [9].

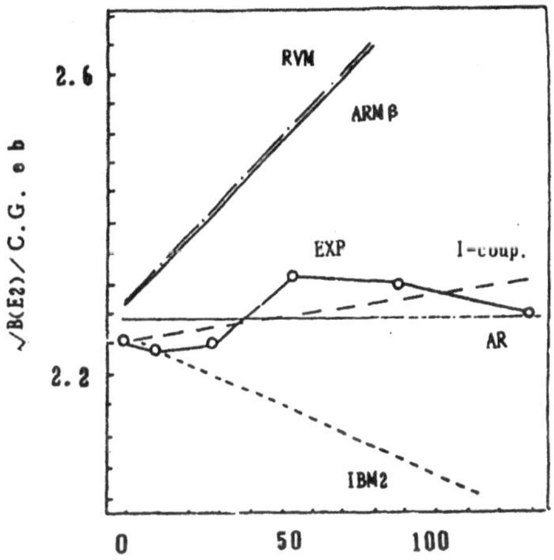

Fig. 1. The B(E2) values within ground state bands

3. Results and discussion

3.1 The B(E2) values for Transitions within a Ground State Band

The change of average B(E2) values of 12 nuclei with $E_{4_g^+}/E_{2_g^+} > 3.20$ for transitions within the ground state band with $I_f(I_f+3)$ have a regular behavior shown in Fig. 1. It is approximently flat from $I_f=0$ to 4, then arises and arrives at maximum at $I_f=6$ and then goes down. It is interpreted as the competetions between deformations and the broken pairs[10], i. e. the centrifugal and coriolis

Table 2 The E2 relative amplitudes of transitions within a ground state band

Nucl.	A	I-coup.		RVM		ARM_β	
		Z	δZ	Z_β	Z_γ	Z_β	Z_γ
Sm	154	0.200	86	1.59	0.14	1.60	0.10
Dy	160	0.113	96	0.87	0.17	0.87	0.14
	162	0.033	138	0.97	0.18	0.97	0.14
	164	-0.302	178	0.52	0.14	0.52	0.17
Er	166	0.042	197	0.89	0.20	0.89	0.18
	168	-0.022	11	1.23	0.21	0.16	0.16
	170	-0.048	145	2.13	0.24	0.22	0.22
Yb	170	-0.393	126	1.74	0.19	1.74	0.18
	172	0.239	172	1.61	0.18	1.61	0.16
	174	0.068	115	0.81	0.08	0.81	0.08
Hf	178	0.009	88	1.67	0.21	1.68	0.22
W	182	0.410	203	2.08	0.25	2.09	0.24
Average		0.028	128	1.34	0.18	1.35	0.17

forces. The theoretical results of AR, RVM, ARM_β and IBM_2 are also shown in Fig. 1. The IBM_2 result is taken from ref. 9. The ones of AR and I-coup. are close to the average values. However the change of the B(E2) values calculated by eq. (3) for the RVM and ARM_β and ones by IBM_2 with $I_f(I_f +3)$ are straight lines with positive and negative slopes, respectively. The amplitudes Z_β and Z_γ calculated by eq. (3) and Z for I-coup. are listed in Table 2, respectively. As seen, the Z_γ of γ vibrations or γ deformations are close to Z values. But the average value of β vibrations are about fifty times as large as the average Z. A conclusion can be drawn that β vibrations make the theoretical B(E2) values being unconsistent with experimental ones.

Table 3 The E2 relative amplitudes for transitions between the γ and g band obtained by RVM, ARM_β and I-coup.

Nucl.	A	I-coup.		RVM			ARM_β		
		Z	δZ	Z_β	Z_γ	$Z_{\beta\gamma}$	Z_β	Z_γ	$Z_{\beta\gamma}$
Sm	154	1.35	125	0.02	1.03	0.59	-0.02	1.04	0.59
Gd	154	3.49	10	-0.09	2.03	2.06	-0.09	2.05	2.02
	158	0.83	23	-0.02	1.24	-1.62	-0.02	1.25	-1.65
	160	1.00	18	-0.01	1.46	-0.01	-0.01	1.47	-0.10
Dy	160	2.09	32	-0.01	1.68	-0.17	-0.01	1.70	-0.17
	162	1.50	26	-0.01	1.70	-0.20	-0.01	1.72	-0.21
	164	3.34	41	-0.01	1.85	-0.08	-0.01	1.87	-0.08
Er	166	1.91	14	-0.01	1.93	-0.15	-0.01	1.96	-0.16
	168	1.59	69	-0.02	1.80	-0.28	-0.02	1.82	-0.28
	170	4.57	9	-0.03	1.52	-5.44	-0.03	1.53	-5.79
Yb	168	1.34	11	-0.02	1.61	-0.64	-0.02	1.63	-0.64
	172	-0.20	61	-0.02	1.30	-6.06	-0.02	1.31	-6.51
Hf	178	0.25	32	-0.902	1.36	-1.47	-0.02	1.37	-1.50
W	182	-4.09	12	-0.02	1.35	-25.09	-0.02	1.36	-36.25
Average		1.36	46		1.56			1.58	

The δZ are errors of Z values

3.2 The B(E2) values for Transitions Between the γ and g Band

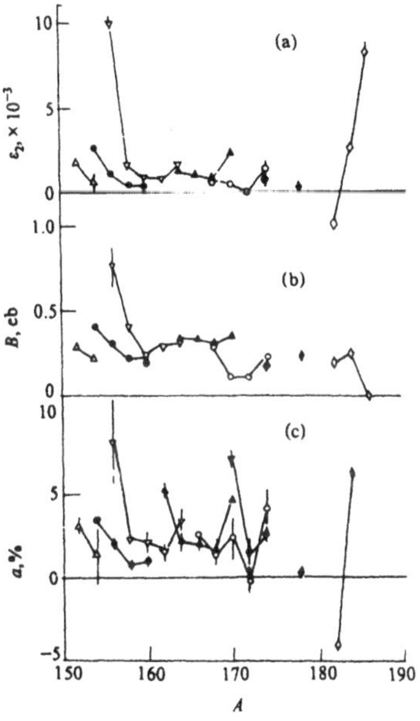

Fig. 2 The ε_2, $<K_f=2|\mu(E2,\nu=2) K_i=0>$ and α values via A for transitions between the γ and g band
(a) ε_2-values
(b) quadrupole matrix elements $B= <K_f=2|\mu(E2,\nu=2)K_i=0>$
(c) α-values for E2; Δ Sm; • Gd; ∇ Dy; Δ Er; ∘ Yb; ♦ Hf; ◇ W

The γ band is the "stable" collective rotational band because the higher spin states in the band for most of the REDN can be found experimentally. Therefore there are more experimental data of B(E2) values. They are subjected to Mikhailov rule quite well. The changes of the coupling strength ε_2 and quadrupole moment component $\mu(\nu = 2)$ deduced from Mikhailov parameters and ratio α of M_2 to M_1 are shown in Fig. 2, respectively. Their changes with mass number A are smooth, firstly decrese and then arise. It is stated that nuclei can be considered as rigidified and then softened again[10]. At some nuclei such as Yb and Hf the α values have obvious fluctuations caused by non-collective motions.

The E2 transition amplitudes obtained by eq. (3) for RVM and ARM_β are listed in Table 3. The ones of β vibrations for transitions between the γ and g band are very small. But the coupling terms between the β vibrations and γ vibrations or deformations are quite large for some nuclei such as ^{154}Gd, ^{170}Er, ^{172}Yb and ^{182}W and their changes with $[I_f(I_f + 1) - I_i(I_i + 1)]$ are non-linear and deviate significantly from Mikhailov rule. The ones of γ vibrations and γ deformations are consistent with the ones of I-coup.The average Z = 1.56 is the same as the Z = 1.36 shown at the bottom in Table 3. The theoretical and experimental B(E2) values versus $[I_f(I_f + 1) - I_i(I_i + 1)]$ are shown in Fig. 3 and 4 for ^{168}Er and ^{170}Er. The M and the relative amplitudes Z_β, Z_γ and $Z_{\beta\gamma}$ against angular momenta are shown in Fig. 5 for ^{170}Er. As seen, the changes of B(E2) values are quite well described by I-coup. The β and γ coupling terms have significantly non-linear changes with $[I_f(I_f + 1) - I_i(I_i + 1)]$ that deviate from the Mikhailov rule.

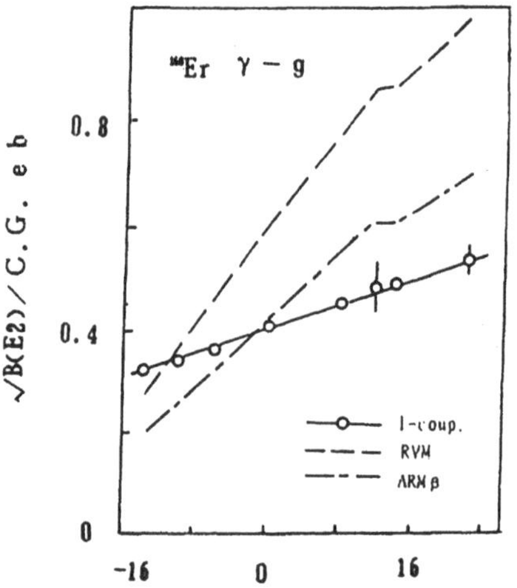

$I_f(I_f + 1) - I_i(I_i + 1)$

Fig. 3. The Mikhailov plot for ^{168}Er

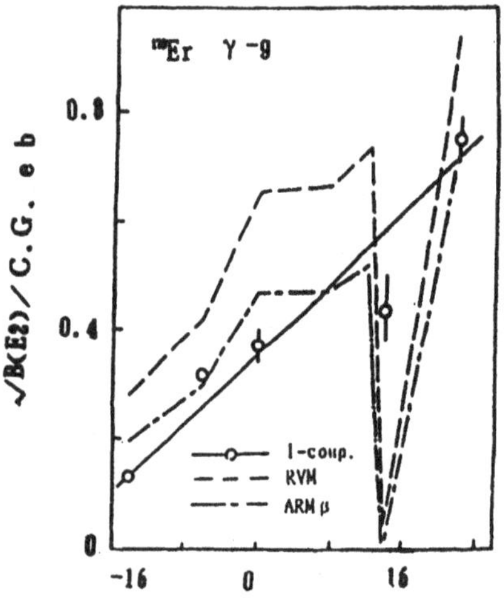

$I_f(I_f + 1) - (I_i + 1)$

Fig. 4. The Mikhailov plot for ^{170}Er

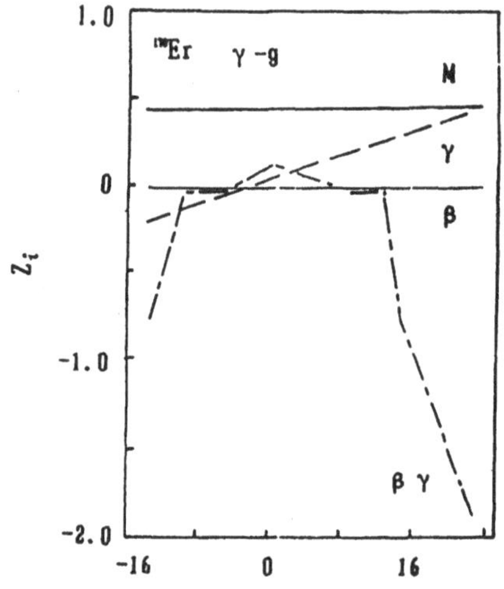

$I_f(I_f + 1)\text{-}I_f(I_f+1)$

Fig. 5. The amplitudes Z_i

3.3 The B(E2) values for transitions between the β and g band

The β band is "unstable "because the higher spin states in it are hardly found for most of the REDN. The changes of B(E2) values with angular momenta are roughly subjected to the Mikhailov rule. The changes of ε_0, $\mu(\nu = 0)$ and α with A, except for the light REDN, have significant fluctuations which are shown in Fig. 6, respectively. It is expressed that the β band intrinsic motions are non-collective. The relative E2 amplitudes Z_β, Z_γ and $Z_{\beta\gamma}$ are listed in Table 4. The average Z_β is 1.7 times as large as Z value shown at the bottom in it. The change of B(E2) values for transitions

Table 4 The E2 relative amplitudes for transitions between the β and γ band obtained by RVM, ARM_β and I-coup.

Nucl.	A	I-coup.		RVM			ARM_β		
		Z	δZ	Z_β	Z_γ	$Z_{\beta\gamma}$	Z_β	Z_γ	$Z_{\beta\gamma}$
Gd	158	0.72	18	1.79	-0.01	1.33	1.80	-0.01	0.69
Er	170	1.35	21	2.31	-0.01	4.40	2.33	-0.01	2.34
Hf	178	3.23	16	2.08	-0.01	1.46	2.10	-0.01	0.75
	180	-0.70	47	2.26	-0.01	1.41	2.28	-0.01	-0.69
W	182	1.72	33	2.36	-0.01	24.5	2.38	-0.01	17.68
Average		1.26	29	2.16			2.18		

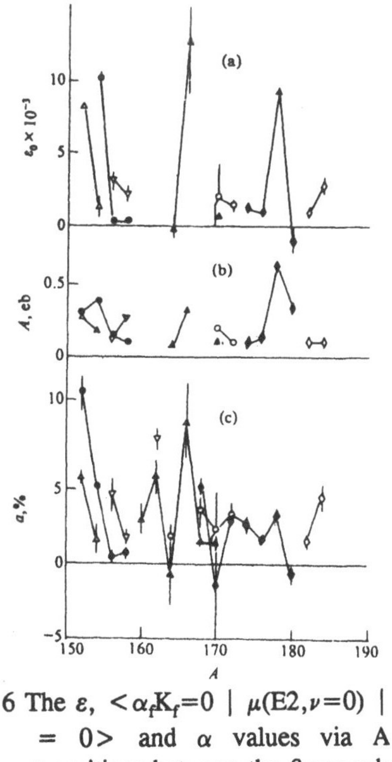

Fig. 6 The ε, $<\alpha_f K_f=0 \mid \mu(E2,\nu=0) \mid \alpha_i K_i = 0>$ and α values via A for transitions between the β ang g band

(a) ε_0-values

(b) quadrupole matrix elements $A=<\alpha_f K_f=0 \mid \mu(E2, \nu=0) \mid \alpha_i K_i=0>$

(c) α-values. Δ Sm; • Gd; ∇ Dy; Δ Er; ∘ Yb; ♦ Hf; ◇ W

between the β and g band with $[I_f(I_f + 1) - I_i(I_i + 1)]$ are shown in Fig. 7 for ^{170}Er. The non-linearity is from the coupling of the β and γ vibrations or deformations.

3.4 *The B(E2) values for transitions between the β and γ band*

The B(E2) values versus $[I_f(I_f + 1) - I_i(I_i + 1)]$ are shown for ^{168}Er in Fig. 8.

3.5 *The B(E2) values are obtained by IBM2 for ^{154}Sm, $^{156, 158}$Gd and ^{158}Dy.*

The B(E2) values calculated by IBM_2 for intraband and interband transitions for ^{154}Sm, $^{156, 158}$Gd and ^{158}Dy four nuclei are obtained and given in detail in ref. 9. The main results about E2 transitions are quoted here. The B(E2) values for

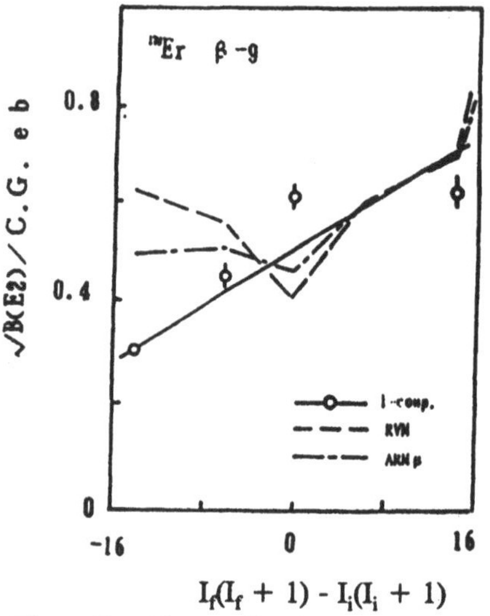

Fig. 7. The Mikhailov plot for transitions between the β and g band for ^{170}Er

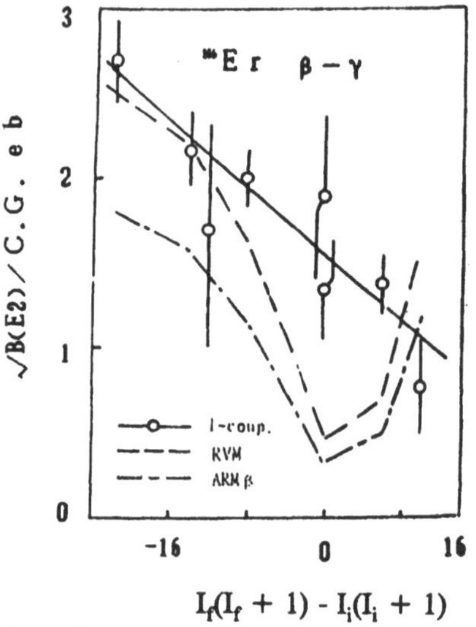

Fig. 8. The Mikhailov plot for transitions between the β and γ band for ^{168}Er

interband transitions are subject to Mikhailov rule except for those of ^{158}Dy. The effective charges e_π and e_ν are difference from those of SU(3) limit and the relative effective charges for g, β and γ band dependes on band intrinsic structures.

4. Conclusion

The B(E2) values for transitions within the ground state band have a regular behavior, which can not be fitted by the models.

The γ band is the better collective rotational band. Its intrinsic structure is γ vibrations or γ deformations that are important for the REDN. The γ vibrations and γ deformations can not be distinguished by B(E2) values. The two quasiparticle configurations in the γ band intrinsic structures should be considered for some nuclei, such as Yb and Hf.

The intrinsic structures of the β band are not the β surface vibrations or the β vibrations are not important for most of the REDN. They can contain the two quasiparticle configurations[10].

If the Mikhailov rule is correct for interband E2 transitions, there are no coupling terms between the β vibrations and γ vibrations or deformations, in other words both of them can not present simultaneously in the collective motion.

The I-coup. is quite well to fit B(E2) values of intraband and interband transitions. The coriolis and centrifugal forces are important for nuclear structure.

The interband B(E2) values calculated by IBM_2 are subjected to the Mikhailov rule for REDN with the headband energy $E_\beta < E_\gamma$, but are not for the REDN with $E_\beta > E_\gamma$, for example ^{158}Dy and ^{162}Dy[11]. The effective charges of the proton and neutron bosons are obviously different from that of SU(3). They can depend on the intrinsic structures of the bands.

Acknowledgement

The author thanks Drs. J. Y. Wang, Y. X. Huang and K. Y. Gao for doing the calculations and Profs. Y. Z. Su, D. Z. Ding, P. K. Weng, C. X. Yang and Y. S. Chen for helpful discussions.

References

1. A. Bohr and B. R. Mottelson, *Nuclear Structure*, **Vol. 2** (Benjamin, New York, 1975)
2. J. M. Eisenberg and W. Greiner, *Nuclear Theory*, **Vol. 1**
3. A. Arima and F. Iachello, *Ann. Phys.* (NY) **99** (1976) 253
4. A. Arima and F. Iachello, *Ann. Phys.* (NY) **111** (1976) 201
5. A. Arima and F. Iachello, *Ann. Phys.* (NY) **123** (1976) 468
6. F. Iachello, *Proc. of the Fourth Int. Sym. on Neutron Capture Gamma-RAY; Spectroscopy and Related Topics*, (Grenoble, France, 7-11 Sep. 1981) p. 15
7. A. Arima, *Proc. Beijing Int. Sym. on Phys.at TANDEM*, (Beijing, China, May 26-30, 1986) p. 3
8. T. Otsuka and N. Yoshida, *JAERI-M85-094*, 185
9. Z. R. Shi, Y. X. Huang and Y. Z. Su, *Chinese J. of Nucl. Phys.* **Vol. 13** (1991) 1
10. Z.R.Shi, K. Y. Gao, J. Y. Wang and Y. Z. SU, *Chinese J. of Nucl. Phys.* **Vol. 13** (1991) 245
11. D. D. Warner, Z. R. Shi, W. Gelletly, H. G. Borner, F. Hoyler, K. Schreckenbach and P. van Isacker, *Proc. Int. Conf. on Capture Gamma-Ray Spectroscopy*, (Institute of Physics Conference Series 88 (1987)) 562.

STUDY OF HIGH ENERGY γ-RAYS IN FAST NEUTRON CAPTURE REACTIONS*

HUANG ZHENGDE, ZHU LIHUA, HOU LONG
and DING DAZHAO

China Institute of Atomic Energy
P. O. Box 275(15), Beijing 102413, China

ABSTRACT

γ-ray spectrum of the $^{12}C(n, \gamma_0)^{13}C$ reaction has been measured and remeasured in pygmy resonance region. The agreement between present results and TUNL's results is quite good. The comparison with DSD model calculation indicates the validation of the inviolation of isospin conservation in the $^{12}C(n, \gamma_0)^{13}C$ reaction. The E1 radiative transition is dominante in the reaction.

1. Introduction

The status of nulear structure study according to the spin and excitation energy can be summarized as follows: LL- low spin and low excited states; LH- low spin and high excited states; HL- high spin and low excited states; and HH- high spin and high excited states. The GDR is studied by measuring high energy γ-rays.

In the case of LH, the GDR built on LH has been studied in fast neutron and proton capture reactions. In the case of HH, the GDR built on HH has been explored with the measurement of high energy γ-rays formed by heavy ion induced fusion reactions. The systematics of giant resonance (GR) indicates that giant resonances are general properties of nuclei. The excitation energy of GR varies smoothly with nuclear mass and GR exhaust an appreciable fraction of the energy weighted sum rule. The shape of the resonance is well described by the Lorentzian distribution. Some of our experimental measurements are listed in table 1.

In this paper I only discuss fast neutron capture reactions. There are three aspects developed for these reactions: (1) capture by light nuclei; (2) capture by deformed nuclei; and (3) search for IVGQR. As an example the $^{12}C(n, \gamma_0)^{13}C$ capture reaction will be discussed here.

*The project partly supported by National Natural Science Foundation of China

Table 1 Exprimental measurements

Type	Reaction	Excitation function	Angular distribution
(LH)	$^{12}C(n, \gamma_0)^{13}C$	7 - 14MeV 13 points	9, 11, 14 MeV; 55°-125°
	$^{12}C(n, \gamma_0)^{13}C$	9.3 - 14MeV 4 points	
	$^{12}C(p, \gamma_0)^{13}N$	7 - 18 MeV	
	$^{40}Ca(n, \gamma)^{41}Ca$	9.3 - 14 MeV	10.8 MeV; 55°-125°
	$^{56}Fe(n, \gamma)^{57}Fe$	9.3 - 14 MeV	9, 15 MeV; 55°, 125°
	$^{238}U(n, \gamma)^{239}U$	9.3 - 14 MeV	
(HH)	$^{124}Sn(^{16}O, \gamma)$	68, 80 MeV	
	$^{116}Sn(^{16}O, \gamma)$	68, 80 MeV	

2. The status of the $^{12}C(n, \gamma_0)^{13}C$ reaction

^{13}C is a (4n+1) nucleus. The feature of GDR for light (4n+1) nuclei is that there is always a "pygmy" resonance at low energy side associated with GDR. The conclusion was obtained clearly from the comparison of $^{13}C(\gamma, n_0)^{12}C$ and $^{12}C(\gamma, n_0)^{11}C$ reactions[1].

About the $^{12}C(n, \gamma_0)^{13}C$ reaction, we found that TLU's[2] cross section of the $^{12}C(n, \gamma_0)^{13}C$ reaction is approxmately a factor of 2 lower than that deduced from the measurement of the $^{13}C(\gamma, n_0)^{12}C$[3] reaction by detailed balance principle,and TUNL's cross section of the $^{12}C(n, \gamma_0)$ reaction. The discrepacies in $^{12}C(n, \gamma_0)$ reaction cross sections and comparison with DSD model calculation which fits the $^{12}C(p, \gamma_0)^{13}N$ reaction cross section can not confirm the violation or inviolation of isospin conservation. That is why we take the measurement of the $^{12}C(n, \gamma_0)^{13}C$ reaction in the pygmy resonance region.

3. Measurement

We have measured and remeasured the differential cross section of the $^{12}C(n, \gamma_0)^{13}C$ reaction at the neutron energy range 7 to 14 MeV independently. The experimental arrangement is shown in Fig. 1.

The neutrons were produced by the D(d, n) reaction. The deuterium gas target was a 2.8 cm long cell with a entrance window of 5.65 mg/cm^2 Havor foil.

The deuteron beam was chopped and bunched into a 1 ns beam burst at a repetition rate of 4 MHz. The ^{12}C sample was a 40 mm by 40 mm or 30 mm by 30 mm cylinder of graphite. The cylinder was oriented with its symmetry axis perpendicular to the reaction plane. The γ-ray detector is a large NaI with plastic anticoincidence and heavy shielding. The sample was in front of 12 cm away from the gas cell.

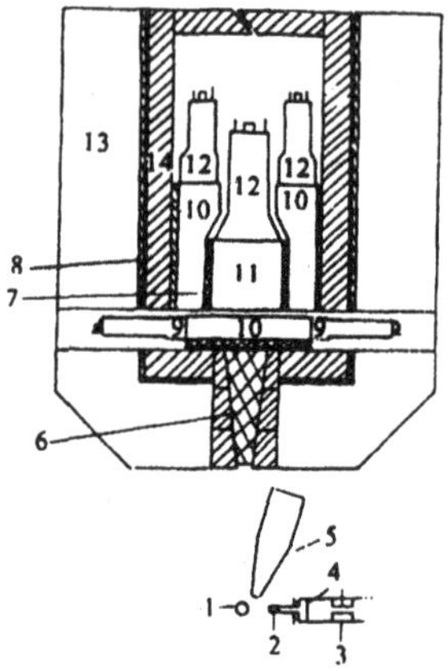

Fig. 1. Experimental arrangement

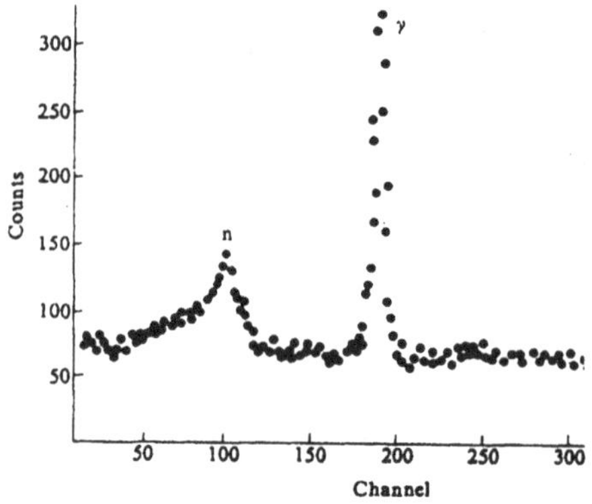

Fig. 2. The TOF spectrum of the $^{12}C(n, \gamma)^{13}C$ reaction $E_n = 9$ MeV, $\theta = 90°$

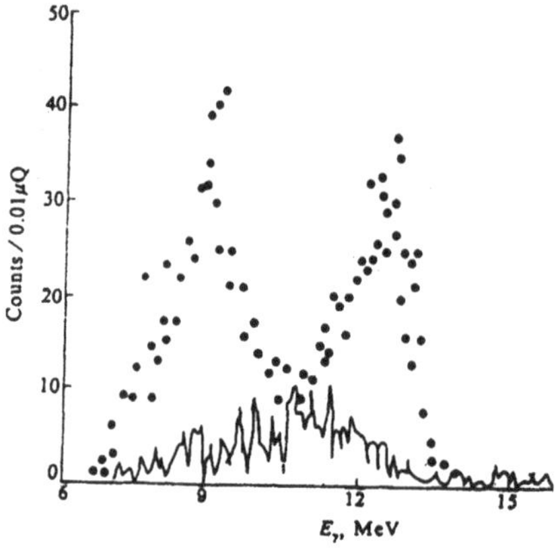

Fig. 3. The accepted spectrum $E_n = 9$ MeV, $\theta = 90°$

The time of flight technique has been used to seperate neutrons and γ-rays. The spectrum of time of flight is shown in Fig. 2. Neutrons and γ-rays can be seperated clearly. The accepted spectrum is shown in Fig. 3 after the subtracting the rejected spectrum and being gated by TAC γ-rays. The efficiency of the NaI detector system was calibrated by 15.1 MeV g-ray produced by the $^{12}C(p, \gamma_0)^{13}N$ resonance reaction at $E_p = 14.23$ MeV.

4. Results and analysis

4.1 The excitation function of the $^{12}C(n, \gamma_0)^{13}C$ reaction

The absolute cross section for a given energy and angle can be written as

$$\sigma(\theta, E_n) = \frac{N_\gamma \cdot r_{gs}^2 \cdot f}{N_A \cdot N_d \cdot n_g \cdot l \cdot \sigma_n(0°) \cdot \varepsilon \cdot \Omega},$$

where N_γ is the number of γ-rays detected by NaI detector system; ε is the effeciency of the NaI detector system; N_d the number of the deutrons incident on the deuterium gas cell; n_g the density. of the deuterium gas in the cell; r_{gs} the distance between the centers of gas cell and sample; l the length of the gas cell; Ω the solid angle sustended by the NaI detector at the sample; $\sigma_n(0°)$ correspond to the 0° differential cross section for the D(d, n) reaction. f is the correction factor of the finite geometry. The Monte-carlo code was used for the correction of γ-ray attenuation, and neutron attenuation and multiple scattering in the sample. The program also takes into account energy dependence in the neutron angular distribution,which is folded in the sample.

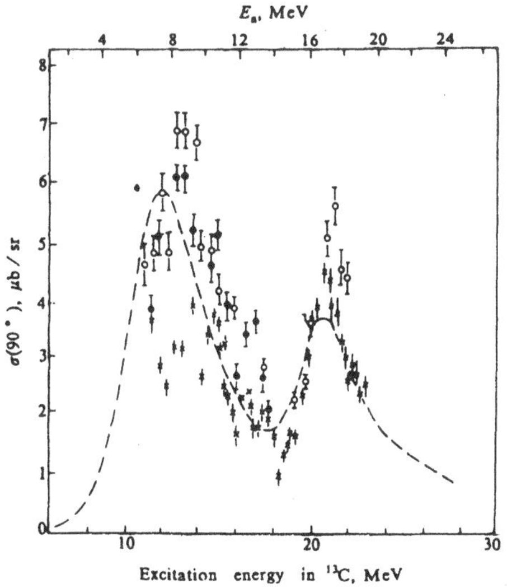

Fig. 4. The 90° absolute differential cross section of $^{12}C(n, \gamma_0)^{13}C$ reaction
• present data, ∘ TUNL data[4],
× TLU data[2], --- DSD calculation[2]

The absolute differential cross section at 90° is shown in Fig. 4 along with previous measured results. The agreement between the present results and TUNL's is quite good in the measured energy region.

DSD calculation for both $^{12}C(p, \gamma_0)^{13}N$ and $^{12}C(n, \gamma_0)^{13}C$ reactions has been made in the same way. The parameters of the model are adjusted to fit proton capture data, Then the neutron capture reaction is predicted by dropping off the coulomb terms from the optical potential. The prediction for the neutron capture reactions compared with the present results and previous data in Fig. 4. The present results are in much better agreement with DSD calculation. According to the DSD calculation the present results and TUNL's data indicate the inviolation of isospin conservation, but TLU's data indicate that the violation of isospin conservation. The discrepancy of the data from different laboratories needs to be clarified in the future.

4.2 Measurement of the angular distribution of the $^{12}C(n, \gamma_0)^{13}C$ reaction.

The angular distribution of the $^{12}C(n, \gamma_0)^{13}C$ reaction has been measured at three energies, 9, 11 and 14.2 Mev and at five angles from 55° to 125° .The angular distribution of the cross section was fitted by using Legendre polynomials

$$\sigma(\theta, E) = A_0(E) [1 + \Sigma a_k \cdot P_k (\cos\theta)].$$

The results indicate that a_2 is dominant. The results are in agreement with the results deduced from the inverse reaction $^{13}C(\gamma, n_0)$[4]. That means E1 radiative transition dominates in the $^{12}(n, \gamma_0)^{13}C$ reaction.

Legendre polynomials can be written as a function of transition amplitude and phase

$$A_0 = {}^2S_{1/2}^2 + 2 \cdot {}^2D_{2/3}^2 \qquad \text{and}$$

$$A_0 \cdot a_2 = -2 \cdot {}^2S_{1/2} \cdot {}^2D_{3/2} \cdot \cos(\phi_d - \phi_s) - {}^2D_{3/2}^2 \cdot$$

If we assume $s = {}^2s_{1/2} / \sqrt{A_0}$ and $d = \sqrt{2}\, {}^2D_{3/2} / \sqrt{A_0}$,

then we get

$$1 = s^2 + d^2 \qquad \text{and}$$

$$a_2 = - d^2 /2 - \sqrt{2} \cdot s \cdot d \cdot \cos(\phi_d - \phi_s).$$

The relation between ratio of transition amplitude, s/d and cosin of phase difference $(\phi_d - \phi_s)$ can be obtained. We have known that the ratio s/d is an approximate constant in the GDR region. If we take the ratio s/d = 1.91 ± 0.34 then the phase difference $(\phi_d - \phi_s)$ =101° ± 10°, 112° ± 10° and 133° ± 15° for E_n = 9 MeV (E_x = 13.25 MeV), 11 Mev (15.09 MeV) and 14.2 MeV (17.9 MeV), respectively.

References

1. J. W. Jury and B. L. Beerman et al., *Phys. Rev.* **C19** (1979) 1684.

2. I. Beergqvist and D. M. Drake et al., *Nucl.Phys.* **A456** (1986) 426; A. Hakansson and A. Lindholm et al., *Phys. Rev.* **C41** (1990) 2556.
3. J. G. Woodworth et al., *Nucl. Phys.* **A327** (1979) 53.
4. R. A. August et al., *Phys. Rev.* **C35** (1987) 393.

THERMAL NEUTRON CAPTURE γ-RAY SPECTROSCOPY IN LIGHT AND MEDIUM NUCLEI

Klaus-Peter Lieb
II. Physikalisches Institut, Universität Göttingen,
D-3400 Göttingen, Fed. Rep. Germany

ABSTRACT

Complete spectroscopy in some selected nuclei with mass number $30 \le A \le 90$ has been carried out using thermal neutron capture reactions at the high flux reactor of the Institut Laue-Langevin/Grenoble. The results are summarized with emphasis on nuclear level density parameters and primary γ-ray strength functions. The novel GRID (n,γ) lifetime technique is described and its potential for measuring nuclear lifetimes and the slowing-down process of low-energy atoms in solids is discussed.

§1. INTRODUCTION

For several decades thermal neutron capture γ-ray spectroscopy has been a well-established field in low-energy nuclear physics and has provided a wealth of information on the properties of particle-bound excited states. As thermal neutrons are generally captured in a non-selective s-wave process and the primary spectrum is dominated by dipole transitions, the spins of intermediate states normally extend over the range $|I_t - 3/2| \le I \le I_t + 3/2$ where I_t denotes the target spin. With the high neutron fluxes and sophisticated γ-ray detector systems available, "almost all" excited states in this spin window and up to about half the neutron binding energy E_B can be identified in light and medium nuclei. Up to 85-98% of the total γ-ray flux through the final nucleus can be established. This not only adds tremendously to our knowledge of low-lying states, concerning their energies, spins, parities, γ-decays, and therefore model interpretations, but also contributes to our understanding of statistical or even chaotic behaviour of nuclei.

Rather than attempting to give a general overview of the field, I have concentrated on some recent results of the Göttingen-Grenoble-Munich collaboration which may illustrate the present state of the art. The topics addressed are γ-ray spectrometers (§2), level density parameters (§3) and primary γ-ray strength functions (§4).

§ 2. SPECTROMETERS AT ILL

Given the high thermal neutron flux of 5 10^{14} /cm^2 s at the target site which is at a distance of some 50 cm from the reactor core, it is mainly the design of the target assembly and the γ-ray spectrometers which determine the sensitivity and resolution with which the capture radiation can be measured. At ILL, the three **curved crystal spectrometers GAMS 1-3** [1] and a Ge-detector which, in connection with an annular two-segment NaI-detector can be operated in **γ-ray singles, pair and Compton suppressed (CS)** mode, are available.

As we are usually interested in resolving the full γ-ray flux through a given final nucleus, with hundreds of discrete cascade transitions connecting the capture state with the ground state, we need both high sensitivity and good energy resolution. In most light and medium $A \leq 100$ nuclei, a good compromise between counting statistics, isotopically enriched target material and measuring time was to scan the whole capture spectrum with the pair/CS detector and to calibrate some strong secondary transitions with GAMS2/3. Alternatively, one may calibrate the capture spectrum relative to the ^{35}Cl(n,γ) ^{36}Cl reaction which has been carefully calibrated relative to some γ-ray standards [2,3].

Recently, the **double flat-crystal spectrometer GAMS4** has been installed at ILL as a joint NIST/ILL instrument [4,5]. It consists of two perfect Si or Ge crystals, 1.5 - 3 mm thick and 50 x 25 mm^2 in area, in which the γ-ray beam from the capture target experiences twice Bragg-Laue diffraction before being recorded in a Ge detector. By means of two sophisticated Michelson interferometers, the angle between the two crystals can be adjusted ("rocked") with a precision of about 10^{-10} radians giving an experimental energy resolution as small as $\Delta E/E \approx 2$ ppm (FWHM). As shown in fig. 1, the spectrometer can be operated in the non-dispersive mode with parallel crystals: in that way the response function of the spectrometer is measured with smallest line width and thus can provide new high precision γ-ray standards, e.g. the deuteron binding energy (and neutron mass [6]) and the ^{36}Cl transitions [3].

In the dispersive mode, the photon beam is diffracted twice into the same direction. This allows one to measure the γ-ray line profile of the source folded with the response function. If the γ-ray emitting nucleus is not at rest, we will notice in the dispersive mode a Doppler broadening of the radiation. In this spirit, Börner [7] has proposed the **GRID** method (Gamma Ray Induced Doppler broadening) by which nuclear lifetimes after neutron capture can be measured. Let us assume for simplicity an isotropic two-step γ-ray cascade with primary radiation E_1 and secondary radiation E_2 involving an intermediate nuclear state with lifetime τ in the range 10^{-15}-10^{-12} s. The Doppler broadened spectral distribution $N(E_2)$ of the secondary radiation reflects the initial recoil velocity v(O)/c which the nucleus obtains during the emisssion of the unobserved radiation E_1, and the subsequent slowing-down process of the nucleus/atom in the lattice. For A=50, E_1=5 MeV and M_N=938 MeV/c^2, the initial velocity is $v(0)/c \approx E_1/AM_Nc^2 \approx 10^{-4}$ and produces a Doppler shift of only 200 eV at $E_2 \approx 2$ MeV. The spectral distribution $N(E_2)$ thus intimately correlates τ with the low-energy atomic stopping power in solids. In the limit $\tau \gg 10^{-12}$ s, $N(E_2)$ approaches the non-dispersive line shape. Two Doppler-broadened lineshapes and the corresponding response functions taken in the reaction Cl(n,γ) are shown in fig. 2.

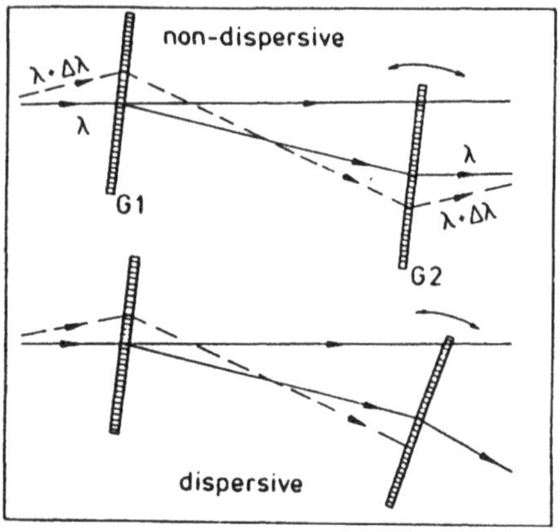

Fig. 1: Non-dispersive and dispersive diffraction modes of GAMS4 [11].

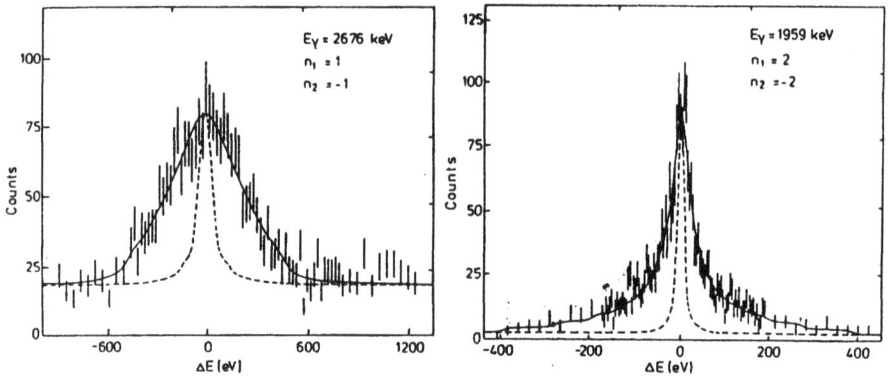

Fig. 2: Doppler-broadened lineshapes of the 2676 and 1959 keV transitions in ^{36}Cl, after the reaction ^{35}Cl (n,γ) [8].

GRID measurements have so far been mostly performed for medium mass targets with high capture cross-sections and simple primary feeding conditions, i.e. ^{35}Cl [8], ^{48}Ti [7,9], ^{53}Cr [10], ^{56}Fe [11] and ^{58,60,62}Ni [12], first results for heavy nuclei are given in [13]. The following two examples [8,12] may illustrate the potential of GRID for measuring **nuclear lifetimes** and for testing the **interatomic potentials** which ultimately determine the details how the excited nucleus looses its initial velocity by atomic collisions. An extensive comparison of nuclear lifetimes in ^{59,61}Ni over quite a broad time range was carried out which, together with our previous results, is presented in fig. 3. The GRID method works best for lifetimes in the range 10^{-14}-$3*10^{-13}$ s, where statistical errors below 5% can be achieved. The conventional Doppler Shift Attenuation (DSA) method after charged particle -induced reactions covers the region from $2*10^{-14}$ to 10^{-12} s. In general, the agreement between GRID and DSA lifetime values turned out quite satisfactory, with the need to extend this comparison to still smaller lifetimes.

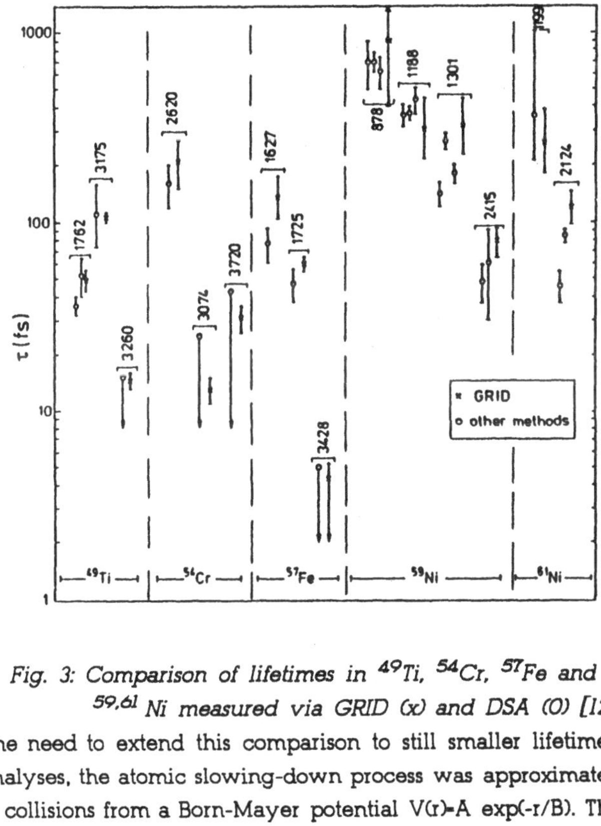

Fig. 3: Comparison of lifetimes in ^{49}Ti, ^{54}Cr, ^{57}Fe and ^{59,61} Ni measured via GRID (x) and DSA (O) [12].

In these GRID analyses, the atomic slowing-down process was approximated via isotropic hard-sphere collisions from a Born-Mayer potential V(r)=A exp(-r/B). The potential parameters A and B were taken from [14]. Recently, Kuronen [15] has approximated the scattering process by molecular dynamics theory. In cases where the lifetime of the intermediate state is accurately known via an independent method, GRID provides a very elegant method for checking the atomic slowing-down process - and one of the very few methods to study sub-keV atomic collisions inside the bulk. Lacking this calibration possibility, consistency checks between different target materials can be carried out. Fig. 4 illustrates the lifetimes of four states in ^{36}Cl which were deduced from GRID lineshape fits measured with four different chloride targets (NaCl, KCl, $SrCl_2$, $BaCl_2$). These compounds differ by their cation masses which leads to quite different momentum transfers of the colliding ^{36}Cl atom. Fig. 4 illustrates the high degree of consistency of the various lifetime values in the 30-100 fs range. For the shortest lifetimes and the two heavy compounds, the line shape fits even turned out to depend on the most probable collision sequence [8].

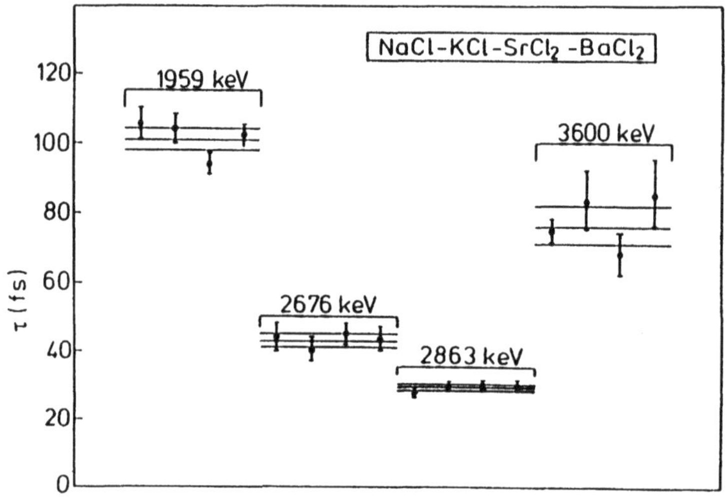

Fig. 4: Comparison of the lifetimes in ^{36}Cl measured in four different chlorides [8].

§ 3. COMPLETE SPECTROSCOPY, LEVEL DENSITIES

Complete spectroscopy aims to localize, in a given spin and energy range, all quantum states of the nucleus. On the one hand, completeness is essential for correlating all experimental states with those which a nuclear model predicts (or is not able to predict). On the other hand, correlations among level energies and decay widths/amplitudes are sensitive to the validity of statistical assumption on the mixing of states with the same I^{π} . One of the goals of **complete spectroscopy** is thus to determine, for given spin and parity, the distribution of nuclear level energies which ultimately may shed some light onto the problem of chaoticity in nuclei [16,17]. Thermal neutron capture proceeds via s-wave neutrons to highly complicated compound (capture) states. Their decays are generally considered to be solely governed by level density and γ-ray phase space arguments, besides angular momentum and parity conservation. In heavy nuclei ($A>100$), the high line density in the γ-ray spectra restricts the range of completeness to 1 - 1.5 MeV excitation. In a number of $A<100$ nuclei we have succeeded to extend the region of completeness up to about half the binding energy E_B.

The construction of a level scheme out of an experimental catalogue of γ-rays is based on some known low-lying transitions and proceeds with the Ritz combination procedure, details are given in [2,18]. After recoil corrections made, the energies of all individual γ-ray cascades are required to add up to the total binding energy E_B. Furthermore, the incoming and outgoing flux through each intermediate level must be balanced. In the many nuclei, we have also taken high resolution (d,p) spectra, by means of the Munich tandem accelerator and Q3D spectrometer. In the case of ^{31}P and ^{89}Y targets, a vector polarized deuteron was used which allowed to make spin-parity assignments to a large number of states in ^{32}P [19] and ^{90}Y [20].

As an example, fig. 5 illustrates the **level densities** in $^{87,88}Sr$ measured by Winter et al. [21]. The solid lines are fits to the data (histograms) with a constant temperature Fermi gas (CTFG) model: $\rho(E_x) = (1/T)\exp\{(E_x-E_0)/T\}$, with the temperature T and the backshift E_0 being adjustable parameters. As usual the level density at spin I was parametrized as $\rho(E_x,I) = \rho(E_x)\, f(I)$ with the spin distribution function $f(I)$ given by $f(I) = \exp\{-I^2/2\sigma^2\} - \exp\{-(I-1)^2/2\sigma^2\}$, σ is the spin cut-off parameter. The fit sensitively depends on the s-wave level density at the neutron binding energy E_B which is extracted from the average low-energy s-wave resonance spacing[22]. The dashed lines in fig. 5 indicate the expected level densities according to an extrapolation into the non-complete region developed by Krusche and Lieb [23]. In the non-complete region more and more primary transitions escape detection because of the limited spectrometer sensitivity and/or their overlap with secondary transitions. The experimentally observed level density ρ_{expt} is related to the "true" level density $\rho(E_x)$ via a weighting factor

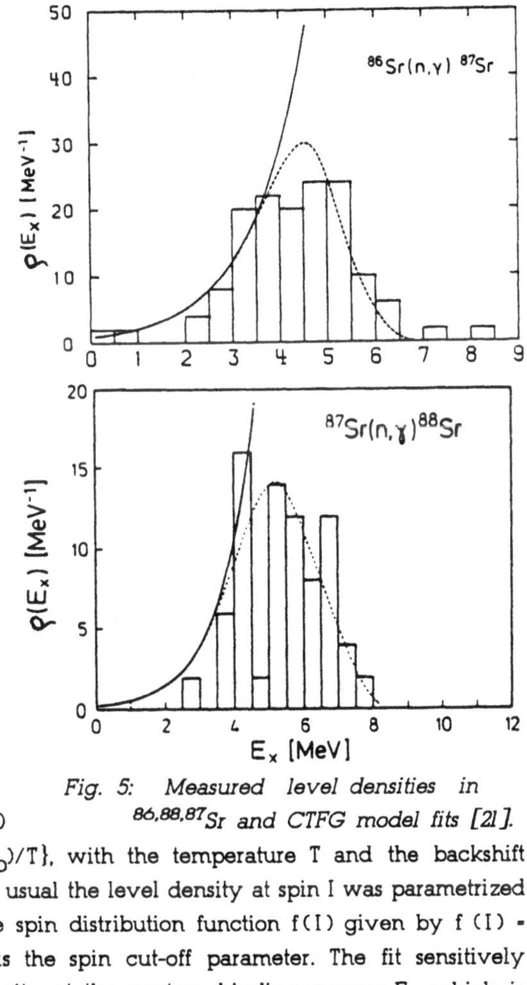

Fig. 5: Measured level densities in $^{86,88,87}Sr$ and CTFG model fits [21].

$$W(E_x) = \rho_{expt}(E_x)/\rho(E_x) = \int_0^{E_x} P_\nu(r)\,\varepsilon(r,E_x)\,dr.$$

$\varepsilon(r, E_x)$ is the sensitivity function and varies between 0 and 1. The intensities of the primary transitions are Porter-Thomas distributed with ν degrees of freedom and scale as E_1^n on the average; n and ν are fitting parameters. When comparing, for a given final nucleus and spin-parity window, the level density deduced from the (n,γ) and (d,p) reactions, the non-selective (n,γ) process usually leads to a somewhat higher level density than the (selective) single-particle transfer reaction. The difference was most pronounced in the case of the neutron-magic ^{88}Sr target [21], but gave in both reactions the same temperature T.

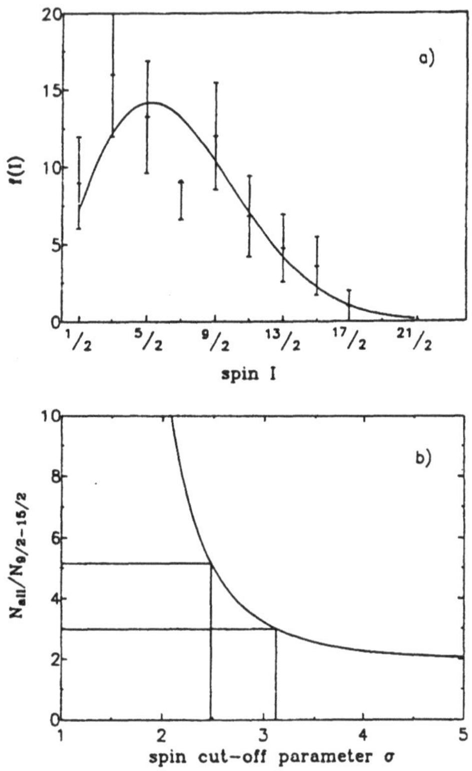

Fig. 6a): Measured and fitted spin distribution f(I) in ^{51}V.

Fig. 6b): The ration $Na_{11}/N_{9/2-15/2}$ sensitively depends on σ and gives σ = 2.8(3) for ^{51}V [21].

Among the three parameters T, E_0 and σ of the constant temperature Fermi gas model, the spin cut-off parameter σ usually has the largest uncertainty, since the narrow (n,γ) spin window generally only covers low spin states. Indeed, a comprehensive fit of many nuclei throughout the Periodic Table gave [24] $\sigma(A) = (0.98 \pm 0.23)\, A^{(0.29 - 0.06)}$. There exist, however, few high spin targets accessible to neutron capture, among them the 6$^+$ ground state of ^{50}V. Michaelsen et al. [25] have recently reported on their detailed ^{50,51}V(n,γ) work in which they identified or confirmed 59 levels with I=9/2-15/2 in ^{51}V. Fig. 6a illustrates the measured **spin distribution** f(I) and a fit for T=1.36 MeV, E_0 =-2.1 MeV and σ =3.1. The experimental ratio N_{all} / $N_{9/2-15/2}$, i.e. the number of all states relative to that with high spins (see fig. 6b), yields a rather precise value of σ = 2.8(3).

§ 4. GAMMA RAY STRENGTH FUNCTIONS

The many decay branches of the neutron capture state(s) constitute a good data base for investigating the **"downward" E1 and M1 strength functions,** they are related to the usual "upwards" strength functions which are a measure of the electric and magnetic polarizability of the nucleus [26,27]. While in heavy nuclei, the primary spectrum is dominated by the E1 giant resonance (GDR), the situation appears to be more complex in light nuclei where E1 and M1 transitions compete with each other. As the the GDR energy scales like $E_O = 80\ A^{-1/3}$, the resonance is closer to the binding energy $E_B \approx$ 7-11 MeV in heavy nuclei than in light nuclei: E_O= 13.5 MeV for A=208 versus E_O= 23.4 MeV for A=40. In addition the GDR peak cross section $\sigma_O \approx$ 9 NZ/A mb decreases in light nuclei. For that and other reasons, the E1 giant resonance affects the primary (n, γ) spectrum much less in light nuclei than in heavy ones.

According to Bartholomew et al. [28], the E1 and M1 strength functions can be expressed as $f(E1) = \Gamma_i\ (E1)/E_i^3\ D_\lambda$, $f(M1) = \Gamma_i\ M1)/E_i^3\ D_\lambda$. Here Γ_i denotes the absolute primary transition width for the transition with energy E_i (in MeV) and D_λ (in eV) the average spacing of accessible (s-wave) neutron resonances at E_B . Fig. 7 illustrates the experimental f(E1) and f(M1) distributions for the Sr isotopes discussed by Winter and Lieb [29]. Inserted are in fig. 7a the predictions of the single-particle model, $f_{sp} = 1.35\ 10^{-6}$ MeV^{-3}, the average E1 strength, $<$(E1)$> = 1.03\ 10^{-7}$ MeV^{-3}, and the GDR prediction evaluated for the GDR parameters E_O=6.84 MeV, Γ_O=4.5 MeV and σ_O= 206 mb. For the distribution of 23 M1 transitions in Sr illustrated in fig. 7b, one

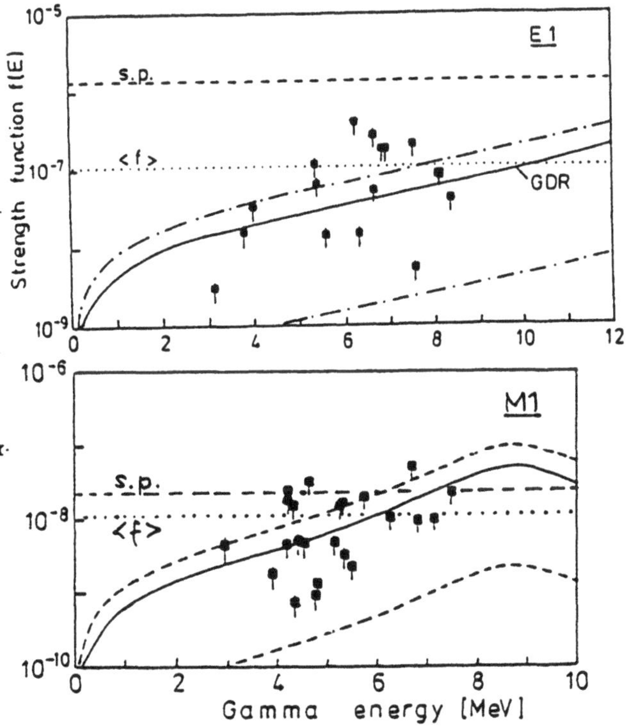

Fig. 7: E1 and M1 strength functions in the Sr isotopes. For details see text [29].

obtains f_{sp} = 2.1 10^{-8} MeV^{-3} and $<f(M1)>$ = 1.2 10^{-8} MeV^{-3}, the "best" fit with a GDR with fixed resonance energy E_0 =8.9 MeV requires the resonance width $\Gamma_0 \approx 3.0$ MeV and the peak cross-section $\sigma_0 \approx 5$ mb (see solid line). The E1 strength is in Sr by an order of magnitude larger than the M1 strength: $<f(E1)>/<f(M1)>$ = 8.6.

Before turning to the sd-shell nuclei, let us shortly discuss the **energy scaling of the strength function** in $^{51,52}V$ [25]. It has been customary to compare the integrated primary flux $Y(E_x)$ = $\Sigma\, b(E_i)$ with $E_i \geq E_B - E_x$, where $b(E_i)$ denotes the branching ratio to the level $E_x = E_B - E_i$, with the theoretical quantity

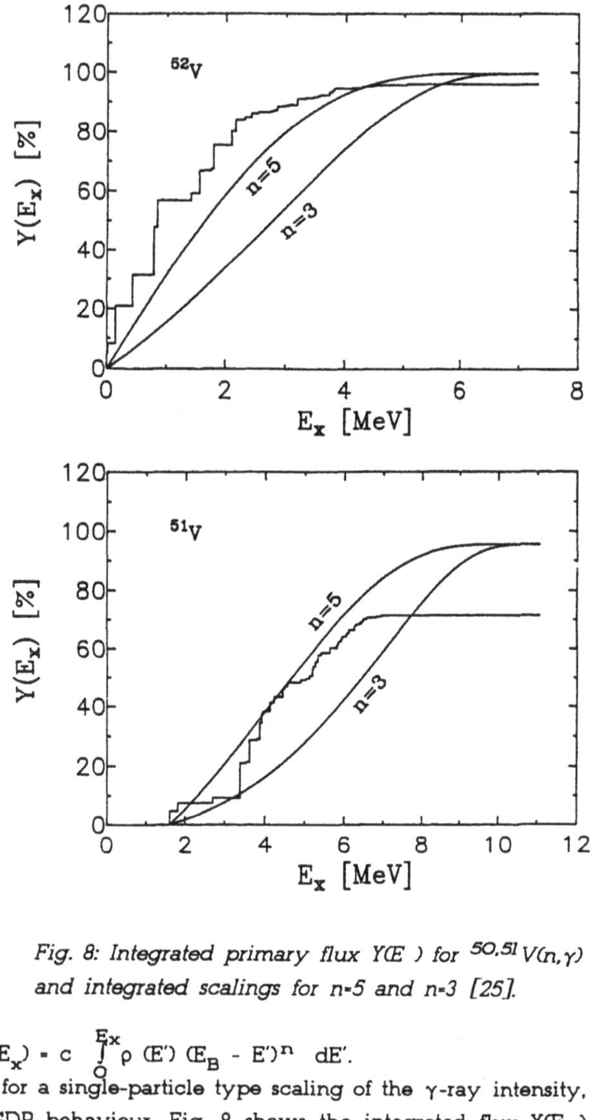

Fig. 8: Integrated primary flux Y(E) for $^{50,51}V(n,\gamma)$ and integrated scalings for n=5 and n=3 [25].

$$Y(E_x) = c \int_0^{E_x} \rho(E')(E_B - E')^n \, dE'.$$

The exponent n=3 is typical for a single-particle type scaling of the γ-ray intensity, while n=5 approximates a GDR behaviour. Fig. 8 shows the integrated flux $Y(E_x)$ in ^{51}V and ^{52}V which in both nuclei follows n=5. As more than 90% of these primary transitions are E1 in nature, we conclude that a simple E1 giant resonance scaling prevails [25].

We finally discuss the El and Ml primary strength functions in some odd-odd nuclei of the (25 ld) shell. As an example, fig. 9 illustrates the integrated primary γ-ray flux of the reaction $^{31}P(n,\gamma)^{32}P$ [18]. The three parts refer to the total dipole flux irrespective of the parity of the final state (a), and the individual El and Ml component (b, c). In this reaction the capture state has $I_c = 0^+, 1^+$ and the spin window accessible via primary dipole transitions only covers I=0-2. It is seen that the scaling exponent dramatically depends on whether one differentiates between electric and magnetic transitions or not. The El strength clearly follows n=3, while the Ml strength even rises more steeply than for n=5. The situation in ^{36}Cl is similar leading to n=3 for the El component, n=5 for the Ml component and n≈3 for the total spectrum [18].

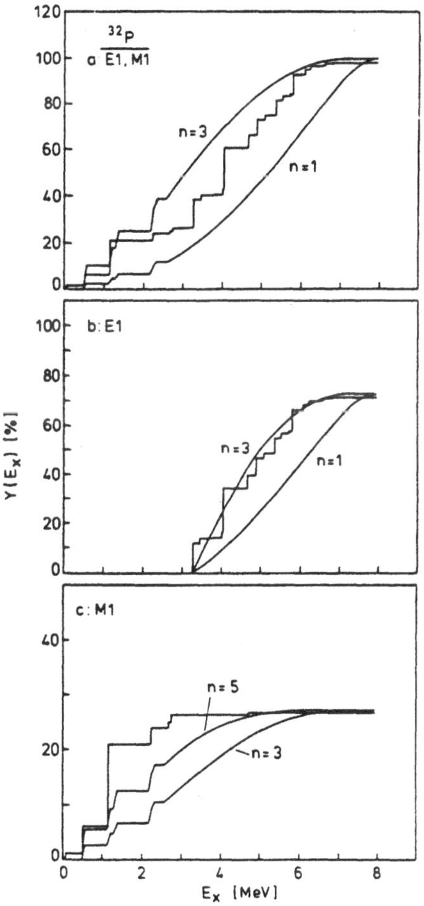

Fig. 9: Integrated primary γ-ray flux $Y(E_x)$ of the reaction $^{31}P(n,\gamma)$ and its El and Ml components [18].

This behaviour is related to two nuclear structure effects:

(i) The low-lying states of these light odd-odd nuclei have positive parity and can be explained via proton and/or neutron excitations within the (2s ld) shell. Those states can only be reached via primary Ml transitions from the positive parity capture state(s). The primary Ml strengths are highly correlated to the s-wave (d,p) spectroscopic factors, as a consequence of a semi-direct capture process proposed by Lane and collaborators [26].

(ii) Primary El transitions require negative parity states and therefore excitation of one nucleon across the Z=N=8 or 20 shell gap. The energy required is ≈3.2 MeV in ^{32}P. It thus appears that these light nuclei exhibit statistical aspects (Fermi gas level densities, statistical El transitions) together with non-statistical features (Ml transitions correlated to the distribution of stripping spectroscopic factors). This is due to the weakness of the El giant resonance contribution which in heavier nuclei tends to dominate other structure effects.

I have shown that the distribution of level energies and primary transition strengths after thermal neutron capture can be well described by the statistical model, provided that some individual parameters (T, E_0, σ, <f(E1)>, <f(M1)>, n) are admitted. As the capture states by no means are singled out in their structures, these parameters should also be valid for the secondary spectrum and in principle should allow us to **predict the full γ-ray flux** through the nucleus. In order to test this conjecture, we have recently calculated the number and intensities of primary and secondary transitions to be observed in the reactions $^{41}K(n,\gamma)^{42}K$ and $^{65}Cu(n,\gamma)^{66}Cu$ for a given finite sensitivity of the spectrometer. The simulated and measured flux patterns agree rather well and indeed confirm the very good consistency of this approach [23].

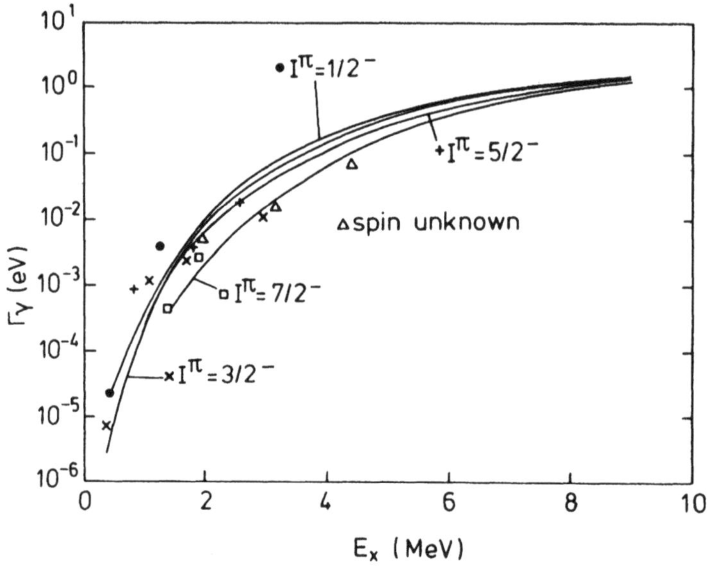

Fig. 10: Gamma decay widths Γ_γ of low-lying states in ^{59}Ni (I^π=1/2⁻-7/2⁻) and statistical model predictions [12].

We can even go one step further and simulate the absolute secondary transition strengths, i.e. the mean lives of these levels. As we measured a large number of lifetimes in the reaction $^{58}Ni(n,\gamma)^{59}Ni$ (see § 2), we have compared the natural line widths Γ_γ of these states with the statistical model predictions. Fig. illustrates the rather good agreement (within a factor of 2) for a nucleus which is magic in the proton number (Z=28) and has only three valence neutrons in the ($1f_{5/2}$ 2p) shell.

The author would like to thank B. Krusche, Ch. Winter and S. Michaelsen who as PhD students have taken the burden of extensive data analysis, and the collaborators at the Institut Laue-Langevin for their invaluable help to accumulate the data. This work has been funded by Deutsches BMFT.

REFERENCES

(1) H.R. Koch, et al., Nucl. Instr. Meth. 175 (1980) 401

(2) B. Krusche, et al., Nucl. Phys. A386 (1982) 245

(3) E.G. Kessler, et al., Phys. Rev. C32 (1985) 374

(4) R.D. Deslattes, et al., Ann. of Phys. (N.Y.) 129 (1980) 378

(5) M.S. Dewey, et al., Nucl. Instr. Meth. A284 (1989) 151

(6) G.L. Greene, et al., Phys. Rev. Lett. 56 (1986) 819

(7) H.G. Börner, J. Phys. G14 (1988) S143, H.G. Börner, et al., Phys. Lett. B215 (1988) 45,

(8) S. Ulbig, doctoral thesis Göttingen (1991) unpubl.,
S. Ulbig, et al., Phys. Lett B259 (1991) 29

(9) J. Jolie, et al., Europhys. Lett. 10 (1989) 231

(10) K.P. Lieb, et al., Phys. Lett. B215 (1988) 50

(11) S. Ulbig, et al., Nucl. Phys. A505 (1989) 193

(12) S. Ulbig, et al., Z. Phys. A338 (1991) 397

(13) H.G. Börner, et al., Asilomar Conference 1990, in press

(14) R.S. Nelson, "The observation of atomic collisions in crystalline solids" (North Holland, Amsterdam, 1968)

(15) A. Kuronen, J. Phys. Condens. Matter 3 (1991) 1363

(16) O. Bohigas, H.A. Weidenmüller, Ann. Rev, Nucl. Part. Sci. 38 (1988) 421

(17) B. Eckhardt, Phys. Rep. 163 (1988) 205

(18) S. Michaelsen, et al., Nucl. Phys. A501 (1989) 437

(19) F.J. Eckle, et al., Nucl. Phys. A501 (1989) 413

(20) S. Michaelsen, et al., in preparation

(21) Ch. Winter et al., Nucl. Phys. A460 (1986) 501, *ibid.* A473 (1987) 129

(22) S.F. Mughabghab, M. Divadeenam, M. Holden, "Neutron Cross Sections" Part I (Academic Press, New York, 1981)

(23) B. Krusche K.P. Lieb, Phys. Rev. C34 (1986) 2103, J. Phys. G14 (1988) S183

(24) T. von Egidy, A.N. Behkami, H.H. Schmidt, Nucl. Phys. A454 (1986) 109, A481 (1988) 189

(25) S. Michaelsen, K.P. Lieb, S.J. Robinson, Z. Phys. A338 (1991) 371

(26) A.M. Lane, in: "Neutron Gamma-Ray Spectroscopy" (Petten, 1974) p.31

(27) C.F. Clement, A.M. Lane, J. Kopecky, Phys. Lett. B71 (1977) 10

(28) G.A. Bartholomew, et al., Adv. Nucl. Phys. 7 (1972/73) 229

(29) Ch. Winter, K.P. Lieb, Z. Phys. A332 (1989) 397

COMMENTS ON APPLICATIONS OF DOPPLER SHIFT ATTENUATION ANALYSIS IN FAST NEUTRON EXPERIMENTS

WEN CHENLIN
Institute of Low Energy Nuclear Physics
Beijing Normal University
Beijing 100875, China

ABSTRACT

The applications of Doppler shift attenuation method in fast neutron experiments were reviewed. Some suggestions on further developments were proposed.

The Doppler shift attenuation method for determining lifetime of excited nuclear states populated in nuclear reactions induced by charged particles (proton, alpha and heavy ions) has been widely used for last two decades[1]. Hundreds papers on this subject were published. Thousands of excited levels for a large variety of nuclei from Li to U have been investigated. Y. G. Kosyak[2] applied this method to neutron inelastic scattering for reactor neutrons. Recently, such study was successfully extended to 14 MeV neutron induced reactions[3]. It was found that Doppler shift attenuation method is promising for fast neutron experiment analyses. From such analyses a lot of valuable information of different kinds, such as reaction mechanism, lifetimes of excited states, stopping power and angular distribution of inelastic scattering of fast neutrons can be derived. For 14 MeV neutron induced reactions on light and medium nuclei the Doppler shift effect is quite clear, many excited states and reactions can be studied.

In the report [3] first of all reaction mechanism of the n + Al reaction was analyzed. The stopping time of recoil nuclei ^{26}Mg in target material Al is about 0.6 ps. When the lifetime of a transition is much less than the stopping time of the recoil, then the shape of gamma line reflects mainly the reaction kinetics and the mechanism of light particle emission. In this manner the shapes of 1129 KeV line of ^{26}Mg resulted from Al(n, npγ) reaction at 55°, and 140° were compared with the calculated curves. The authors claimed that the contribution of direct mechanism is less than 5%. Assuming pure evaporation mechanism through compound nucleus for Al(n, npγ) and lifetime of 0.20 ps, a nice reproduction of experimental data was achieved. The adequate description of experimental data was considered as an evidence of correctness of the assumptions.

Precision of determining lifetime of a nuclear state by the Doppler shift attenuation method depends on stopping power of recoils in target to a great extent. Experimental studies on electron stopping power for different materials showed large deviations of experimental data from theoretical calculation. So in general the uncertainty of lifetime measurement was about 20% or worse by Doppler method, unless reliable experimental stopping power had been taken. The Doppler effect of 1808.7 KeV γ line of ^{26}Mg from Al(n, pnγ) reaction is very suitable for determining stopping power because of its well-known life-time (0.70 + 0.03 ps determined by Coulomb excitation), high sensitivity to stopping power, correct knowledge of recoil kinetics, absence of disturbing lines and good statistics. In the same work the stopping power of Mg in Al in energy region 5 - 500 KeV was determined. The results (correction coefficients f_e = 1.27[6], f_n = 0.77[7] and ϕ_n = 0.90(15)) differed from LSS theory(where $f_e = f_n = \phi_n = 0$).

Using the experimental stopping power lifetimes of nine γ lines of ^{26}Mg, ^{27}Mg and ^{24}Na populated in reactions Al(n, npγ), (n, pγ) and (n, $\alpha'\gamma$) were determined. Most of them were in good agreement with previous data acquired by other methods except for 3476 KeV and 1341 KeV γ-lines of ^{27}Mg for which the life-time of Ref. [3] is 1.05 ps respectively, while pevious value is < 0.01 ps and 0.09 ps[4].

Other application of Doppler shift attenuation analysis consists in determination of angular distribution of inelastically scattered neutrons. In fast neutron DDX measurements on ^{7}Li at 14 MeV, it is difficult to separate the inelastic neutrons from the elastic neutrons because of insufficient energy resolution. In other words, the "elastic neutron peak" contains inelastic neutrons of first excited state. However, the angular distribution of Li recoil nuclei can be derived from the shape of Doppler-broadened gamma line of 478 KeV. Such a work was completed by H. Liskien in the incident neutron energy below 8 MeV[5]. A. A. Pasternak obtained the angular distribution of pure elastic scattering for 14 MeV neutrons in a similar way independently. The forward part of the angular distribution is several times greater than that at neutron energy lower 8 MeV. The asymmetric part amounts to 51 percent of total neutron yield. In the authors' opinion, a new reaction mechanism was involved with increasing the incident neutron energy from 8 to 15 MeV. It can be explained in terms of quasidirect process.

The applications of Doppler shift attenuation method in 14 MeV neutron experiment showed its powerful potential for getting some important information. The procedure for charged particle induced reactions became nearly perfect long ago. But for 14 MeV neutron experiments the procedure must be used with modifications. It is necessary to consider some effects in a different way. The preequilibrium and direct processes might play more important part[6]. Finite size effects of scattering sample, particularly neutron multiple scattering should be taken into account. Side feeding may disturb determination of short lifetime quite seriously. In deducing angular distribution of inelastic scattering it is desirable to use nuclear model calculation.

Reference

1. I. H. Lemberg, A. A Pasternak and A. E. Zobov, *Izvestya AN SSSR (ser.fiz.)* **V.54(5)** (1990) 998
2. Y. G. Kosyak, D. K. Kaypov and L. V. Chekushina, *Izvestya AN SSSR (ser.fiz.)*, **V.46(11)**, (1982) 2257
3. A. A Pasternak and Zhou Hongyu, *41th conf. Nucl. Spect. and Nucl. Struct.* (Nauka, Leningrad, 1991) p.239-241
4. P. M. Endt and C. Van der Leun, *Nucl. Phys.* **A310** (1978) 1.
5. H. Liskien, Proc. *Int. Conf. on Nucl. Data for Bas. and Appl. Sc.*, (St. Pr. 1985) p.1277
6. Zhao Zhixiang and Zhou Delin *INDC(CPR)-8/L*, (IAEA, 1986)

MEASUREMENTS OF NEUTRON CAPTURE CROSS SECTIONS OF WOLFRAM AND THULIUM

XIA YIJUN, WANG CHUNHAO, YANG JINGFU,
YANG ZHIHUA and LUO XIAOBING
Institute of Nuclear Science and Technology,
Sichuan University, P. O. Box 390-1, Chengdu 610064,
P. R. of China

and

LIU JIANFENG
Physics Department, Zhengzhou University,
Zhengzhou 450002, P. R. of China

ABSTRACT

The neutron capture cross sections of wolfram and thulium were measured in the energy range from 10 to 100 KeV using gold as the standard. Kinematically collimated neutrons were produced via the $^7Li(p, n)^7Be$ reaction with a 2.5 MV pulsed Van de Graaff accelerator at Sichuan University. The capture events were detected by a pair of Moxon-Rae detectors. Time-of-flight technique was used to improve the signal-background ratio. The present results are compared with data by other authors.

The capture cross sections were calculated from 10 to 100 KeV for two nuclides by the Hauser-Feshbach statistical theory with width fluctuation correction. The nonstatistical effects such as potential capture and radiative capture in elastic and inelastic channels of a compound nucleus were included in the calculations. The calculated results show that the nonstatistical contribution to the capture cross sections is negligible compared with that of the statistical effects.

1. Introduction

The knowledge of neutron capture cross section in KeV region is of fundamental importance for designing reactors, for nuclear reaction theories and for astrophysical theories. To wolfram, as an important structural material of reactor (especially for fast reactor), its capture cross sections are required in reactor development. And ^{169}Tm is a rare earth nuclide with a natural abundance of 100%. The neutron capture cross section of ^{169}Tm is important for reactor designing and radiochemical diagnostic studies.

So, in this work, the neutron capture cross sections of wlofram and ^{169}Tm

have been measured in the energy range from 10 to 100 KeV with ^{197}Au as the standard.

2. Experiment

The $^{7}Li(p, n)^{7}Be$ reaction was used for neutron production in the experiment. When the energy of proton is just above the reaction threshold, the neutrons are kinematically collimated. All neutrons are emitted within a cone in a forward direction, the opening angle of which is determined by the proton energy. The capture events can be observed by detector placed at backward angles completely outside the neutron cone.

The detailed description of the experimental set-up, block diagram of the electronics and data analysis is given in Refs. [1] and [2]. Therefore, only a brief description is given here.

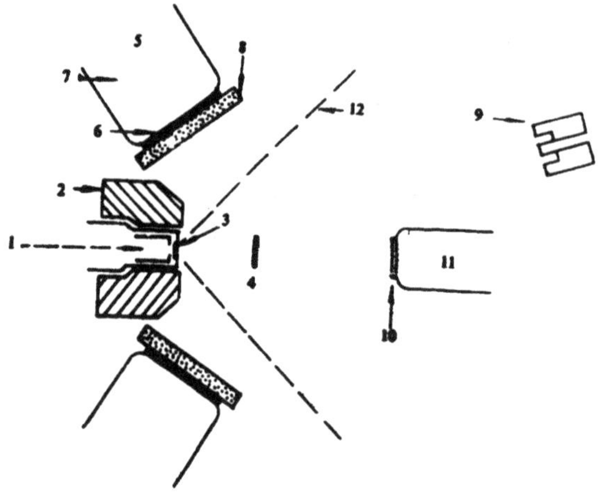

Fig. 1. Experimental setup. 1. proton beam, 2. lead shielding, 3. Li target, 4. sample, 5. Moxon-Rae detector, 6. plastic scintillator. 7. 58AVP photomultiplier, 8. graphite converter, 9. long counter, 10. ^{6}Li glass scintillator, 11. 56AVP photomultiplier, 12. neutron cone.

The scheme of the experimental set-up is shown in Fig. 1. The pulsed proton beam hits a water-cooled metallic lithium target producing kinematically collimated neutrons. The proton energy was adjusted only 20 KeV above the reaction threshold (1.881 MeV) in order to obtain kinematically collimated neutrons which had a half maximum opening angle of 45.5°. By using a thick lithium target, the continuous energy neutron beam between 10 and

100 KeV was obtained.

The capture events produced in the sample were detected by a pair of Moxon-Rae detectors symmetrically located outside the neutron cone at an angle of 120° with respect to the incident proton beam axis. The neutron energy was determined by time-of-flight (TOF). With a flight path of only 7cm, the overall experimental time resolution was 1.7 ns and, consequently, the energy resolution was 24 ns/m.

A ^{6}Li-glass detector was placed at 0° with respect to the proton beam axis as a transmission detector. It was used to measure transmitted neutron spectra and to monitor the changes of incident neutron spectra during the experiment.

Neutron fluence was monitored with a long counter placed at an angle of 20 degrees with respect to the beam direction and at a distance of 1.7m from the target.

The purity of Au and W samples (metal disks) is 99.99% and 99%, respectively. The sample of ^{169}Tm was prepared by pressing 99.99% pure Tm_2O_2 powder into thin disk. In order to subtract the background produced due to the neutron elastic scattering on the sample, graphite was used as pure scattering sample. All samples were 4cm in diameter. The samples of W, Tm, Au and C were mounted on three pieces of steel wires 0.15 mm in diameter controlled by a step motor.

The data acquisition was carried out by an on-line microcomputer. The TOF spectra recorded by the Moxon-Rae detector and ^{6}Li-glass detector were all stored on magnetic disks.

3.Data Analysis and Results

The data analysis has been described in detail in Ref [2]. The ratios of the capture cross sections of wolfram and ^{169}Tm to that of gold were determined according to the relation:

$$\frac{\sigma_x}{\sigma_{Au}} = \frac{C_x}{C_{Au}} \cdot \frac{N_{Au}}{N_x} \cdot \frac{B_{Au}}{B_x} \cdot \frac{(MS \cdot SS)_{Au}}{(MS \cdot SS)_x} \cdot \frac{S_{Au}}{S_x} \cdot \Delta\varepsilon,$$

where C is the background corrected counting rate in the TOF spectrum. N is the sample thickness. B is the neutron binding energy. MS, SS and S are the correction factors for multiple scattering, self-shielding and gamma-ray self-absorption, respectively. $\Delta\varepsilon$ is the correction factor for the deviation of the detector efficiency from a linear relation with gamma-ray energies. X is the index for wolfram and ^{169}Tm.

The correction factors MS and SS are calculated by the "SESH" Monte-Carlo code[3]. The correction factor S is calculated by using the mass absorption coefficient given by Hubbell[4]. The correction factor $\Delta\varepsilon$ is not calculated and an uncertainty of 2% is added to the systematic error to account for it.

The experimental results for the capture cross section of wolfram and ^{169}Tm are compiled in table 1 and table 2. To compare our results with recent measurements of other authors, the experimental ratios have been converted to absolute cross sections by using the gold cross section evaluated by Ref [5].

Table 1 Experimental results of neutron capture cross sections of W

En, KeV	σ(w)/σ(Au)	σ(w), mb	Error(%)		
			statistical	systematic	total
11.2 ± 0.3	0.370	430	8.2	5.5	9.9
12.8 ± 0.4	0.385	387	7.4	4.4	8.6
14.3 ± 0.5	0.423	397	7.3	5.1	8.9
15.4 ± 0.5	0.429	385	7.0	5.2	8.7
16.6 ± 0.6	0.395	336	6.6	4.7	8.1
18.0 ± 0.7	0.403	322	6.5	4.5	7.9
19.6 ± 0.8	0.415	315	6.2	4.6	7.7
21.4 ± 0.9	0.444	326	5.6	5.7	8.0
23.4 ± 1.0	0.451	316	5.3	4.6	7.0
25.8 ± 1.2	0.455	300	4.3	5.0	6.6
28.5 ± 1.4	0.464	284	4.2	5.4	6.8
31.7 ± 1.6	0.483	277	3.9	4.3	5.8
35.5 ± 1.8	0.475	257	3.3	4.5	5.6
40.0 ± 2.3	0.465	235	2.7	4.6	5.3
45.4 ± 2.7	0.498	240	2.6	4.4	5.1
51.9 ± 3.4	0.478	215	2.5	4.6	5.1
60.0 ± 4.2	0.503	206	2.0	4.3	5.0
70.2 ± 5.3	0.524	199	1.8	4.5	4.7
83.1 ± 6.8	0.551	195	1.6	4.5	4.8
99.9 ± 9.0	0.538	173	1.9	8.5	8.7

For wolfram, the present results are in good agreement with those of Macklin[6]. They also agree with the absolute measurements of Fricker[4] and Bartolome[8] within the uncertainties in the energy range from 10 to 80 KeV, but they are higher than that of Bartolome in the 80 to 100 KeV range. The data of relative measurement given by Kompe[9] are systematically higher than ours.

Our values for the capture cross sections of ^{169}Tm are in good agreement with results of Macklin[10], while the results of Gibbons[11] are systematically higher.

By using the optical model and statistical theory, the capture cross sections were calculated for wolfram and ^{169}Tm[12]. In the energy range 10 to 100 KeV, the values of calculation are in agreement with those of our measurement within the uncertainties.

Table 2 Experimental results of neutron capture cross sections of ^{169}Tm

En ± ΔEn (KeV)	$\sigma_r(^{169}Tm)/\sigma_r(^{197}Au)$	$\sigma_r(^{169}Tm)$ mb	uncertainty(%) statistical	systematic	total
11.2 ± 0.36	2.26	2509	14	7.2	15.7
12.9 ± 0.45	2.13	2132	12	5.0	13
14.0 ± 0.5	1.82	1713	12	5.6	13.2
15.1 ± 0.56	1.91	1749	10	5.0	11
16.4 ± 0.64	1.82	1593	8.3	5.4	9.9
17.9 ± 0.73	1.66	1328	6.5	5.3	8.4
19.6 ± 0.84	1.65	1244	5.6	5.3	7.9
21.6 ± 0.96	1.82	1291	5.3	4.4	6.9
23.9 ± 1.1	1.80	1236	4.3	4.3	6.1
26.5 ± 1.3	1.69	1062	3.8	4.3	5.7
29.6 ± 1.6	1.71	1010	3.4	4.3	5.5
33.3 ± 1.9	1.77	979	3.0	4.2	5.2
37.8 ± 2.2	1.79	925	2.9	4.4	5.3
43.2 ± 2.7	1.72	830	2.6	4.9	5.5
49.8 ± 3.4	1.83	802	2.5	4.3	5.0
58.1 ± 4.3	1.76	732	2.2	4.3	4.8
68.8 ± 5.6	1.65	632	2.0	4.2	4.8
82.3 ± 7.2	1.71	609	2.2	5.1	5.6
100 ± 9.8	1.72	552	3.2	11.3	11.7

References

1. Xia Yijun et al., *Chin. Jour. of Nucl. Phys.* **10** (1988) 102
2. Xia Yijun et al., *Chin. Jour. of Atomic Energy Science and Technology* **22** (1988) 316
3. F. H. Frohner, **GA-8380** (1968)
4. J. H. Hubbell, *Int. J. Appl. Rad. Isot.* **33** (1982) 126
5. Jiang Sonsheng, "*Evaluated neutron capture cross section for* ^{197}Au" **hsj-77141** (1978) (bp)
6. Macklin R. L., *Nucl. Sci. Eng.* **84** (1983) 98
7. Fricker M. P. et al., *2nd Inter. Conf. Nucl. Data for reactors*, (Helsinki, 1970), **Vol. II**, P.265

8. Bartolome Z. M. et al., *Nucl. Sci. Eng.* **37** (1969) 137
9. Kompe D. et al., *Nucl. Phys.* **A133** (1969) 513
10. Macklin R. L., *Nucl. Sci. Eng.* **82** (1982) 143
11. Y. H. Gibbons et al., *Phys. Rev.* **122** (1961) 182
12. Yunshan Mu et al. *Nucl. Sci. Eng.* **108** (1991) 302

NUCLEAR THEORY

CALCULATION OF THE NUCLEAR REACTIONS IN TERMS OF THE TWO-COMPONENT EXCITON MODEL WITH REALISTIC STATE DENSITIES

G.REFFO, GE ZHIGANG①, M.HERMAN② and C.COSTA③
ENEA,Inn. Svil. Divisione di Calcolo
Viale Ercolani 8, Bologna 40138, Italy

ABSTRACT

The two-component exciton model is applied for the calculation of neutron and proton emission in neutron induced reactions. Microscopic exciton level densities, accounting for neutron-proton distinction and pairing effects are used throughout the calculations. The results are compared against experimental data.

1. Introduction

The problem of neutron-proton distinction in the equilibration of the composite nuclear was addressed already in the early stages of preequilibrium[1,2], receiving little attention however. Even after Betak and Dobes[3] and later Kalbach[4] had formulated and applied the two-component version of the exciton model, one-component approaches have been exclusively used in reaction calculations. This may be related to the fact that, on the one hand, two-component calculations are much more involved, and on the other, the results reported in Refs.3 and 4 did not seem to present a convincing improvement over the one-component version of the model. A consistent derivation of the effective one-component exciton model, from the two-component formulation, was given by Gupta[5]. He has shown that, using a one-component model, transitions inconsistent with the assumption of the two-body nature of the equilibration process aer included, and that this can be compensated by modification of the averaged matrix element $|M|^2$ without affecting the cross section and spectra predictions. The result by Gupta provides the justification of the one-component model which is only valid, however, if the densities for equivalent proton and neutron configurations are the same. Our recent shell-model calculations[6-8] indicate that the assumption of an equal state

---

① On leave from China Institute of Atomic Energy, P.O.Box 275-41, Beijing, P.R.China.
② Under ENEA contract
③ Guest researcher

density for neutron and proton configurations underlying the model by Gupta, usually dose not hold. The thresholds for excitation of the given excition configurations may be very different as well as may be the structure in the energy distribution of the states. We stress that accounting for these differences is only possible if the explicit two–component formulation of the exciton model is used.

In this paper, we present the first application of the two–gas formulation of the exciton model as propoesd in Ref.8 Internal transition and emission rates are expressed in terms of the state densities, calculated within the microscopic approach of Ref.6.

2. Two–component exciton model

The model, we apply here in, is a natural extension of the exciton model to a system of two distinguishable compoenets, as proposed by Dobes and Betak and reformulated in Ref.8 Actually, in this paper we go beyond the formulation of Ref.8 considering also conservation of the angular momentum and following the linear momentum of the unbound nucleon to obtain angular distributions.

In Ref.8 the equilibration of the composite system(see Fig.1) is described through the master equation which takes the form

$$
\begin{aligned}
\frac{dP(N,h_\pi,t)}{dt} &= [\lambda_{\nu\pi}^{++}(E,N-1,h_\pi-1)+\lambda_{\pi\pi}^{++}(E,N-1,h_\pi-1)]P(N-1,h_\pi-1,t) \\
&+[\lambda_{\pi\nu}^{+0}(E,N-1,h_\pi)+\lambda_{\nu\nu}^{+0}(E,N-1,h_\pi)]P(N-1,h_\pi,t) \\
&+[\lambda_{\nu\nu}^{-0}(E,N+1,h_\pi)+\lambda_{\pi\nu}^{-0}(E,N+1,h_\pi)]P(N+.1,h_\pi,t) \\
&+[\lambda_{\pi\pi}^{--}(E,N+1,h_\pi+1)+\lambda_{\pi\nu}^{--}(E,N+1,h_\pi+1)]P(N+1,h_\pi+1,t) \qquad (1) \\
&+\lambda_{\nu\pi}^{0+}(E,N,h_\pi-1)P(N,h_\pi-1,t)+\lambda_{\pi\nu}^{0-}(E,N,h_\pi+1)P(N,h_\pi+1,t) \\
&-[\lambda_{\nu\nu}^{+0}(E,N,h_\pi)+\lambda_{\pi\nu}^{+0}(E,N,h_\pi)+\lambda_{\nu\pi}^{++}(E,N,h_\pi)+\lambda_{\nu\nu}^{++}(E,N,h_\pi) \\
&+\lambda_{\pi\nu}^{--}(E,N,h_\pi)+\lambda_{\pi\pi}^{--}(E,N,h_\pi)+\lambda_{\nu\pi}^{-0}(E,N,h_\pi)+\lambda_{\nu\nu}^{-0}(E,N,h_\pi) \\
&+\lambda_{\pi\nu}^{0-}(E,N,h_\pi)+\lambda_{\nu\pi}^{0+}(E,N,h_\pi)+W(E,N,h_\pi)]P(N,h_\pi,t).
\end{aligned}
$$

For the explanation of the symbols appearing in the equations of this section we refer to Table 1. The internal transition rates λ are defined as

$$
\lambda=\frac{2\pi}{\hbar}|M|^2\,\omega_f, \qquad (2)
$$

where the average squared matrix element $|M|^2$ is, as usual, a parameter of the model, and ω_f stands for the density of accessible final states. The emission rates $W(E,N,h_\pi)$ are written in analogy to the one–component exciton model, having one–gas state densities replaced by the two–gas quantities

$$W_\beta(\varepsilon,N,h_\pi)=\frac{2s_\beta+1}{\pi^2 h^3}\mu_\beta\varepsilon\sigma_\beta(\varepsilon)\frac{\omega(E-\varepsilon,p_\nu-p_{\nu\beta},h_\nu,p_\pi-p_{\pi\beta},h_\pi)}{\omega(E,p_\nu,h_\nu,p_\pi,h_\pi)}, \tag{3a}$$

$$W(N,h_\pi)=\sum_\beta\int W_\beta(\varepsilon,N,h_\pi)d\varepsilon. \tag{3b}$$

In comparison with the one–component master equation appears to be appreciably more complicated as the population probability is a two–dimensional matrix instead of a vector. To facilitate the solution of the master equation we have proposed in Ref.8 to transform the two–dimensional population matrix into a vector. In order to do so we ascribe a running index j to each substage which is related to N and h_π by a simple relation

$$j=N(N-1)/2+h_\pi+1. \tag{4}$$

This way, the Eq.1 is transformed into the standard system of linear differential equations. In Fig.2 we show the structure of the master equation after the transformation. Using $P(t)$ for the population probability vector, we can rewrite Eq.1 in the matrix form

$$\dot{P}(t)=AP(t). \tag{5}$$

The formal solution of Eq.5 reads

$$P(t)=e^{At}P(0). \tag{6}$$

In most of the applications, we are interested in the time integral of $P(t)$, which gives average life–times τ of the substages. Because all eigenvalues of A have nonvanishing negative real part, Eq.5 can be integrated from zero to infinity to give

$$A=-P(0). \tag{7}$$

Similar to the one–component model, we have to solve a set of linear equations. This set is appreciably larger, and more complicated (as A is no longer tri–diagona) than in the one–component case but, due to the negative diagonal dominance of a, a unique solution exists, and can be obtained using standard numerical procedures.

The differential cross section is given finally by

$$\frac{d\sigma_{\alpha\beta}(\varepsilon)}{d\varepsilon}=\sigma_\alpha\sum_j W_\beta(\varepsilon,j)\tau(j). \tag{8}$$

The conservation of the angular momentum in this formalism means that the master equation has to be formulated and solved for each spin of the composite nuvleus, and the summation over channels is performed observing angular momentum coupling. As far as the angular distribution of the emitted nucleons is considered we follow the approach of Ref.9 based on the leading particle model by Mantzouranis and Weidenmüeller[10]. We factorize the generalized internal transition rates from substage (N,h_π,Ω) to substage (N',h'_π,Ω') into the usual transition rates between different substages and an angle–dependent part that is supposed to be proportional to the differential free nucleon–nucleon scattering cross section;

$$\lambda_{N,h_\pi\to N',h'_\pi}(\Omega\to\Omega')=\lambda_{N,h_\pi\to N',h'_\pi}G(\Omega\to\Omega')=\lambda_{N,h_\pi\to N',h'_\pi}\left(\int d\Omega\frac{d\Omega'}{d\Omega}\right)^{-1}\frac{d\Omega'}{d\Omega}. \tag{9}$$

Let us investigate the eigenfunctions of the integral operator v defind by

$$Vf(\Omega)=\int G(\Omega,\Omega')f(\Omega')d\Omega', \tag{10}$$

where f is any quadratically integrable function on surface of the unit ball and kernel G is given by Eq.(9).

As a consequence of Eq.(9) the integral kernel G only depends on the angle alpha between the ingoing and outgoing direction Ω and Ω'. Hence it is possible to expand the kernal G into a Legendre polynomial series. Due to the axial symmetry of the solutions of generalized master equation with respect to the incident direction of the projectile(Ref.10, Appendix A), one can obtain

$$\int G(\Omega,\Omega')L_l(cos\theta)=\mu_l L_l(cos\theta), \tag{11}$$

where L_l is the Legendre polynomials, μ_l is the eigenvalue of V corresponding to the Legendre polynomial of order l and we take μ_l from Ref.9.

$$\mu_l=2\int_{-\beta}^{1}\frac{(x+\beta)(1+\beta x)}{(1+2\beta+\beta^2)^2}L_l(x)dx, \tag{12}$$

where bets stands for the ratio of the masses of the projectile and the target nucleus.

Next we expand the generalized occupation probability into a Legendre polynomial series,

$$p(n,h_\pi,\Omega,t)=\sum_l P_l(N,h_\pi,t)L_l(cos\theta), \tag{13}$$

and we obtain a set of linear differential equations similar to the Eq.(1). We adopt as initial condition according to Ref.9:

$$P(0)=\delta_{nn_0}(4\pi)^{-1}(2l+1)\mu_l. \tag{14}$$

3. Internal transition rates.

Assuming two–body interaction, there are 18 diagrams that contribute to internal transitions. Fig.3 shows how particular transition rates are constructed. In the two–component formulation one can distinguish between the $|M|^2$

corresponding to neutron–proton, proton–proton, and neutron–neutron interaction. In practice, it is presumably sufficient to deal with two matrix elements, one for like and one for unlike interactions.

To obtain the density of accessible states, entering Eq.2, we split the initial configuration into the interacting and passive part. The probability of finding the required interacting part with energy ε is multiplied by the density of ststes for the interacting part before and after the transition and integrated over ε from 0 to E to give the desired result. This simple procedure requires performing one numerical integration of the product of the tabulated states densities.

The transition rates are obtained by applying the Golden Rule(Eq.2) and summing the appropriate diagrams of Fig.3. In the case of λ^{++} we get for example

$$\lambda^{++} = \frac{2\pi}{h}(|M_{\pi\pi}|^2\,\omega_{f1} + |M_{\pi\pi}|^2\,\omega_{f2} + |M_{\nu\pi}|^2\,\omega_{f3} + |M_{\nu\pi}|^2\,\omega_{f4}), \quad (15)$$

where fi index at ω indicates accessible state densities for the transition corresponding to the i–th diagram of Fig.3.

4. Microscopic state densities

The state densities used in the present work were calculated according to the formalism described in Ref.6 where details of numerical procedure can be found. Here we restrict our discussion to the physical aspects of the method.

Assuming that the Hilbert space may be splitted into classes m of increasing complexity the state density in class m may be written as

$$\omega_m(E) = \sum_k \langle k|P_m|k\rangle \delta(E - E_k), \quad (16)$$

where

$$(H - E_k)|k\rangle = 0, \quad (17)$$

and P_m is a projector onto class m, being subject to the condition $\langle k|P_m|k\rangle = 0,1$ since no mixing between classes is assumed. In non–interacting particle model one assumes that exists one–fermion model space given by the single particle orbitals $|\nu\rangle$ being eigenfunctions of the shell–model Hamiltonian H_0

$$(H_0 - \varepsilon_\nu)|\nu\rangle = 0, \quad (18)$$

where each orbital $|\nu\rangle$ is characterized by its angular momentum J_ν, magnetic number M_ν, and parity π_ν. In the second quantization formalism H_0 may be written as:

$$H_0 = \sum_\nu \varepsilon_\nu a_\nu^+ a_\nu, \quad (19)$$

and N–fermion wave functions are given by

$$|N\rangle = \prod_{i=1}^{N} a_{\nu_i}^{+} |-\rangle, \tag{20}$$

with

$$H_0|N\rangle = \left(\sum_{i=1}^{N} \varepsilon_{\nu_i}\right) |N\rangle, \tag{21}$$

where $|-\rangle$ stands for the particle vacuum. The N–fermion ground state wave function $|\ \rangle$ is subject to the condition

$$\langle|H_0|\rangle = min. \tag{22}$$

Next, we define Ω_k^m operator which acting on a ground state $|\ \rangle$ creates excited state k belonging to class m

$$\Omega_k^m = \prod_{i=1}^{m} a_{\nu_i}^{+} a_{\mu_i} . \tag{23}$$

Here ν stands for an orbital above the Fermi energy(particle), while μ denotes an orbital below the Fermi energy(hole). Thus each state k in class m may be written as

$$|m,k\rangle = \Omega_k^m|\ \rangle \equiv |\nu_1,...,\nu_m,\mu_1,...,\mu_m\rangle. \tag{24}$$

Pauli principle requires that $\nu_{i+1} > \nu_i$ and $\mu_{i+1} > \mu_i$. This formulation essentially provides states with equal number of particles and holes $p = h = m$. States with $p \neq h$ may, however, be obtained by selecting subsets of Ω_k^m operators having some particles or holes fixed at the orbital closest to the Fermi level.

Pairing interaction is introduced through the BCS Hamiltonian

$$H = H_0 - \frac{1}{2} G \sum_{\nu,\mu} a_\nu^{+} a_{\bar{\nu}}^{+} a_{\bar{\mu}} a_\mu, \tag{25}$$

where $\bar{\nu} = (J_\nu, -M_\nu, \pi_\nu)$. Extending the blocking method of Wahlborn[11], this Hamiltonian is diagonalized by the following trial wave function for a state k in class m

$$|m,k\rangle = \prod_{i=1}^{m} a_{\nu_i}^{+} \prod_{\nu_i \neq \eta \neq \mu_i} (u_{k\eta} + v_{k\eta} a_{\bar{\eta}}^{+} a_\eta^{+})|0\rangle. \tag{26}$$

Index ν_i runs over orbitals occupied by the unpaired nucleons above the Fermi energy (particle excitons), while hole excitations are introduced by excluding μ_i orbitals from the second product. In such representation it is assumed that no residual interaction is effective between excitons. According to the BCS theory the energy of state k is given by

$$E_k = \sum_{i=1}^{m}(\varepsilon_{\nu_i} - \varepsilon_{\mu_i}) + \sum_{\nu_i \neq \eta \neq \mu_i} 2v_{k\eta}^2 \varepsilon_\eta - G(\sum_{\nu_i \neq \eta \neq \mu_i} v_{k\eta} u_{k\eta})^2 , \tag{27}$$

here

$$v_{k\eta}^2 = \frac{1}{2}\{1 - \frac{\varepsilon_\eta - \lambda_k}{[(\varepsilon_\eta - \lambda_k)^2 + \Delta_k^2]^{1/2}}\}, \tag{28}$$

and

$$u_{k\eta}^2 = \frac{1}{2}\{1 + \frac{\varepsilon_\eta - \lambda_k}{[(\varepsilon_\eta - \lambda_k)^2 + \Delta_k^2]^{1/2}}\}, \tag{29}$$

λ and Δ are solution of the two BCS equations

$$\sum_{\nu_i \neq \eta \neq \mu_i} 2v_{k\eta}^2 = \frac{N}{2} - m, \tag{30}$$

$$\sum_{\nu_i \neq \eta \neq \mu_i} [(\varepsilon_\eta - \lambda_k)^2 + \Delta_k^2]^{-1/2} = \frac{2}{G}. \tag{31}$$

In the microscopic approach a new solution has to be found for each state k.

Calculation of the state density consists in:

(i) generating all Ω_k^m for class m

(ii) solving BCS equations to determine state energy E_k

(iii) sorting states according to their energy

The above is done separately for neutrons and protons, as no interaction between two types of nucleons is assumed. The densities of mixed states are obtained folding neutron and proton state densities.

For the purpose of the present calculations, the spin distributions were parametrized using spin cut–off parameter $\sigma^2 = 0.28nA^{2/3}$ (See Ref.6), and both parities have been assumed equally probable.

5. Results

The calculations were performed using the appropriately modified version of the code PENELOPE[12]. The angular momentum and parity conservation was taken into account. The two–component preequilibrium emission was considered from the second and the third stage of the equilibration chain, i.e. exciton numbers up to 5 were considered. There results were suplemented with the Hauser–Feshbach calculations to account for the equilibrium contribution and second chance emission. This way total emission spectra and angular distributions for neutrons and protons as well as various reaction cross sections were obtained. Double–differential cross sections were calculated in term of the two–component exciton model while compound nucleus contribution was assumed to be istropic. The calculations of state densities were carried out in the space of single particle orbitals by Nix and Moller[13]. Values of the pairing strength parameter G were determined from the mass differences and taken configuration independent.

In Fig.4 we compare experimental neutron spectra emitted in the interaction of the 14 MeV neutrons with Al, Zr and Nb with the predictions of the one–gas and two–gas model. In the calculations $|M|^2 = K/A^3 E$ with $K = 190$ has been used. In the two–gas case , following Gupta, the $|M|^2$ has been increased by a factor 8 / 3. Apparently, a two–gas approach presents a clear improvement over the one–gas model. This can be traced back to the use of the microscopic state densities in the two–gas approach while the William's formula based on the equidistant spectrum of the single particle levels was used in the one–gas calculations. We note that in the case of Al the two–gas calculations approximately reproduce the structure in the spectrum. There is an overall factor about 1.3 by which our calculations overestimate experimental results. This can be related to the neglect of the direct contribution which is relatively large in the Al(n,n′) reaction and leads to the reduction of the statistical processes. A very good agreement is achieved in the case of Nb where two–gas calculations fit experimental data perfectly except of the high energy region dominated by the elastic pick which was not included in our calculations. Good agreement is obtained also in the case of Zr, however , we note an underestimation of the high energy data in the backward angle 120°.

In Fig.5 and 6 we show the effect of the pairing interaction and of the assumption on the strength of the like and unlike interactions on the ^{93}Nb(n,n′) neutron spectrum. We present the results with the same $|M|^2$ for $\nu\nu, \pi\pi, \nu\pi$ and those obtained assuming different values for $|M|^2$ according to Ref.3, i.e.:

$$|M|^2_{\nu\pi} = \frac{K_T}{ANZE}, \tag{32a}$$

$$|M|^2_{\nu\nu} = \frac{K_T}{AN^2 ER}, \tag{32b}$$

$$|M|^2_{\pi\pi} = \frac{K_T}{AZ^2 ER}, \tag{32c}$$

$$K_T = 3 \times \frac{8R(R+1)}{3(2R+1)^2} \times K, \tag{32d}$$

where we take $R=2.89$ and K is a parameter used in the one–gas model. We introduce a factor of 3 in equation (32d) to reproduce $\Gamma\downarrow$ as obtained from the imaginary part of the optical model potential. In the case of the neutron spectrum both pairing and like–unlike interaction strength ratio have rather limited influence, involve only the high energy end of the calculated spectrum, and comparison with experimental data doesn't allow for a definite conclusion. The same analysis performed in the proton emission channel(Fig.7 and 8) appears to be more significant. One observes a strong increase of the emission when pairing is neglected, and at the same time a substantial decrease of the emission when like and unlike nucleon interactions are taken the same. Unfortunately, comparison of the calculated results with the experimental data is not satisfactory. This discrepancy can be explained observing that experimental proton spectrum at 160° (Fig.8) multiplied by 4π is very close to the total spectrum (Fig.7), indicating rather istropic angular distribution. This suggest that the proton emission goes mostly through the MSC mechanism, which yields angular distribution symmetric around 90° rather than through MSD mechanism. It follows that our exciton model which, for a time being, does not distinguish between MSC and MSD mechanisms overestimates MSD contribution in the proton channel. This is not so evident in the case of the neutron channel where the angular distributions are generally better reproduced. This is confirmed by the Fig.9 where neutron spectra at various angles are presented for the case of Zr. Also here the effect of changing K and like–unlike interaction strength ratio are shown. There is a slight prference for $K=380$ value and $|M_{nn}| = |M_{np}|$.

In Fig.10 and 11 total neutron spectrum and double–differential neutron spectra at different angles are presented for two different values of K in the case of Al. Also here a good description of the angle dependence is obtained and again $K=380$ seems to be somewhat better.

6. Conclusions

In all nuclei considered the calculations performed in the frame of the two–gas model present a significant improvement with respect to the previous one–gas calculations. It turns out that the effects of the two–gas distinction manifest themselves mostly in the proton channel. Unfortunately, it seems that proton channel is dominated by the MSC process and therefore exciton model fails to reproduce experimental data. The situation is much better in the neutron channel, where exciton model provides good prediction of the angular distribution, except very high energy end of the spectra and very backward angles. In this respect exciton model could be judged as containing to high MSD contribution and underestimating the MSC one.

Our preliminary calculations indicate that in the two–gas model the value of the residual interaction matrix element $|M|^2$, which is responsable for the composite nucleus equilibration, should be reconsidered. This is mostly due to the use of the microscopic state densities and goes beyond the factor of $8/3$ pointed out by Gupta. We also note, that our results do not provide an undoubtful evidence for the necessity of using different $|M|^2$ for like and unlike interaction.

As far as the angular distributions are concerned we mention that in the present calculations we negected λ^{00} transitions, that are irrelevant for the total cross section estimate, but can affect an angular distribution, increaseing it at backward angles.

7. References

1. E.Betak and J.Dobes, Acta Phy. Slov. 29(1979)76.
2. E.Betak, P.Oblozinsky and I.Ribansky, Nukleonika 19(1974)687.
3. J.Dobes and E.Betak, Z.Phys. A310(1983)329.
4. C.Kalbach, Phys.Rev. C33(1986)818.
5. S.K.Gupta, Z.Phys. A303(1981)329.
6. M.Herman and G.Reffo, Phys.Rev. C36(1987)1546.
7. M.Herman, G.Reffo and R.A.Rego, Phys.Rev. C36(1988)797.
8. M.Herman, G.Reffo and C.Costa, Phys.Rev. C39(1989)1269.
9. J.M.Akkermans, H.Gruppelaar and G.Reffo, Phys.Rev. C22(1980)73.
10. G.Mantzouranis, D.Agassi and H.A.Weidenmüller, Phys.Lett. 57B(1975)220 and G.Mantzouranis, H.A,Weidenmüller, and D.Agassi, Z.Phys. a276(1976)154.
11. S.Wahlborn, Nucl.Phys. 37(1962)554.
12. F.Fabri and G.Reffo, code PENELOPE, unpublished.
13. J.R.Nix and P.Moller(private communication).
14. A.Takahashi, J.Yamamoto, T.Murakami, K.Oshima, H.Oda, K.Fujimoto, M.Ueda, M.Fukazawa, Y.Yanagi, J.Miyaguchi and K.Sumita, OKTAVIAN report A–83–01, Osaka University (1983).
15. A.Takahashi, J.Yamamoto, K.Ohshima, M.Fukazawa, Y.Yanagi, M.Ueda, J.Miyaguchi, S.Kohno, K.Yugami, H.Nonaka, E.Ichimura, H.Sugimoto and K.Sumita, OKTAVIAN report A–87–01, Osaka University (1987).

16. A.Takahashi, E.Ichimura, Y.Sasaki and H.Sugimoto,
OKTAVIAN report A–87–03, Osaka University (1987).
17. A.Takahashi, Y.Sasaki and H.Sugimoto, R,INDC(JAP)–118 / L,8907.
18. R.Fischer, M.Uhl and H.Vonach, J,PR / C,37,(20)578,8802.

TABLE I. Definition of symbols.

ν,π	Subscript denoting neutrons or protons
β	Subscripts denoting emitted particle
p, h	When used as a subscript relate a quantity to particle or hole degrees of freedom respectively
o	When used as a subscript relates exciton numbers to the initial configuration($N=1$)
E	Excitation energy of the composite nucleus
U	Excitation energy of the residual nucleus
g	Single particle state density
p	Number of particle excitons
h	Number of hole excitons
n	Total number of excitons
N	Stage of the reaction
j	Running index enummerating substages
$P(N, h_\pi, t)$	Population probability of the substage denoted by N and h_π at time t
$P(t)$	Population probability vector (components being $P(j,t)$) at time t
τ	Vector of the average liftime of the substages (integral of $P(t)$)
A	Matrix of trarnsition and emission rates in the master equation
$\omega(E, p_\nu, h_\nu, p_\pi, h_\pi)$	Density of states with specified exciton numbers at excitation energy E
$\|M\|$	Average transition matrix element
$w_\beta(\varepsilon, N, h_\pi)$	Emission rate of a particle β into the continuum with channel energy ε from the configuration denoted by N and h_π
$W(E, N, h_\pi)$	Total emission of particles into continuum from the configuration denoted by N and h_π at energy E
σ_α	Reaction cross section
σ_β	Inverse reaction cross section
s_β	Spin of emitted particle
μ_β	Reduced mass of emitted particle

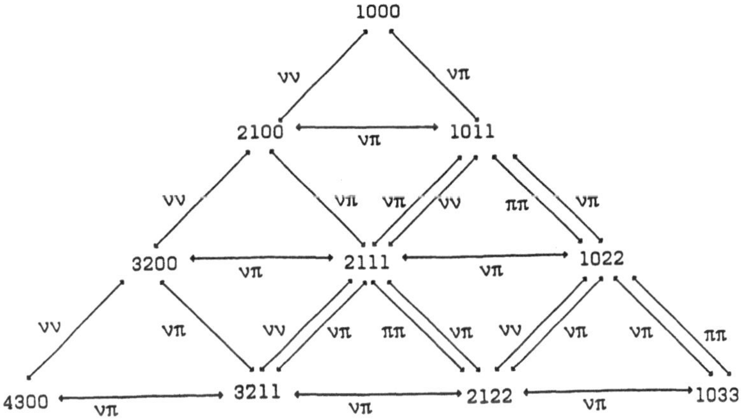

Fig.1 Schematic diagram of the equilibration of the composite nucleus formed in the neutron induced reaction. Different substages are denoted with p_ν, h_ν, p_π, and h_π. The arrows show the internal transitions consistent with the two-body assumption for the nuclear forces, and are labeled with the type of the interacting nucleons.

N		1	2		3			4			
	j	1	2	3	4	5	6	7	8	9	10
1	1	L	λ^{-0}	λ^{--}							
2	2	λ^{+0}	L	λ^{0-}	λ^{-0}	λ^{--}					
	3	λ^{++}	λ^{0+}	L		λ^{-0}	λ^{--}				
3	4		λ^{+0}		L	λ^{0-}		λ^{-0}	λ^{--}		
	5		λ^{++}	λ^{+0}	λ^{0+}	L	λ^{0-}		λ^{-0}	λ^{--}	
	6			λ^{++}		λ^{0+}	L			λ^{-0}	λ^{--}
4	7				λ^{+0}			L	λ^{0-}		
	8				λ^{++}	λ^{+0}		λ^{0+}	L	λ^{0-}	
	9					λ^{++}	λ^{+0}		λ^{0+}	L	λ^{0-}
	10						λ^{++}			λ^{0+}	L

Fig.2 Schematic representation of the j-indexed master equation. Thick lines separate different stages of the equilibration process. Off-diagonal elements represent gain of the flux from other substages, while diagonal elements L are responsible for the loss of the flux due to the coupling to other substages and to the open channels.

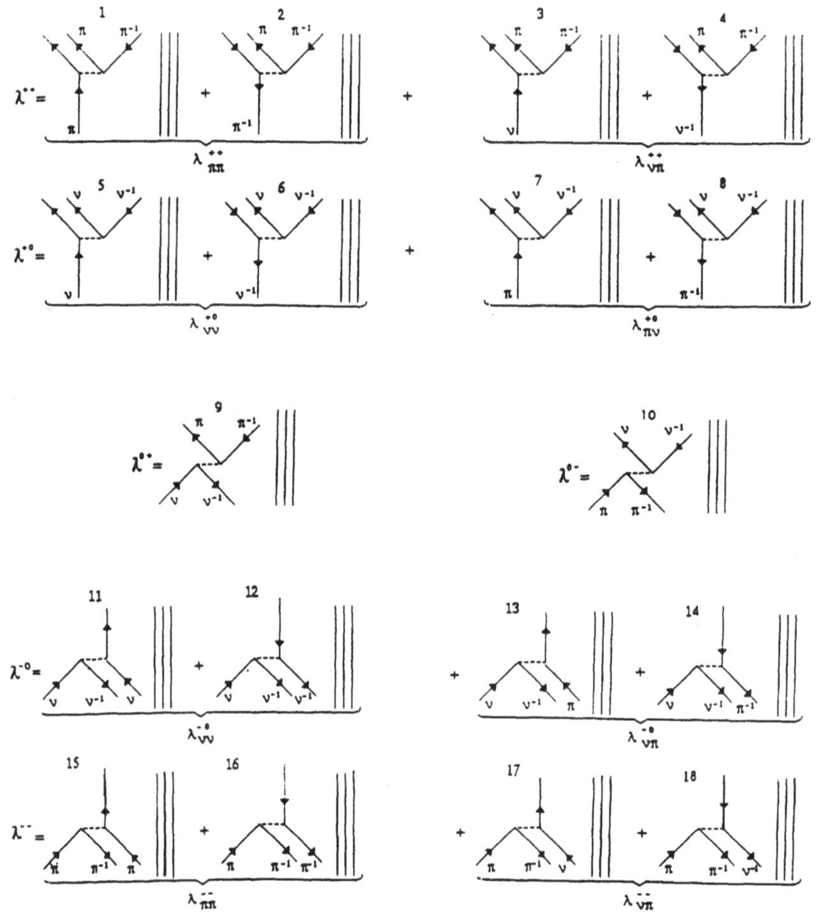

Fig.3 Diagrams illustrating internuclear transitions in the two-component exciton model. Interacting particles (holes) are represented by upward (downward) arrows, and their nucleon type is indicated. Vertical lines stand for the passive excitons. The diagrams are enumerated to facilitate reference.

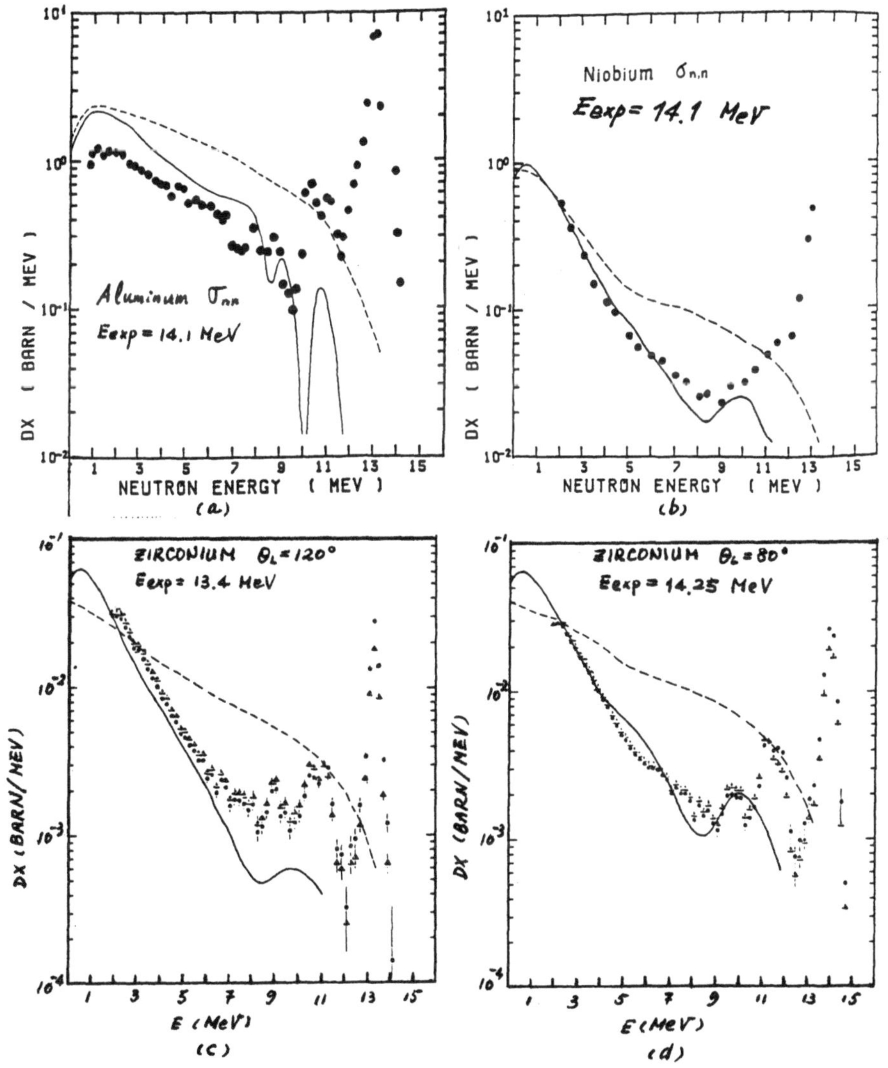

Fig.4 Comparison of the two-component (full line) and one-component (dashed line) calculations for 27Al(n,n′) total neutron spectrum at 14.1 Mev (a); 93Nb(n,n′) total neutron spectrum at 14.1 Mev (b) and 91Zr(n,n′) double differential cross section at angles 120° (c) and 80° (d).

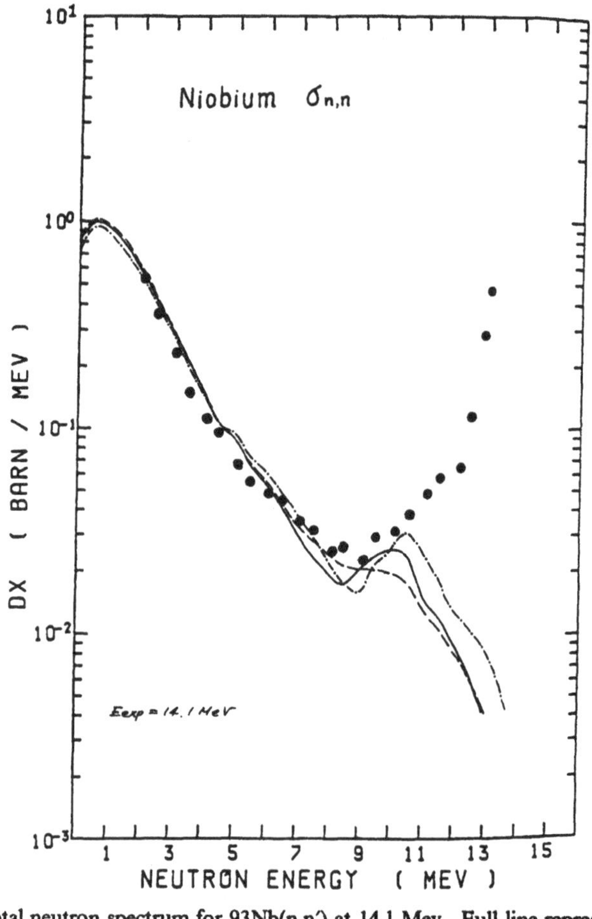

Fig.5 Total neutron spectrum for 93Nb(n,n′) at 14.1 Mev. Full line represents the results obtained including pairing correction and assuming $|M|^2$ given by the formula (32); dot-dashed line corresponds to the same calculation but neglecting pairing interaction; dashed line is obtained including pairing and assuming $|M|^2_{\nu\nu} = |M|^2_{\nu\pi}$.

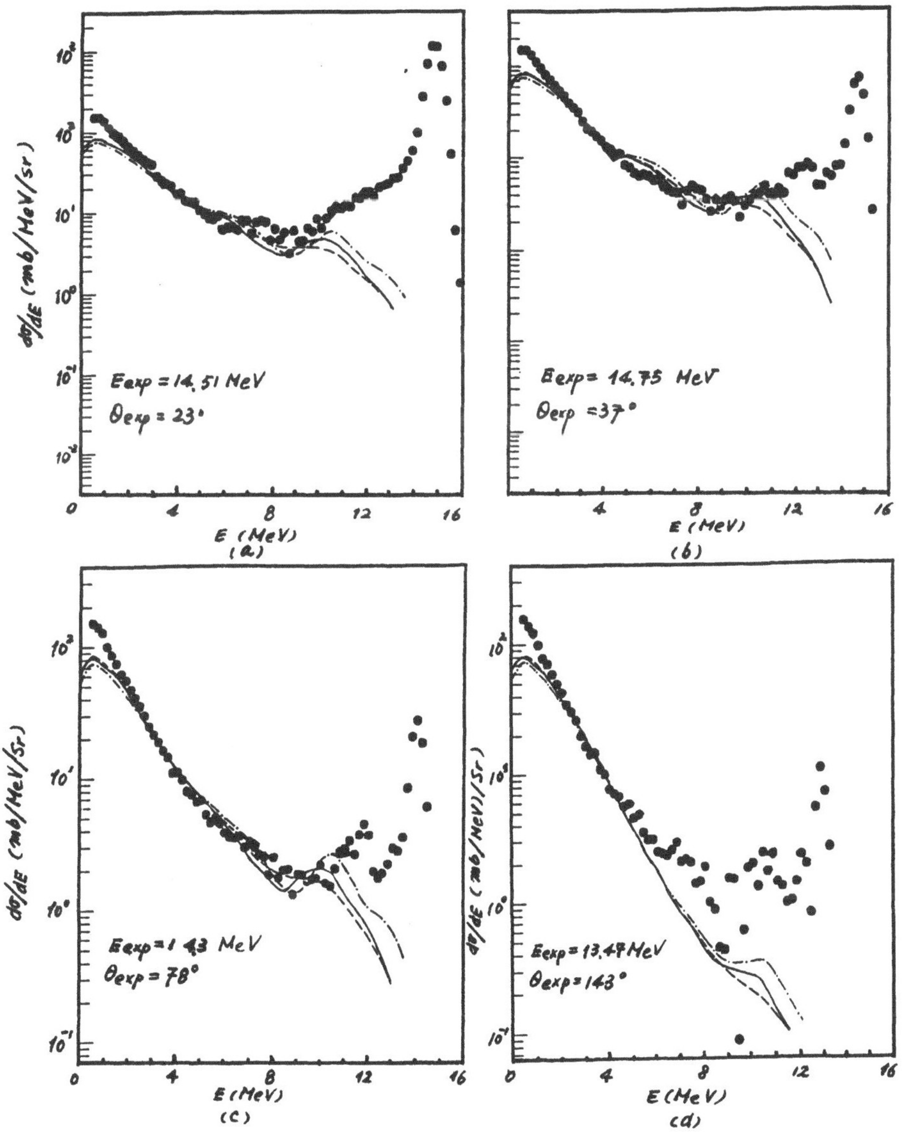

Fig.6 Double differential cross section at angles 23°,37°, 78° and 150° for 93Nb(n,n′) at 14.1 MeV. Full line represents the results obtained including pairing interaction and assuming $|M|^2$ given by the formula (32); dot-dashed line corresponds to the same calculation but neglecting pairing; dashed line is obtained including pairing and assuming $|M|^2_{\nu\nu} = |M|^2_{\nu\pi}$.

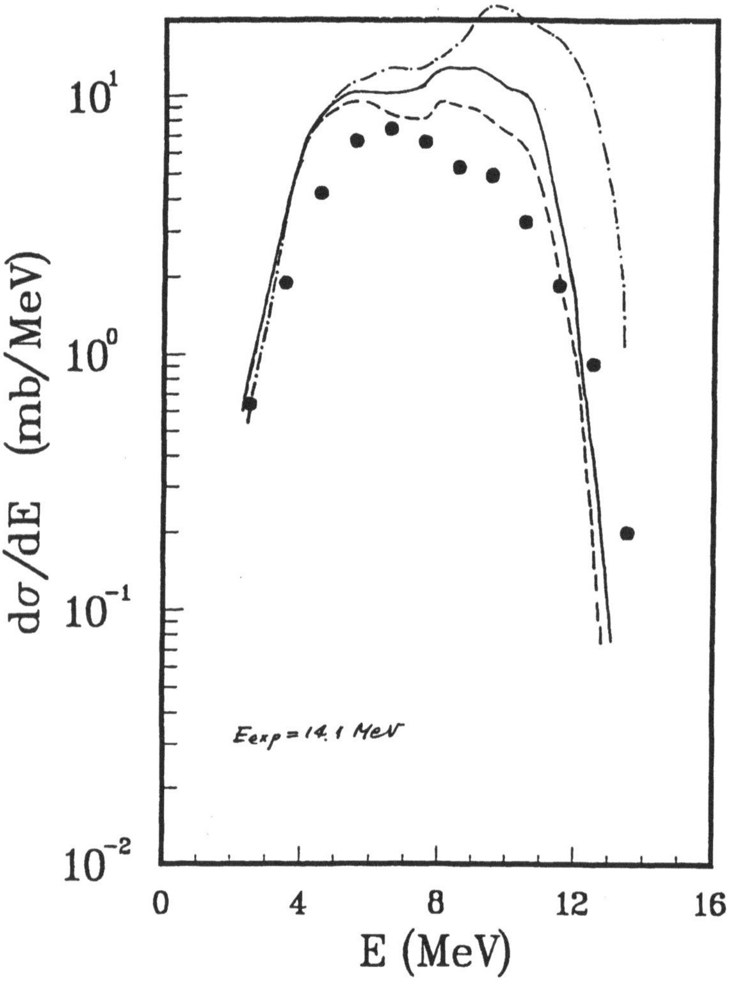

Fig.7 Total proton spectrum for 93Nb(n,p) at 14.1 MeV. Full line represents the results obtained including pairing and assuming $|M|^2$ given by the formula (32); dot-dashed line corresponds to the same calculation but neglecting pairing; dashed line is obtained including pairing and assuming $|M|^2_{\nu\nu} = |M|^2_{\nu\pi}$.

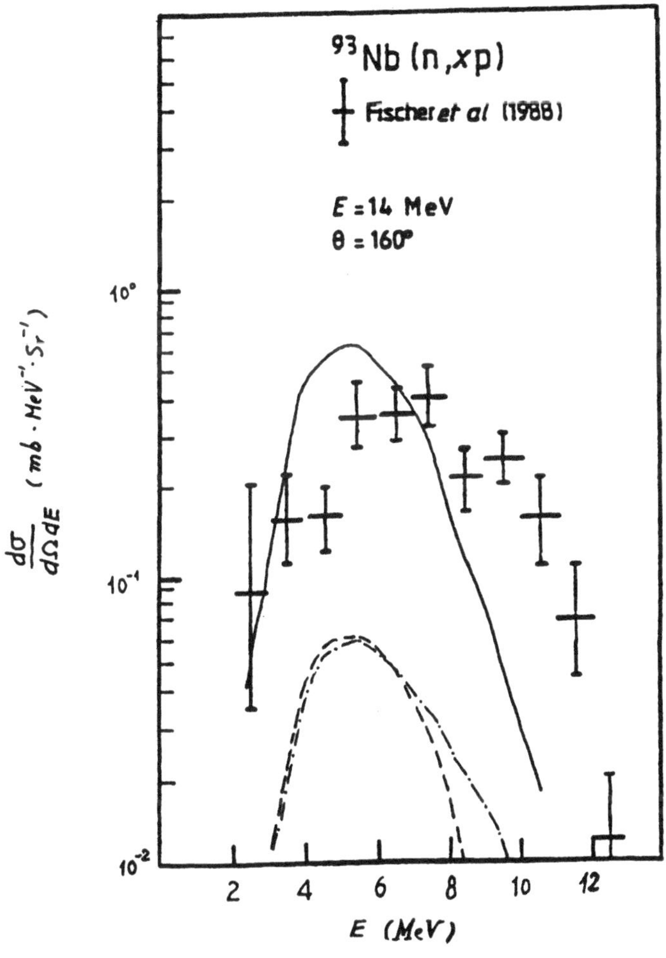

Fig.8 Double differential cross section for 93Nb(n,p) at 14.1 MeV at angle 160°. Full line represents the results obtained including pairing and assuming $|M|^2$ given by the formula (32); dot-dashed line corresponds to the same calculation but neglecting pairing; dashed line is obtained including pairing and assuming $|M|^2_{\nu\nu} = |M|^2_{\nu\pi}$.

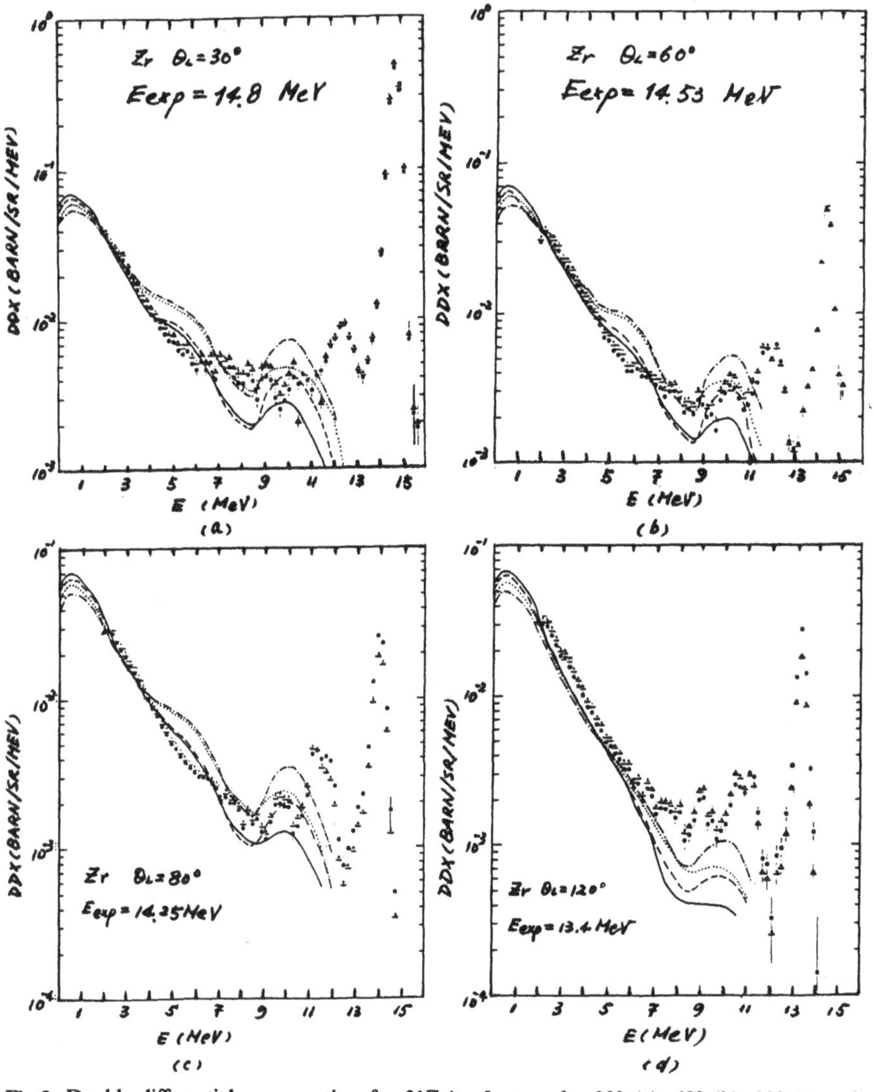

Fig.9 Double differential cross section for 91Zr(n,n′) at angles 30° (a), 60° (b), 80° (c) and 120° (d). The full lines and the dotted lines refer to the assumption of $|M|^2_{\nu\nu} = |M|^2_{\nu\pi}$ the first one with K = 380 and the second with K = 190. The dashed and dashed-dotted lines are obtained assuming $|M|^2_{\nu\nu} \neq |M|^2_{\nu\pi}$ according to formula (32) with K = 380 and K = 190.

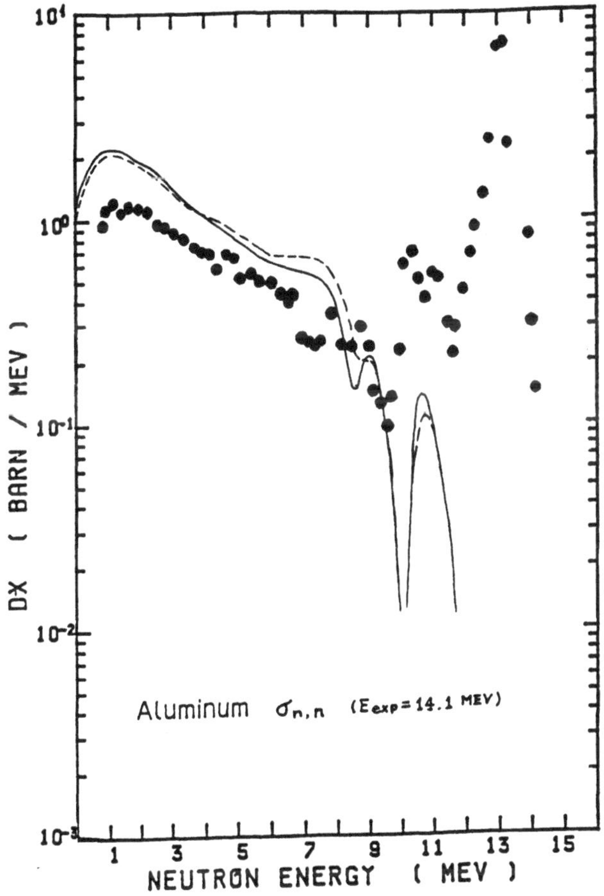

Fig.10 Total neutron spectrum for 27Al(n,n′) at 14.1 MeV. The full line corresponds to K = 380 and the dashed one to K = 190. Both curves are obtained including pairing interaction and assuming $|M|^2$ according to formula (32);

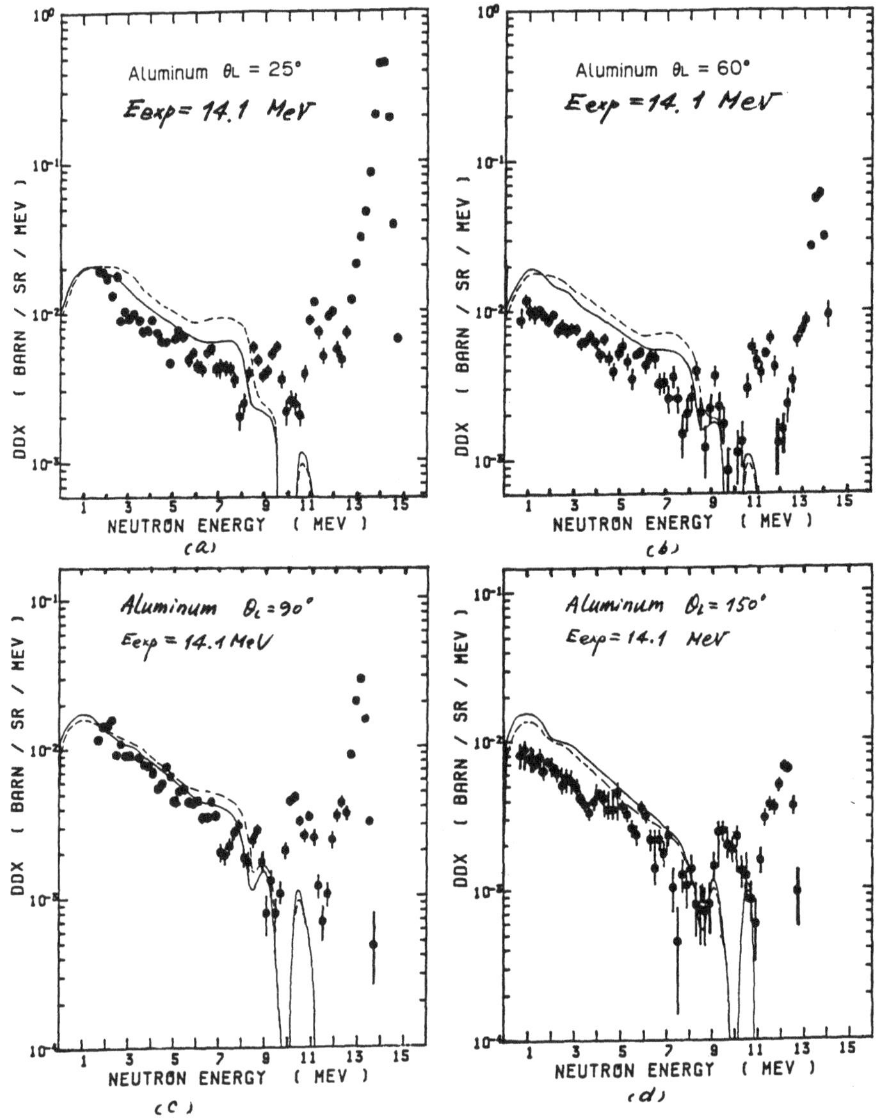

Fig.11 Double differential cross section for 27Al(n,n′) at 14.1 MeV at angles 25° (a), 60° (b), 90° (c) and 150° (d). The full line corresponds to K = 380 and the dashed one to K = 190. Both curves are obtained including pairing interaction and assuming $|M|^2$ according to formula (32).

COMPARISONS OF NON–RELATIVISTIC AND RELATIVISTIC PHENOMENOLOGICAL AND MICROSCOPIC OPTICAL POTENTIALS FOR NEUTRON INDUCED REACTIONS AT LOW AND MEDIUM ENERGIES

Shen Qingbiao and Zhuo Yizhong
China Institute of Atomic Energy, Beijing 102413, China

and

Cai Chonghai
Nankai University, Tianjin 300071, China

ABSTRACT

Non–relativistic and relativistic phenomenological and microscopic optical potentials are studied and comparied for neutron induced reactions at low and medium energies. The calculated results show that the microscopic optical potential based on extended Skyrme force, which has an analytical formalism without any free parameters, is useful in nuclear data calculation and evaluation. The global neutron relativistic phenomenological optical potential based on the available experimental data for various nuclei ranging from ^{12}C to ^{238}U with incident energies 20–1000 MeV has been obtained through automatic search of the best parameters by computer. The nucleon relativistic microscopic optical potential is studied by utilizing effective lagrangian based on popular Walecka model. Through extensive comparisons we can understand the neutron optical potential in large range of energy quite well.

I. Introduction

The optical model is one of the most important theoretical approaches in nuclear data analyses and evaluations. The optical potential can be divided into two classes, the phenomenological and the microscopic optical potentials.

The phenomenological optical potential (POP) contains many adjustable parameters. For those elements for which there are enough experimental data, we take the energy–dependent Woods–Saxon potential and adjust the potential parameters to minimize the χ^2 for each element. Obviously, the cross sections calculated by this set of parameters will be in the best agreement with the experimental values. We call this set of parameters the optimal optical potential (OOP).

For those elements lack of experimental data, systematics are generally used to obtain optical potential parameters. Becchetti and Greenlees gave a set of global optical potential parameters that is quite popular; we call it BGP[1]. Based on more and newer experimental data, Varner et al. provided a set of global optical potential parameters call CH86[2]. Because these parameter sets were obtained by fitting the experimental data of many elements, the cross sections of a specific element calculated using BGP or CH86 are not as good as those using OOP in comparison with experimental data.

It is well known that the elements for which there are enough experimental data are mostly the ones near the beta–stability line. Therefore, it is not appropriate to extrapolate this set of parameters to the elements far from the beta–stability line, such as the fission products. Even for stable elements for which there exist plenty of experimental data, one cannot always be certain that all of them are included in the range of adjusting parameters. Therefore, an extrapolation to those elements and / or energy regions that are far away from the domain being included in the parameter adjusting process is obviously also unreliable.

The microscopic optical potential based on nuclear forces can be derived from the nuclear many–body theory. To find a useful theory of microscopic optical potential for nuclear data evaluation, we propose to calculate the mass operator based on effective force only up to second order in nuclear matter with local density approximation (LDA) to obtain the microscopic optical potential for finite nuclei. We call it MOP method[3–6].

The MOP method has an analytical formalism for calculating the optical potentials; hence, it is convenient to cross–section evaluations. Because Skyrme force parameters are fitted from the properties of nuclear matter and nuclear structure, there are no free parameters in the MOP optical model calculation. If we use MOP for cross–section calculations of those elements which have a large amount of experimental data and obtain good theoretical results in agreement with experimental values, with confidence one can extend it to those elements for which the experimental data are lacking.

In the medium energy domain, the nucleon relativistic optical model phenomenology based on Dirac equation with a mixture of the Lorentz scalar potential and the time–like component of the Lorentz four–vector potential developed by Arnold and Clark[7,8] has been used to analyze the medium energy proton elastic scattering data with great success.

In the past, the form and parameters of the relativistic phenomenological optical potential (RPOP) have been extensively investigated. They, however, were restricted to fit the experimental data for some specific targets and in a certain range of incident energies. Only very recently the global Dirac optical potentials for proton elastic scattering from heavy nuclei at energies between 65 and 1040 MeV have been published[9]. They are functions of both energy and target mass number. The number of parameters for set 1 and set 2 is 77 and 84, respectively.

The studies on neutron RPOP so far are rather limited due to scarcity of experimental data. Recently more and more medium energy data have become available. The availability of new neutron experimental data provides the opportunity for studying the global neutron RPOP as well as its microscopic foundation.

In this talk, the global neutron relativistic phenomenological optical potential for target nuclei ranging from ^{12}C to ^{238}U at incident energies $E_n = 20–1000$ MeV have been obtained through automatic search of the best–fit parameters by computer[10]. Then the relativistic nucleon–nucleus microscopic optical potential (RMOP) is studied with the effective lagrangian based on Walecka model including only nucleon, σ and ω meson with two adjustable isoscalar meson coupling constants g_σ and g_ω chosen to reproduce the nuclear matter saturation properties[11–13].

In Sec. II, we present the basic theory of the microscopic optical potential based on skyrme interaction and some calculated results. The comparisons of phenomenological and microscopic optical potential for low energy neutron induced reactions are given in Sec. III. In Sec. IV, the relativistic phenomenological and microscopic optical potentials for medium energy range are studied. Finally, a brief summary is given in Sec. V.

II. Microscopic Optical Potential Based on Skyrme Interaction

The generalized Skyrme force and the modified Skyrme force can be written in an unified form as follows[6]:

$$\begin{aligned} V_{12}(\vec{R},\vec{r}) = {} & t_0(1+X_0P_\sigma)\delta(\vec{r}) + \frac{1}{6}t_3(1+X_3P_\sigma)\rho^\alpha(\vec{R})\delta(\vec{r}) \\ & + \frac{1}{2}t_1(1+X_1P_\sigma)(\vec{k}'^2\delta(\vec{r}) + \delta(\vec{r})\vec{k}^2) + \frac{1}{2}t_4(1+X_4P_\sigma)(\vec{k}'^2\rho(\vec{R})\delta(\vec{r}) \\ & + \delta(\vec{r})\rho(\vec{R})\vec{k}^2) + t_2(1+X_2P_\sigma)\vec{k}'.\delta(\vec{r})\vec{k} + t_5(1+X_5P_\sigma)\vec{k}'.\rho(\vec{R})\delta(\vec{r})\vec{k} \\ & + iW_0(\vec{\sigma}_1+\vec{\sigma}_2)\vec{k}' \times \delta(\vec{r})\vec{k} , \end{aligned} \tag{1}$$

where

$$\vec{r} = \vec{r}_1 - \vec{r}_2, \quad \vec{R} = \frac{1}{2}(\vec{r}_1 + \vec{r}_2) . \tag{2}$$

The relative momentum operators are

$$\vec{k} = \frac{1}{2i}(\nabla_1 - \nabla_2) \qquad \textit{acting on the right,}$$

$$\vec{k}' = -\frac{1}{2i}(\nabla_1 - \nabla_2) \qquad \textit{acting on the left.} \tag{3}$$

p_σ is the spin exchange operator and $\vec{\sigma}$ are the Pauli spin matrices. We call Eq.(1) the extended Skyrme force.

The single–particle Green function satisfies the Dyson equation

$$G_{\alpha\beta}(\omega) = \delta_{\alpha\beta}G^{(0)}_\alpha(\omega) + G^{(0)}_\alpha \sum_\gamma [U_{\alpha\gamma} - M_{\alpha\gamma}(\omega)]G_{\gamma\beta}(\omega), \tag{4}$$

where $U_{\alpha\gamma}$ is the mean field, $M_{\alpha\gamma}(\omega)$ is the mass operator, and

$$M_{\alpha\gamma}(\omega) = M^{(1)}_{\alpha\gamma} + M^{(2)}_{\alpha\gamma}(\omega) + \ldots\ldots \quad . \tag{5}$$

The mass operator $M_{\alpha\alpha}(E)$ can be identified with the optical model potential for scattering process with the energy E.

For the extended Skyrme force which involves only two–body interaction, the mass operators $M^{(1)}_{\alpha\alpha}$ and $M^{(2)}_{\alpha\alpha}(E)$ are

$$M^{(1)}_{\alpha\alpha} = \sum_{\rho} V_{\alpha\rho,\alpha\rho} n_{\rho} , \tag{6}$$

$$M^{(2)}_{\alpha\alpha}(E) = \frac{1}{2} \sum_{\mu\lambda\nu} \frac{V_{\alpha\mu,\lambda\nu} V_{\lambda\nu,\alpha\mu}}{E + \varepsilon_{\mu} - \varepsilon_{\lambda} - \varepsilon_{\nu} + i\eta} n_{\mu}(1 - n_{\lambda})(1 - n_{\nu}) . \tag{7}$$

According to the formula of the principal value integral

$$\frac{1}{X + i\eta} = \mathcal{P}\frac{1}{X} - i\pi\delta(X) , \tag{8}$$

one can obtain the imaginary part of the second–order mass operator $M^{(2)}_{\alpha\alpha}(E)$ from Eq.(7) as follows :

$$ImM^{(2)}_{\alpha\alpha}(E) = -\frac{\pi}{2} \sum_{\mu\lambda\nu} V_{\alpha\mu,\lambda\nu} V_{\lambda\nu,\alpha\mu} n_{\mu}(1 - n_{\lambda})(1 - n_{\nu})\delta(E + \varepsilon_{\mu} - \varepsilon_{\lambda} - \varepsilon_{\nu}) \tag{9}$$

The real part of MOP obtained from Eq.(6) is as follows :

$$\begin{aligned}
V_{\tau_\alpha} = \frac{m^*_{\tau_\alpha}}{m_{\tau_\alpha}} & \{ t_0[(1 + \frac{X_0}{2})\rho - (X_0 + \frac{1}{2})\rho_{\tau_\alpha}] + \frac{t_3}{6}\rho^{\alpha}[(1 + \frac{X_3}{2})\rho - (X_3 \\
& + \frac{1}{2})\rho_{\tau_\alpha}] + \frac{1}{4} \{ t_1[(1 + \frac{X_1}{2})\rho - (X_1 + \frac{1}{2})\rho_{\tau_\alpha}] + t_4\rho[(1 + \frac{X_4}{2})\rho - (X_4 \\
& + \frac{1}{2})\rho_{\tau_\alpha}] + t_2[(1 + \frac{X_2}{2})\rho + (X_2 + \frac{1}{2})\rho_{\tau_\alpha}] + t_5\rho[(1 + \frac{X_5}{2})\rho + (X_5 \\
& + \frac{1}{2})\rho_{\tau_\alpha}]\} \\
& * \frac{2m_{\tau_\alpha}}{\hbar^2}(\frac{M}{M + m_{\tau_\alpha}} E_L - V_C) + \frac{1}{40\pi^2}[t_1(1 - X_1) + t_4\rho(1 - X_4) \\
& + 3t_2(1 + X_2) + 3t_5\rho(1 + X_5)]k^5_{\tau_\alpha} + \frac{1}{20\pi^2}[t_1(1 + \frac{X_1}{2}) \\
& + t_4\rho(1 + \frac{X_4}{2}) + t_2(1 + \frac{X_2}{2}) + t_5\rho(1 + \frac{X_5}{2})](2k^3_F - k^3_{\tau_\alpha})^{5/3}\} ,
\end{aligned} \tag{10}$$

where the effective mass is

$$\begin{aligned}
\frac{m^*_{\tau_\alpha}}{m_{\tau_\alpha}} = \{ 1 + & \frac{2m_{\tau_\alpha}}{\hbar^2}\frac{1}{4} \{ t_1[(1 + \frac{X_1}{2})\rho - (X_1 + \frac{1}{2})\rho_{\tau_\alpha}] \\
& + t_4\rho[(1 + \frac{X_4}{2})\rho - (X_4 + \frac{1}{2})\rho_{\tau_\alpha}] + t_2[(1 + \frac{X_2}{2})\rho - (X_2 + \frac{1}{2})\rho_{\tau_\alpha}] \\
& + t_5\rho[(1 + \frac{X_5}{2})\rho + (X_5 + \frac{1}{2})\rho_{\tau_\alpha}]\} \}^{-1} .
\end{aligned} \tag{11}$$

The imaginary part of MOP obtained from Eq.(9) is

$$W_{\tau_\alpha} = -\frac{1}{64\pi^5} \sum_{i=1} W_i \tag{12}$$

The expressions of W_i can be found in Ref. [6].

The simplest way to obtain MOP for a finite nucleus is to use the LDA. We assume that the densities of the neutrons and protons in a spherical nucleus have the same geometrical distributions and are expressed by Negele's empirical formula[14]

$$\rho_K(r) = \frac{\rho_{oK}}{1 + exp[(r - C)/a]} , \qquad K = N \ or \ Z \tag{13}$$

where

$$\rho_{oK} = \frac{3K}{4\pi C^3 (1 + \pi^2 a^2 / C^2)} , \qquad K = N \ or \ Z \tag{14}$$

$$C = (0.978 + 0.0206 A^{\frac{1}{3}}) A^{\frac{1}{3}}, \qquad a = 0.54 . \tag{15}$$

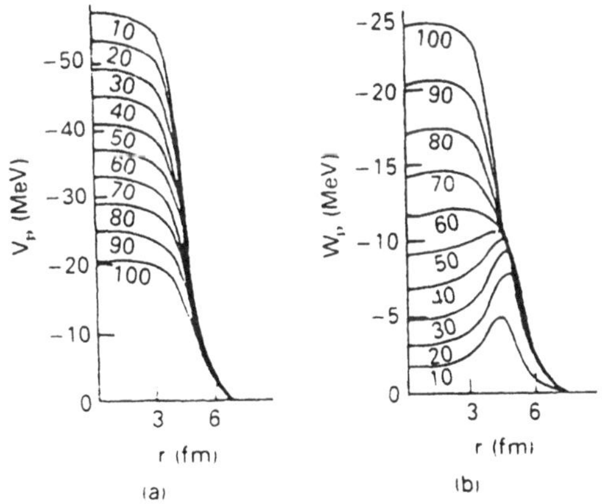

Fig. 1 Radial dependence of the proton MOP for ^{58}Ni at energies 10–100 MeV. The curves are the theoretical values computed with GS2. (a) The real parts. (b) The imaginary parts.

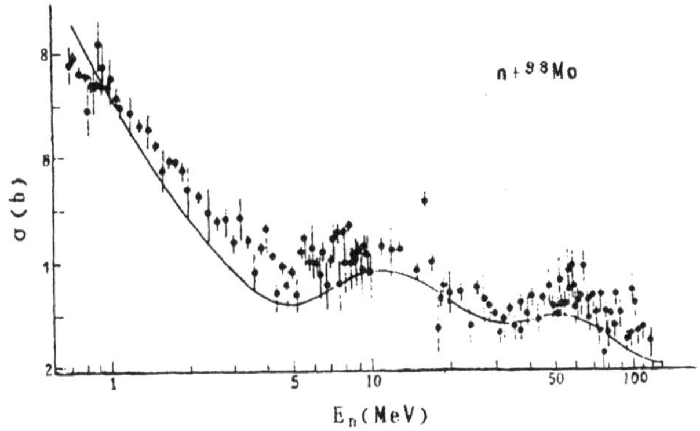

Fig. 2 Comparison of the theoretical values of the neutron total cross sections for ^{98}Mo with experiments.

In Fig. 1 the real and imaginary parts of the proton MOP for ^{58}Ni in the energy range of E = 10–100 MeV computed with GS2 are illustrated. The imaginary part of the MOP changes from the dominant surface absorption into the volume absorption with increase of energy as expected from the POP. Fig. 2 shows the neutron total cross section for ^{98}Mo calculated with MOP. They are in good agreement with the experimental data. Fig. 3 shows that the caculated (n,2n) cross sections of ^{100}Mo are also in good agreement with experimental data. Fig. 4 shows the calculated neutron elastic scattering angular distributions for ^{98}Mo at $E_n = 0.5–26.0$ MeV. The tendency of angular distribution, position and absolute values of peaks and valleys all fit experiments quite well.

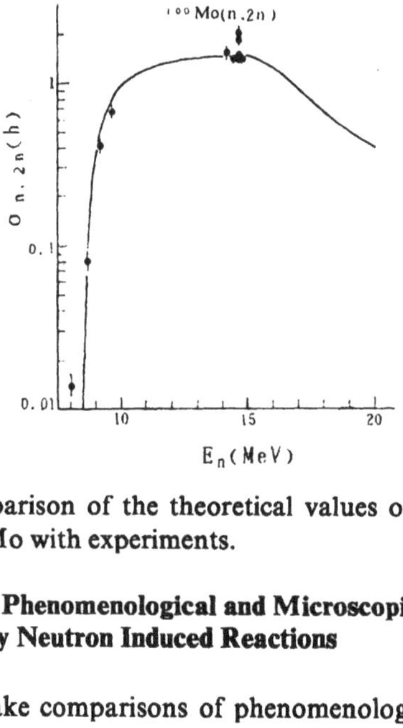

Fig. 3 Comparison of the theoretical values of the (n,2n) cross sections for ^{100}Mo with experiments.

Ⅲ. Comparisons of Phenomenological and Microscopic Optical Potential for Low Energy Neutron Induced Reactions

In order to make comparisons of phenomenological and microscopic optical potential for low energy neutron induced reactions, we calculate and evaluate nuclear data for 14 natural elements using OOP, BGP, CH86, and MOP[15]. For MOP we chose four sets of Skyrme force parameters: GS2, SKa, SⅡ, and SⅢ. In all of our calculations, the compound elastic scattering cross section and angular distribution are calculated by Hauser–Feshbach theory with width fluctuation, in which the transmission coefficients are computed using the same optical potential as in the optical model calculation.

For a certain element, χ^2 represents the deviation of the calculated nuclear data from the experimental values; it is defined as

$$\chi^2_{\sigma_t} = \frac{1}{N_t}\sum_{j=1}^{N_t}\left[\frac{\sigma_t^T(j)-\sigma_t^E(j)}{\triangle\sigma_t^E(j)}\right]^2 , \tag{16}$$

$$\chi^2_{\sigma_{non}} = \frac{1}{N_{non}}\sum_{j=1}^{N_{non}}\left[\frac{\sigma_{non}^T(j)-\sigma_{non}^E(j)}{\triangle\sigma_{non}^E(j)}\right]^2 , \tag{17}$$

$$\chi^2_{\sigma_{el}(\theta)} = \frac{1}{N_{el}}\sum_{j=1}^{N_{el}}\frac{1}{n_j}\sum_{i=1}^{n_j}\left[\frac{\sigma_{el}^T(\theta_{ji})-\sigma_{el}^E(\theta_{ji})}{\triangle\sigma_{el}^E(\theta_{ji})}\right]^2 , \tag{18}$$

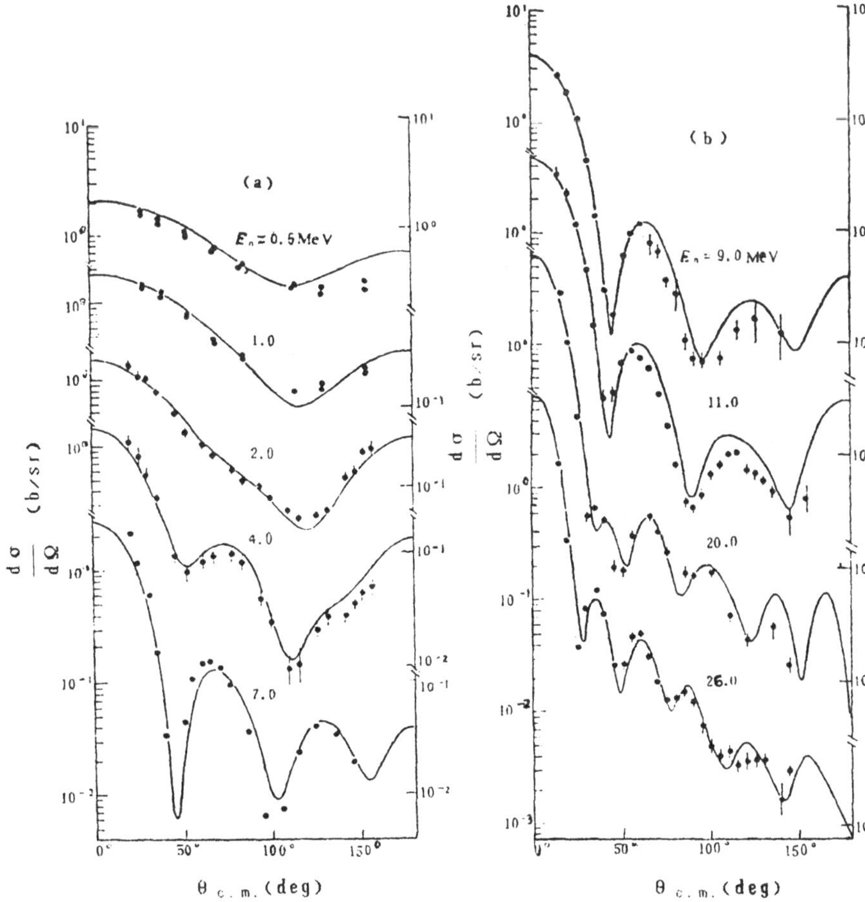

Fig. 4 Comparison of the theoretical values of the neutron elastic scattering angular distribution for ^{98}Mo at $E_n = 0.5$–26.0 MeV with experiments.

where the superscripts T and E represent the theoretical and experimental values, respectively. The values of N_t , N_{non} , and N_{el} are the numbers of energy points of the experimental data, corresponding to σ_t , σ_{non} , and $\sigma_{el}(\theta)$, respectively. Where n_j is the number of angles of the experimental differential cross section corresponding to the j'th energy point, and $\theta_{j\,i}$ is the angle value of the i'th angle for the j'th energy point. The total χ^2 of a certain element is defined as

$$\chi^2 = \frac{1}{3}[\chi^2_{\sigma_t} + \chi^2_{\sigma_{non}} + \chi^2_{\sigma_{el}(\theta)}] , \tag{19}$$

which has equal weight for σ_t , σ_{non} , and $\sigma_{el}(\theta)$.

We have collected the optimal neutron optical potential parameters of 14 elements from Z=13 to 82 and A=27 to 208. These OOP parameters are adjusted in a neutron nuclear data evaluation. Table 1 lists the energy points used in the calculation corresponding to each element, which contains one to four isotopes.

Table 1. The incident neutron energy taken in the calculations

Elements	Number of Energy	Energy, MeV
$_{13}$Al	14	1.5,2.47,3.85,4.56,7.05,9,11,13,14.1,15,17,18.4,20,24
$_{15}$P	13	0.2,0.45,1.09,2.5,3.5,4.8,5.95,7.79,9.05,11,12.5,14.2,14.5
$_{16}$S	15	0.6,1.05,2.47,3.49,4.48,4.56,5.5,7.05,7.97,8.52, 9.8,11.01,14,14.83,20
$_{17}$Cl	12	1.5,2.5,3.6,4.1,5.2,7,8,10,12,14,18.34,20
$_{19}$K	16	1.5,2.06,2.38,3.1,3.66,3.76,4.33,5.5,6.5,7.91,9, 10.2,11.15,14,15.15,20
$_{20}$Ca	13	1,2.06,2.5,3.5,4.1,5.5,6.37,7.05,7.6,8.05,11,14,20
$_{26}$Fe	15	2.33,3.4,4.6,5.05,5.56,6.12,7.55,7.96,9.94,11.93, 13.92,14.7,16.1,17.9,19.7
$_{27}$Co	20	0.5,0.9,1.46,2,2.47,3,3.5,4,4.56,6,7,8,9,10, 11,12,14.7,16,18,20
$_{29}$Cu	18	0.083,0.09,0.1,0.15,0.3,0.39,0.64,0.8,0.966,1.46,2.47, 3.5,4.56,6.09,8.05,11,14.6,20
$_{47}$Ag	14	0.4,1,2.05,3,3.6,4,5,6.7,9,10.5,13,15,17,18.5
$_{49}$In	12	0.8,1,2.03,3,4,4.56,5.34,6.47,8.05,11.01,14,18.4
$_{51}$Sb	15	1,1.6,2.5,3.3,4.1,5,6,7,8,9,10,12,14,17.5,21.4
$_{74}$W	14	0.35,0.79,1.1,2.6,2.9,4.34,4.92,6.44,7.54, 8.56,11,14.1,16,20
$_{82}$Pb	19	0.1,0.3,0.5,1,1.5,2.2,3.3,4,5,6,7,8.05, 10,11,14,14.6,15,17.5,20

Table 2. The total relative average χ^2 for 14 elements.

Potential:	OOP	GS2	SKa	CH86	BGP	SIII	SII
χ^2 :	0.309	0.600	0.646	0.684	0.807	1.368	2.587

We define the relative total average χ^2 for the 14 elements as

$$\chi^2_p = \frac{7}{14}\sum_{k=1}^{14}\frac{\chi^2_{p,k}}{\sum_{p'}\chi^2_{p',k}} , \tag{20}$$

where

p,p′ = OOP,BGP,CH86,GS2,SKa,SII,and SIII,

k = various elements.

The calculated results using Eq. (20) are listed in Table 2. One can clearly see that, on the average, OOP has the smallest deviation, SII has the largest deviation, and SIII has the next largest deviation. The deviation of GS2, SKa, and CH86 are between those of OOP and BGP. It is clear that the agreements of the calculated total, nonelastic, and differential elastic cross sections by CH86, GS2, and SKa with the experimental data are comparable. In some cases GS2 and SKa are better than CH86. Based on fundamental physics and without adjustable parameters, MOP is expected to have a stronger predictive power. We can conclude that the microscopic optical potential based on Skyrme force proposed here has great usefulness in neutron data calculation and evaluation below 20 MeV.

IV. Neutron Relativistic Phenomenological and Microscopic Optical Potential in Medium Energy Range (20–1000 MeV)

In relativistic optical model analyses of medium energy scattering experiments we use the Dirac equation given by

$$[\vec{\alpha}\cdot\vec{P} + r^{0}(M + U_{s}(r)) + U_{0}(r) + V_{C}(r)]\Psi(\vec{r}) = E\Psi(\vec{r}) \ , \tag{21}$$

where U_S is a Lorentz scalar potential, U_0 is the time–like component of a four–vector potential, and V_C is the Coulomb potential for proton, M the nucleon mass and E the nucleon total energy in the C.M. frame. The restriction to local scalar and vector potentials is motivated by conservation laws as well as by meson exchange considerations which suggest that these potentials represent the most important interactions.

The potentials U_0 and U_S are assumed to be complex and dependent on energy and mass number of target nuclei, and they can be treated either as a strictly phenomenological model with a number of adjustable parameter (RPOP) or as a microscopic model derived from some more basic theory without any free parameters (RMOP).

The Dirac equation (21) for the 4–component spinor $\Psi(\vec{r})$ is equivalent to two coupled equations for the large (upper) and small (lower) 2–component spinors. One can eliminate the small component of the Dirac spinor in the standard way. One then obtains a Schrödinger–like equation for the large component of the Dirac spinor:

$$[\frac{P^{2}}{2E} + U_{eff}(r) + V_{C}(r) + U_{so}(r)\vec{\sigma}\cdot\vec{L}]\varphi(r) = \frac{E^{2} - M^{2}}{2E}\varphi(r) \ , \tag{22}$$

where the Schrödinger equivalent potentials U_{eff} and U_{SO} are the central and spin–orbit ones, respectively, and

$$U_{eff}(r) = U_{0} + \frac{1}{2E}[U_{s}(2M + U_{s}) - (U_{0} + V_{C})^{2}] \ , \tag{23}$$

$$U_{so}(r) = -\frac{1}{2ErD(r)}\frac{dD(r)}{dr} \ , \tag{24}$$

$$D(r) = M + E + U_S - U_0 - V_C \ , \tag{25}$$

here a small Darwin term is neglected.The Schrödinger equivalent potential yields exactly the same scattering phase shifts as the original 4×4 relativistic potential in the 4–component Dirac equation.Thus, we call Eq.(22) as relativistic optical model equation which is used to calculate the scattering amplitude.

In the following part, we are aiming to obtain a global neutron relativistic optical potential whose parameters are functions of both energy and target mass number. The experimental data used for this purpose consist of ten nuclei from ^{12}C to ^{238}U over a wide energy region from 20 up to 1000 MeV. The total cross section data taken from Refs.[16–19] are quite complete which can cover the whole energy range for all ten nuclei considered in this paper. The nonelastic cross section data available only below 200 MeV for some specific nuclei[18], the elastic scattering angular distribution data at energies between 20 and 40 MeV for ^{12}C, ^{16}O, ^{27}Al, ^{56}Fe, ^{208}Pb and ^{209}Bi[20–25] as well as at 96 MeV for ^{12}C in small angle region[26] and at 155 MeV for ^{12}C, ^{27}Al, ^{63}Cu and ^{208}Pb also within the small angle range[27] are all neutron data accessible to us in addition to the rather complete set of total cross section. Our global neutron relativistic phenomenological optical potential is parameterized on the basis of above mentioned data set.

We write the scalar and vector optical potentials with Woods–Saxon form, whose parameters depend on energy E, mass number A and charge number Z. Through automatic search of the best parameters in fitting the experimental data σ_t , σ_{non} and $\sigma_{el}(\theta)$ by computer, a global neutron RPOP, which contains 38 paramcters, has been constructed as follows:

$$U_0(r) = V_0 f_0(r) + iW_0 g(r), \tag{26}$$

$$U_S(r) = V_S f_S(r) + iW_S g(r), \tag{27}$$

$$V_0 = 304.08 - 0.1103E^{0.8451 + 0.000634A} - 0.0000922E^{1.968 - 0.000787A} - 2.295\alpha - 5.666\beta, \tag{28}$$

$$V_S = -379.66 - (0.05492 + 0.0000241A)E - 3.637\alpha + 13.328\beta, \tag{29}$$

$$W_0 = -5.090 - (0.2146 + 0.0000464A)E^{0.8945 - 0.000283A} + 0.00000562\beta E^2 + 23.692\alpha + 1.982\beta, \tag{30}$$

$$W_S = -14.21 + (0.2820 - 0.000395A)E^{0.8934 - 0.000285A} + 18.895\alpha - 1.189\beta \tag{31}$$

$$\alpha = \frac{N - Z}{A}, \qquad \beta = \left|\frac{A - 64}{A + 16}\right|^{1.278}, \tag{32}$$

$$f_i(r) = \{1 + exp[(r - r_i A^{\frac{1}{3}}) / a_i]\}^{-1}, \qquad i = 0, s \tag{33}$$

$$g(r) = \{1 + exp[(r - r_W A^{\frac{1}{3}}) / a_W]\}^{-1}, \tag{34}$$

$$r_0 = 1.164, \qquad r_S = 1.159, \qquad r_W = 1.239$$

$$a_0 = 0.5746, \qquad a_S = 0.5923, \qquad a_W = 0.4378$$

For the RMOP, we adopted an approach to the relativistic many–body theory based on effective lagrangian density of Walecka model, which allows one to perform the calculation for nucleon–nucleus interactions in the lowest order approximation. The merit of this approach is obvious that it is very simple, and can easily be applied to the scattering problem.

We start from an effective lagrangian with two adjustable isoscalar meson coupling constants g_σ and g_ω

$$\pounds = \bar{\psi}(i\gamma^\mu \partial_\mu - M)\psi + \frac{1}{2}(\partial^\mu \sigma \partial_\mu \sigma - m_\sigma^2 \sigma^2) + \frac{1}{2} m_\omega^2 \omega^\mu \omega_\mu - \frac{1}{4} F^{\mu\nu} F_{\mu\nu} - g_\sigma \bar{\psi}\sigma\psi$$
$$- g_\omega \bar{\psi}\gamma^\mu \omega_\mu \psi, \qquad (35)$$

with $F_{\mu\nu} = \partial_\mu \omega_\nu - \partial_\nu \omega_\mu$, where ψ, σ and ω_μ are the nucleon, σ and ω–meson field operators, respectively. The values of nucleon and ω meson mass are taken from the experiments, M = 938.9 MeV and M_ω = 783 MeV. The mass of the hypothetical σ meson is fixed to M_σ = 550 MeV, which is commonly used in the NN interaction to simulate the two π exchange.

In the next part, we are going to study the nucleon RMOP. It is known that the self–energy of a nucleon in the nuclear medium is identified with the effective interaction of the nucleon with the nuclear medium, i.e. nuclear optical potential. Based on the Feynman diagram rules one could perturbatively derive the nucleon self–energy in nuclear matter. We let only the second–order (Hartree–Fock) self–energy of a nucleon in the nuclear medium represent the real part of the optical potential and consider the imaginary part of the fouth–order self–energy as the imaginary part of the optical potential.

The effective coupling constants g_σ and g_ω are adjusted by requiring that the empirical saturation properties of the nuclear matter are reproduced, i.e. the binding energy per nucleon $\varepsilon / \rho_B - M = -15.75$ MeV and pressure P = 0 at the normal density, corresponding to $k_F = 1.42$ fm^{-1}.

The calculations of the self–energy of a nucleon are first carried out in the nuclear matter up the fouth–order as mentioned above. The optical potential for finite nuclei is then obtained within the LDA. The density distribution is taken from Negele's empirical formulas[14], except for light nuclei such as ^{12}C and ^{16}O. The Gaussian–type distribution are chosen for ^{12}C and ^{16}O[28]:

$$\rho(r) = \rho_0[1 + \alpha r^2 / a^2] exp(-r^2 / a^2), \qquad (36)$$

$$\rho_0 = A / [4\pi a^3 \sqrt{\pi}(\frac{1}{4} + \frac{3}{8}\alpha)], \qquad (37)$$

where α, a are 4 / 3, 1.65 fm for ^{12}C and 2.0, 1.76 fm for ^{16}O. Thus we could obtain a RMOP without any free parameters.

The real part of the central Schrödinger equivalent potentials of neutron RPOP and RMOP for ^{56}Fe at various energies are calculated. For energy $E_n < 500$ MeV the real potentials of both RPOP and RMOP are rather close and the so–called "bottom of wine bottle" shape is found in both of them in 200–300 MeV energy region. For energy $E_n > 500$ MeV the V_{eff} of RMOP continues to go up rapidly as energy increases, while the one of RPOP first increases slowly with energy and then even comes down as energy goes up . The imaginary potentials always take negative values. The value of imaginary potentials of RMOP increase much more rapidly with the energy as compared with those of RPOP. The spin–orbit Schrödinger equivalent potentials of neutron RPOP and RMOP in the same energy region are also calculated. One can see that the real spin–orbit potentials V_{SO} are negative whereas the imaginary ones W_{SO} are positive. The absolute value of V_{SO} decreases as energy increases , but that of W_{SO} increases as energy increases. The imaginary part W_{SO} can be neglected at low energies but both the real and imaginary spin–orbit potentials should be considered simultaneously at high energies.

The neutron total cross section σ_t, nonelastic cross section σ_{non} and elastic scattering angular distribution $\sigma_{el}(\theta)$ for ten target nuclei ranging from ^{12}C to ^{238}U

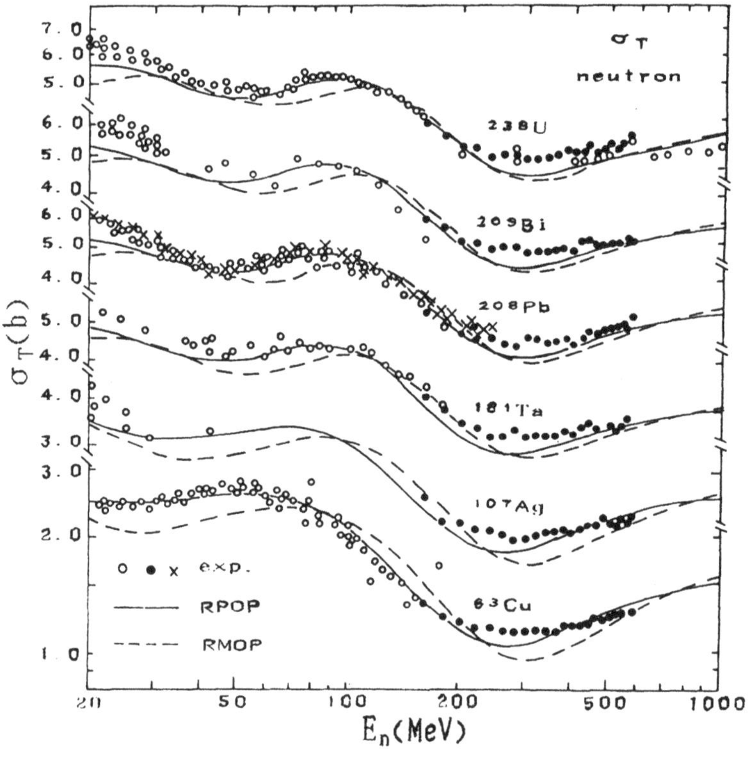

Fig. 5 Neutron total cross sections for ^{63}Cu, ^{107}Ag, ^{181}Ta, ^{208}Pb, ^{209}Bi, and ^{238}U at energies 20–1000 MeV calculated by RPOP (solid line) and RMOP (dashed line). The experimental data taken from Refs.[16–19].

with incident energies $E_n = 20-1000$ MeV have been calculated by both RPOP and RMOP.

Fig. 5 shows the comparison of neutron total cross sections for 6 target nuclei between calculated results of RPOP and RMOP with the experimental data. Generally speaking, the values and tendency of the total cross sections calculated by RMOP agree with the experiments pretty well and the results of RPOP are rather close to the experiments.

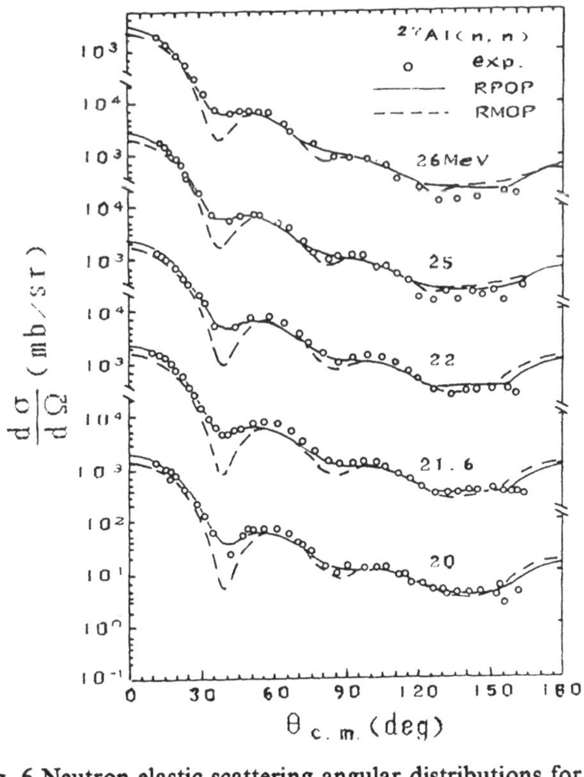

Fig. 6 Neutron elastic scattering angular distributions for ^{27}Al at energies 20,21.6,22,25,and 26 MeV calculated by RPOP (solid line) and RMOP (dashed line). The experimental data taken from Refs.[20,22].

Fig.6 shows the neutron elastic scattering angular distributions for ^{27}Al at energy region of 20–26 MeV calculated by RPOP and RMOP. The experimental data are taken from Refs[20,22]. For most cases the calculated results by both RPOP and RMOP are in pretty good agreement with the experiments, but the valleys in the results by RMOP are too deep as compared with experiments. In Fig.7 the elastic scattering angular distributions for ^{12}C at energy 96 MeV and for ^{12}C, ^{27}Al, ^{63}Cu and ^{208}Pb at 155 MeV calculated by RPOP and RMOP are illustrated. The expermental data are taken from Refs.[26,27]. The overall good agreement with experiments for both RPOP and RMOP is seen except at some specific points.

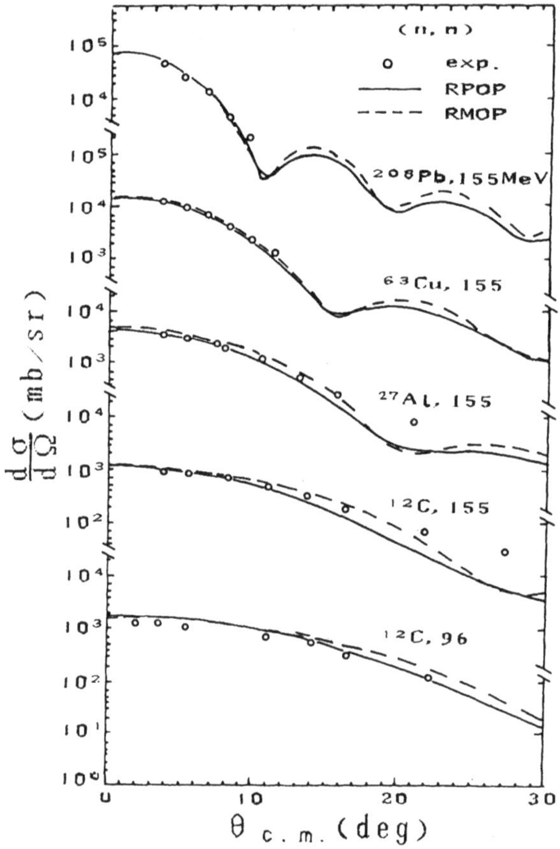

Fig. 7 The same as Fig. 6 except for ^{12}C at energy 96 MeV and for ^{12}C, ^{27}Al, ^{63}Cu, and ^{208}Pb at 155 MeV. The experimental data taken from Refs.[26,27].

V. Summary

Through systematic studies and a large amount of calculations, it is shown that the calculated neutron cross sections and angular distributions by using the MOP based on Skyrme interaction below 20 MeV are better than the global optical potential BGP which is widely used and at the same level compared with the improved global optical potential CH86 in agreement with the experimental data. Considered the MOP has a reliable theoretical basis and there are no free parameters in MOP, we can conclude that the MOP based on Skyrme force has great usefulness in nuclear data calculation and evaluation.

We also made calculations for 20–1000 MeV neutron induced reactions by using the relativistic microscopic optical potential and obtained a global neutron relativistic phenomenological optical potential based the available experimental data for the same energy region. The calculated cross sections and angular distributions by these two kinds of potentials agree with the experimental data pretty well. These rsults are useful for the medium energy nuclear reaction study in the future.

ACKNOWLEDGMENTS

Zhang Jingshang, Tian Ye, Han Yinlu, and Feng Dachun participated in part of this work. We would like to express our appreciation to Ma Zhongyu, Gu Yingqi, and Zhu Ping for their help to part of our work presented in this talk.

References

1. F. D. Becchetti and G. W. Greenlees, *Phys. Rev.* **182** (1969) 1190.
2. R. L. Varner, T. B. Clegg, T. L. Mcabee, and W. J. Thompson, *Phys. Lett.* **B185** (1987) 6.
3. Shen Qingbiao, Zhang Jingshang, Tian Ye, Ma Zhongyu, and Zhuo Yizhong, *Z. Physik* **A303** (1981) 69.
4. Shen Qingbiao, Tian Ye, Zhang Jingshang, and Zhuo Yizhong,*Nuclear Data for Science and Technology, Proc. Int. Conf.*, Antwerp, 6–10 Sept., 1982, p. 565.
5. Shen Qingbiao, Li Zhuxia, Shi Xiangjun, Tian Ye, and Zhuo Yizhong, *Proc. Int. Summer School on Nucleon–Nucleon Interaction and Nuclear Many–Body Problems*, Chongchun, China, 1983, p. 618.
6. Zhuo Yizhong, Shen Qingbiao, and Tian Ye, *Advances in Science of China Physics* Vol. 1, 1985, p. 231.
7. L. G. Arnold and B. C. Clark, *Phys. lett.* **84B** (1979) 46.
8. L. G. Arnold, B. C. Clark, and R. L. Mercer, *Phys. Rev.* **C19** (1979) 917.
9. S. Hama, B. C. Clark, E. D. Cooper, H. S. Sherif, and R. L. Mercer, *Phys. Rev.* **C41** (1990) 2737.
10. Shen Qing–biao, Feng Da–chun, and Zhuo Yi–zhong, *Phys. Rev.* **C43** (1991) 2773.
11. C. J. Horowitz and B. D. Serot, *Nucl. Phys.* **A399** (1983) 529.
12. C. J. Horowitz, *Nucl. Phys.* **A412** (1984) 228.
13. Ma Zhong–yu, Zhu Ping, Gu Ying–qi, and Zhuo Yi–zhong, *Nucl. Phys.* **A490** (1988) 619.
14. J. W. Negele, *Phys. Rev.* **C1** (1970) 1260.
15. C. H. Cai, Q. B. Shen, and Y. Z. Zhuo, *Nucl. Sci. Eng.* **109** (1991) 142.
16. J. Franz, H. P. Grotz, L. Lehmann, E. Rossle, H. Schmitt, and L. Schmitt, *Nucl. Phys.* **A490** (1988) 667.
17. S. Pearlstein, *Nuclear Data for Science and Technology, proc. Inter. conf.*, 1988, Mito, Japan p. 1115.
18. V. McLane, C. L. Dunford, and P. F. Rose, *Neutron Cross Sections*, BNL–325, Vol.2, Academic Press, 1988.

19. R. L. Schutt, R. E. Shamu, P. W. Lisowski, M. S. Moore, and G. L. Morgan, *Phys. Lett.* **B203** (1988) 22.
20. J. S. Petler M. S. Islam, R. W. Finlay, and F. S. Dietrich, *Phys. Rev.* **C32** (1985) 673.
21. J. P. Delaroche, M. S. Islam, and R. W. Finlay, *Phys. Rev.* **C33** (1986) 1826.
22. N. Olsson, B. Trostell, E. Ramström, and B. Holmqvist, *Nuclear Data for Basic and Applied Science, Proc. Inter. Conf.*, Santa Fe, New Mexico, USA, 1985, p. 963.
23. S. Mellema, R. W. Finlay, F. S. Dietrich, and F. Petrovich, *Phys. Rev.* **C28** (1983) 2267.
24. C. H. Johnson, D. J. Horen, and C. Mahaux, *Phys. Rev.* **C36** (1987) 2252.
25. R. L. Walter and P. P. Guss, *Nuclear Data for Basic and Applied Science, Proc. Inter. Conf.*, Santa Fe, New Mexico, USA, 1985, p. 1079.
26. G. L. Salmon, *Nucl. Phys.* **21** (1960) 15.
27. R. S. Harding, *Phys. Rev.* **111** (1958) 1164.
28. H. F. Ehrenberg, R. Hofstadter, U. Meyer–Berkhout, D. G. Ravenhall, and S. E. Sobottka, *Phys. Rev.* **113** (1959) 666.

THEORETICAL RESEARCH ON 13 MeV N-D BREAK-UP PROCESSES

ZHANG TIANYUAN
*Institute of Theoretical Physics, Academia Sinica, Beijing, China and Institute of Applied Physics and Computational Mathematics, P. O. Box 8009, Beijing 100088, China**

and

MO JUNYONG
Institute of Applied physics and Computational Mathematics, P. O. Box 8009, Beijing, China

ABSTRACT

Faddeev equations with local potentials in the momentum space are two-variable integral equations whose kernels have singularities in the case of break-up processes and three-body scattering. From the singular integral equation theory authors proposed a numerical technique for solving the two-variable equations. The convergent numerical solutions can be obtained. Differential cross sections for complete kinematics are given. It is clear that agreement between the theoretical and the experimental is very well.

1. Introduction

Discussing the properties of three-body systems in 1960s, L. D. Faddeev[1] proposed a set of equations as a basic tool for research on the three-body problems. Large-scale computer systems make the great progress so that the three-nucleon scattering problems with complicated realistic nuclear potentials will be able to be solved.

In the momentum representation the Faddeev equations consists of a set of integral equations. If the two-body nuclear interaction is separable, the kernels of the equations will also be separable. In this case the integral equations are reduced to a system of linear algebraic equations[2]. However the realistic nuclear potentials are not separable and their structure is complicated. In order to solve the equations with the separable potentials, first of all, we have to make approximation for the realistic potentials by using functions with separable variables, for example, rational algebraic fraction. The calculations need the

---

* permanent address

separable potentials with higher partial waves, and should spend much computer CPU time.

At present, Paris[3] and Bonn[4] groups have obtained the important realistic potentials from mesonic exchange theory. Reducing angular mometa, the Faddeev equations with the realistic nuclear forces turn to two-variable integral equations whose kernels have singularities in the case of three-body scattering and break-up processes.

In this paper authors develop a numerical technique for solving the equations on the basis of the singular integral equation theory. The paper consists of four sections. In the second section theoretical formulation is discussed for three-body break-up processes with the realistic nuclear potential. We will devote to the numerical technique for two-variable integral equations in the third section. At last, we will analyze some calculated results, and the conclusions are drawn from them.

2. Theoretical Formulation

In general, t-matrix formalism is convenient to discuss the three-body processes[5]. t-matrix of the system makes decomposition

$$T = T^{(1)} + T^{(2)} + T^{(3)}, \tag{2.1}$$

where each of the components $T^{(i)}(i = 1, 2, 3)$ satisfies Faddeev equation as follows

$$T^{(i)}(s) = T_i(s) + \sum_{j \neq i} T_i(s)G_0(s)T^{(j)}(s). \qquad i = 1, 2, 3 \tag{2.2}$$

In the Eq. (2.2) $G_0(s)$ is free three-body Green function

$$G_0(s) = (S - H_0)^{-1}. \tag{2.3}$$

The total kinetic energy H_0 is defined by the kinetic energies of these three particles

$$H_0 = \frac{k_1^2}{2m_1} + \frac{k_2^2}{2m_2} + \frac{k_3^2}{2m_3}. \tag{2.4}$$

$S = (E+i\varepsilon)$ is regarded as a complex energy parameter, where ε is a positive number less than one. T_i $(i = 1, 2, 3)$ in Eq. (2.2) are the two-body off-shell t-matrices, which satisfy the Lippmann-Schwinger equations:

$$T_i(s) = V_i + V_i g_0(s) T_i(s), \qquad (i = 1, 2, 3) \tag{2.5}$$

where V_i is a two-body interaction between the particles j and k($i \neq j \neq k$). $g_0(s)$ is free two-body Green function.

Expanding the Faddeev operator equations into basic functions, the t-matrix element integral equations are derived as

$$\tilde{\Psi}_s = (p,q,\alpha;\ p_0,q_0,\alpha_0) = \tilde{\phi}_s(p,q,\alpha;\ p_0,q_0,\alpha_0) - \sum_{\alpha_2}\int_0^{\infty} dq_2^2 \int_{A(q,q_2)}^{B(q,q_2)} dp_2^2 \tilde{K}_s(p,q,\alpha;\ p_2,q_2,\alpha_2)\tilde{\Psi}_s(p_2,q_2,\alpha_2;\ p_0,q_0,\alpha_0). \tag{2.6}$$

3. The Numerical Technique of Singular Integral Equations

Converting the variables, the equation (2.6) turns to a standard form

$$\Psi(t,\tau,\alpha;\ t_0,\tau_0,a_0) + \sum_{\alpha_2}\frac{B(t,\tau)}{\pi}\int_{-1}^{1} d\tau_2 \int_{-1}^{1} dt_2 \frac{\Psi(t_2,\tau_2,\alpha_2;\ t_0,\tau_0,\alpha_0)}{t_2 + \tau_2 - t_s - \tau_s}$$
$$+ \sum_{\alpha_2}\frac{1}{\pi}\int_{-1}^{1} d\tau_2 \int_{-1}^{1} dt_2 K_0(t,\tau,\alpha;\ t_2,\tau_2,\alpha_2)\Psi(t_2,\tau_2,\alpha_2;\ t_0,\tau_0,\alpha_0)$$
$$= f(t,\tau,\alpha;\ t_0,\tau_0,\alpha_0). \tag{3.1}$$

The solution of Eq.(3.1) has the following form

$$\Psi(t,\tau,\alpha) = W(t)\phi(t,\tau,\alpha), \tag{3.2}$$

where

$$W(t) = (1-t)^{\alpha}(1+t)^{\beta}. \tag{3.3}$$

According to the Hunter quadrature formula[6] the first integration over the variable t_2 in the second term on the left - hand side of Eq. (3.1) can be done. The interval of the variable t is discretized into (m+1) pieces by the zeros $\zeta_{i,m}$ of the Jacobi polynomial $P_m^{(-\alpha,-\beta)}$ of the degree m, and for the variable τ Gauss nodes are used. The integration over τ_2 is performed by the Gauss quadrature. Finally we lead a set of the equations as follows:

$$W(\zeta_{im})f(\zeta_{im},\tau_j,\alpha) + \sum_{\alpha_2}\frac{B(\zeta_{i,m},\tau_j)}{\pi}\sum_{l=1}^{N} w_l I_n^{(\alpha,\beta)}(\tau_l,\zeta_{i,m},\tau_j)\cdot\frac{\phi(\zeta_{i,m}+\tau_j-\tau_l,\alpha_2)}{P_n^{(\alpha,\beta)}(\zeta_{i,m}+\tau_j-\tau_l)} +$$

$$\sum_{\alpha_2}\frac{B(\zeta_{i,m},\tau_j)}{\pi}\sum_{k=1}^{n}\sum_{l=1}^{N} w_l\mu_{k,n}\cdot\frac{\phi(\zeta_{k,m},\tau_l,\alpha_2)}{(\zeta_{i,m},+\tau_j-\zeta_{i,m}-\tau_l)} +$$

$$\frac{1}{\pi}\sum_{\alpha_2}\sum_{k=1}^{n}\sum_{l=1}^{N} w_l\mu_{k,n}\cdot K_0(\zeta_{i,m},\tau_j,\alpha;\ \xi_{k,n},\tau_l,\alpha_2)\phi(\xi_{k,n},\tau_l,\alpha_2) = f(\zeta_{i,m},\tau_j,\alpha), \tag{3.3}$$

where $\xi_{k,n}$ is a zero point of the polynomial $P_n^{(\alpha,\beta)}(x)$, τ_j stands for the jth zero point of the Legendre polynomial $P_n(x)$, w_l is the Gauss weight for the corresponding node.

From the theory of singular integral equations[7], we know that Eq. (3.1) is solvable for any function f of the right- hand side, if the index $k > 0$. If $k < 0$, then necessary and sufficient conditions for the equation being solvable are that the function f must satisfy consistency conditions.

4. Result and Conclusion

In 1961 L.D. Faddeev[1] proved that if two-body potentials in the momentum space content the following conditions:
1. boundness and sufficiently fast falling off

$$|V(k)| < C(1+|k|)^{-1-\theta_0},\ (\theta_0 > 0), \tag{4.1}$$

2. smoothness

$$|V(k) - V(k+h) \le C(1 + |k|)^{-1-\theta_0}|h|^{\mu_0},\quad (|h| \le 1,\ \mu_0 > 0) \tag{4.2}$$

3. real-valuedness

$$V(-k) = \overline{V(k)}\ , \tag{4.3}$$

and $\omega_0 \in B(\theta, \mu)$, where $B(\theta, \mu)$ is a Banach space with parameters θ, μ, then the Faddeev equation

$$\omega(z) = w_0(z) + A(z)\omega(z) \tag{4.4}$$

has a unique solution $\omega(z) \in B(\theta, \mu)$. As is well known, all the realistic nuclear forces satisfy the conditions (4.1), (4.2) and (4.3). So the Faddeev equation (2.6) is uniquely solvable with respect to any given function $\omega_0 \in B(\theta, \mu)$. According to the singular integral equation theory[8], the index of the equation (2.6) is zero.

As the first step, we consider the Herzfeld potential[9]

$$V(r) = \begin{cases} \frac{h^2}{2\mu}U_0, & r < r_0 \\ \frac{h^2}{2\mu}\tilde{U}(r) = \frac{h^2}{2\mu}U_1, & r_1 < r < r_0 \\ \frac{h^2}{2\mu}U_2 = 0. & r < r_0 \end{cases} \tag{4.7}$$

The two-body off-shell t-matrices with this potential have analytic solutions.

For the sake of comparison between calculated results and experimental data, we have to consider complete kinematics of the system[10]. Assume that the energies of the outgoing particles 1 and 2 in laboratory coordinates are E_1^l and E_2^l, respectively. If these two particles are observed and recorded by two coincidence detectors, then their energies will be determined at all. It is easy to deduce allowed kinematical curves of the particles in laboratory coordinates from energy and momentum conservations of the three outgoing particles. We immediately find the differential cross-sections

$$\frac{d^3\sigma}{d\Omega_1 d\Omega_2 dE_s} = \frac{1}{(2\pi)^3}\frac{1}{k_0}\frac{m_p m_i}{m_p + m_i}\rho_2(E_s)\,|T(pq)|^2 \times 0.050667(\text{mb}\cdot\text{MeV}^{-1}\cdot\text{sr}^{-2}) \tag{4.8}$$

where $d\Omega_1$ and $d\Omega_2$ represent solid angle elements spanned by the momenta of outgoing particles 1 and 2, and dE_s is the differential on the kinematical curve. The arc length origin is at a point with the minimum E_1^l and the larger one of the two possible E_2^l. The positive direction of the arc is in the clockwise sense (Fig. 1).

For exploring precision of the numerical technique we compared the results in the case of different numbers of quadrature nodes. Because of the restricted computer storage the number of quadrature points can not be taken too large. The practice shows that the results are stable for various numbers of partition points. In figure 2 the values of the amplitude square in 13 MeV n-d break-up process are sketched for three cases I) 6 × 10, II) 14 × 18, III) 18 × 22, respectively. It can be found out that the largest deviation of case I) from case III) does not exceed 14%, and the other case is less than 5%. Therefore the numerical solution for 18 × 22 nodes approximates the true solution of the

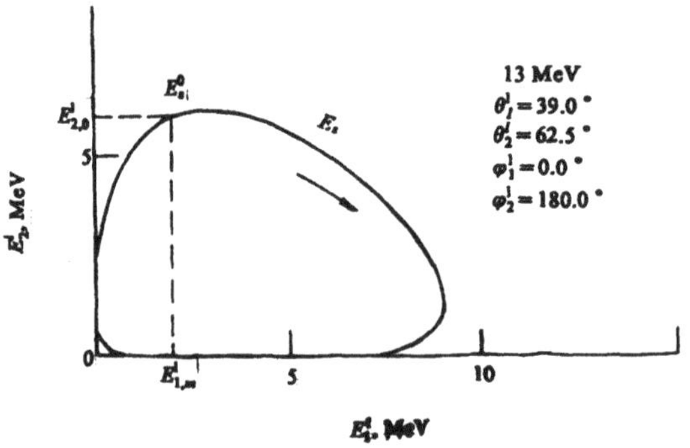

Fig. 1. An allowed kinematical curve. The arc length origin is at a point with the minimum E_1^l and the larger of two possible E_2^l, and the clockwise is in positive direction of the arc

equation. The calculation did not spend too much computer CPU time, for example, 1 hour.

The differential cross sections for three kinematical configurations are drawn in figures 3, 4 and 5. It is easy to see that the agreement between the calculated results and the data of Ref. [11] is quite well.

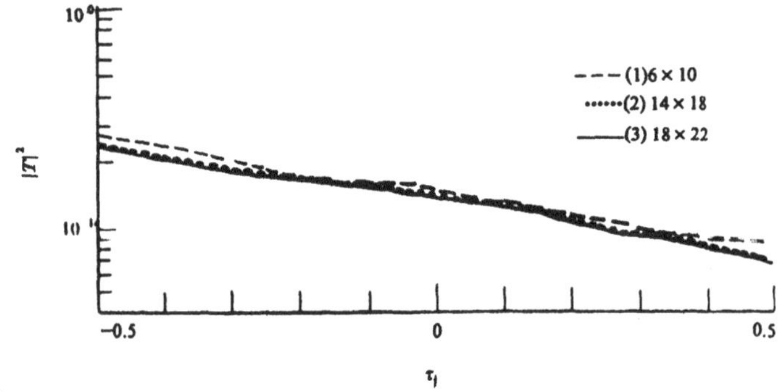

Fig. 2. Dependence of calculated amplitude square upon various nodes

On the whole, the index of two-variable Faddeev equations is zero. The numerical technique in the work is appropriate, and their numerical solution converges. Moreover the technique is also able to discuss the break-up processes and three-body scattering of p-d system with Coulomb modification. And the problems for realistic nuclear forces are being studied.

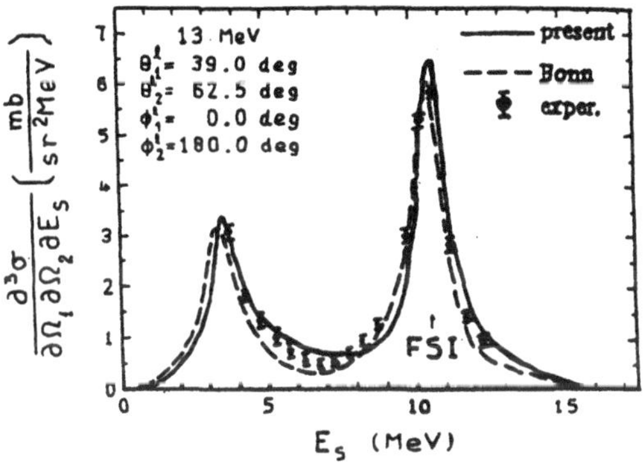

Fig. 3.
Differential cross sections of n-d break-up processes for incident neutron energy $E_p^1 = 13$ MeV. The kinematical configuration is given in the figure.

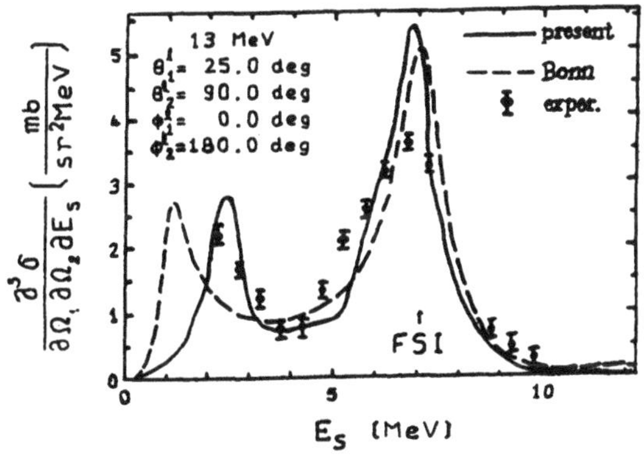

Fig. 4. Same as Fig. 3. The kinematical configuration is given in the figure.

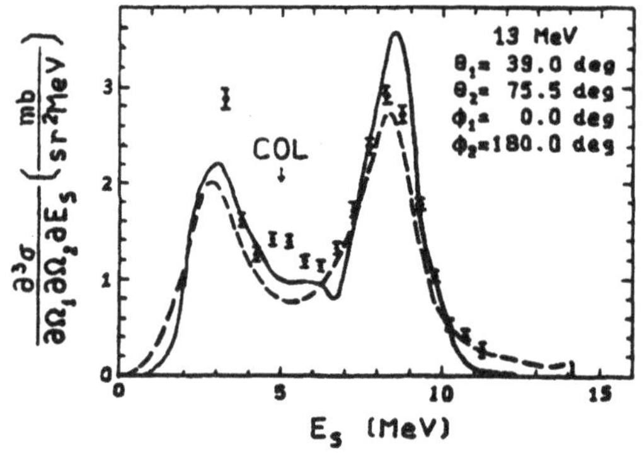

Fig. 5.
Same as Fig. 3. The kinematical configuration is given in the figure.
— present work,
---- Bonn group,
ɸ experimental data

References

1. L. D, Faddeev, Mathematical problems of quantum scattering theory for three-particle system *JETP* **12** (1961) 1014 (Izdt. Akad. Nauk, SSSR, Moscow, 1963).
2. E. O. Alt, P. Grassberger and W. Sandhas, *Nucl. Phys.* **B2** (1967) 167.
3. M. Lacombe, B. Loiseau, J. M. Richard, R. Vinh--Mau J. Cote, P. Pires and R. de Tourreil, *Phys, Rev.* **C21** (1980) 861; *Phys. Rev.* **D12** (1975) 1495.
4. R. Machleidt, R. Holinde and Ch. Elster, *Phys. Rep.* **149** (1987) 1.
5. W. Glockle, *The Quantum Mechanical Few-body Problem*, (Springer-Verlag, 1983).
6. D. Elliott, *SLAM J. Numer. Anal.* **19** (1982) 816; M. L. Dow and D. Elliott, *SIAM J. Numer. Anal.* **16** (1979) 115; Zhang Tianyuan and Mo Junyong, *in Few-body Methods: Principles and Applications*, ed. T. K. Lim, C. G. Bao, D. P. Hou and H. S. Huber (World Scien., Singapore, 1986) P. 847.
7. N. I. Mushkelishvili, *Singular Integral Equations* (Fizmatgiz, Moscow, 1962).
8. S. G. Mikhlin, *Multidimensional Singular Integrals and Integral Equations* (Fizmatgiz, Moscow, 1962); S. G. Mikhlin and S. Prossdrof, *Singular Integral Operators* (Springer-Verlag, 1986).
9. E. P. Harper, Y. E. Kim and A. Tubis, *Phys. Rev.* **C2** (1970) 877.
10. G. G. Ohlsen, *Nucl. Instr. Methods* **37** (1965) 240.
11. T. N. Frank, H. Haberzettl, Th. Januschke, U. Kerwath and W. Sandhas, *Phys. Rev.* **C38** (1988) 1112.

A Semi–Classical Theory of Multi–Step Nuclear Reaction Processes

ZHANG JINGSHANG
China Institute of Atomic Energy
P. O. Box 275–41, Beijing 102413, China

ABSTRACT

The master equation theory of precompound and compound nuclear reaction has been generalized to the inclusion of the conservation of angular momentum and parity. This improved semi–classical theory has been extended for application as an evaluation tool of the calculations of nucleon induced reaction cross sections and double differential cross sections. For structural materials at incident neutron energies below 20 MeV, it is demonstrated that the constructed model contains the Hauser–Feshbach, Weisskopf–Ewing as well as the exciton models as limiting cases. The unified treatment of pre–equilibrium processes includes a number of interesting features, such as the exciton state densities with the exact Pauli exclusion correction which are renormalized to the back–shifted Fermi–gas formula; the introduction of formation factors of composite particle in calculations of pick–up type composite particle emission and the double differential cross sections for all kinds of particles in terms of the leading particle model.

1. Introduction

In recent years pre–equilibrium nuclear reaction theories have been developed with great successful for the description of the double differential cross sections based on the exciton model. This success was mainly due to introducing a leading particle into the generalized exciton model[1] and application of the pick–up mechanism for composite particle emissions [2–3]. For the purpose to calculate the complete set of fast neutron data, a new semi–classical model of multi–step nuclear reaction processes has been proposed, in which the discrete level effect in multi–particle emissions was included as well as the pre–equilibrium phenomenon combining with the parity conservation and angular momentum conservation.

In section 2 the basic formulation of the semi–classical theory of multi–step reaction processes are given. Section 3 is devoted to the description of the double differential cross sections and some related contents. Some calculation results are shown in the last section. The calculation for n+ ^{56}Fe shows that the semi–classical model can rather nicely reproduce the experimental data of the total neutron spectra and the double differential cross sections. The calculation results of the spectra for (n,α),(n,d) and (n,t) indicate that the pick–up mechanism plays an important role in the pre–equilibrium composite particle emissions.

With the increasing of incident energies the components of the pick–up picture increase obviously. In particular to the composite particle with large binding energies, like d,t and ^{3}He, the pick–up components are always the dominant parts. Therefore, the pick–up mechanism must be taken into account in the composite particle emission processes.

To keep the energy conservation the recoil nucleus effects are also taken into account.

2. Basic formulation of the semi–classical theory

Since the Hauser–Feshbach model is successful in the equilibrium reaction processes, but the pre–equilibrium effects are exclusive. The exciton model is a useful tool to describe the pre–equilibrium phenomenon but it is unable to account the parity and angular momentum conservations. For this reason, to develop a new theory which includes both the success of these two models, a semi–classical model has been proposed.

The energy spectrum formula reads

$$\frac{d\sigma}{d\varepsilon} = \sum_{J\pi} \sigma_a^{J\pi} \sum_n P^{J\pi}(n) \frac{T_b^{J\pi}(n,E,\varepsilon)}{T_t^{J\pi}(n,E)} \quad , \tag{2.1}$$

where E is the excitation energy,

ε is the energy of emitted particle,

$\sigma_a^{J\pi}$ stands for the absorption cross section in $J\pi$ channel

$T_b^{J\pi}(n,E,\varepsilon)$ is the T factor of emitted particle b with excitation energy E at n exciton state in $J\pi$ channel,

$T_t^{J\pi}(n,E)$ is the total T factor,which is given by

$$T_t^{J\pi}(n,E) = \sum_b \int T_b^{J\pi}(n,E,\varepsilon) d\varepsilon. \tag{2.2}$$

$P^{J\pi}(n)$ stands for the $J\pi$ channel occupation probability of n exciton state.

Obviously, if we do not consider the parity and angular momentum effects, the eq.(2.1) is reduced to the exciton model, while if the pre–equilibrium effect is omitted, it is reduced to the Hauser–Feshbach model.

The approach to account the exact Pauli exclusion effect in the exciton state densities has been found [4]. For the nucleon induced nuclear reaction processes, the exciton state density was proposed as

$$\omega^{J\pi}(n,E) = P(\pi)\omega(n,E)R_n(J), \tag{2.3}$$

where $P(\pi) = \frac{1}{2}$ *for parity factor,*

$$R_n(J) = \frac{(2J+1)^2}{2\sqrt{2\pi}\,\sigma_n^3} \exp\left\{\frac{-(J+1/2)^2}{2\sigma_n^2}\right\} ,$$

$$\omega(n,E) = g\frac{(gU)^h (gU - A(p,h))^{p-1}}{p!h!(n-1)!} , \tag{2.4}$$

$$n = p + h.$$

$g = \frac{6a}{\pi^2}$ is the single particle level density around the Fermi surface, a is the level density parameter.

$U = E - \Delta$,Δ is pair correction. The Pauli exclusion correction A(p,h) is approximately equal to

$$A(p,h) = \frac{1}{2}p(p-1) + \frac{1}{2}h(h-1). \tag{2.5}$$

To show the correctness of eq.(2.4), we sum up eq. (2.4) over all exciton states and rewrite

$$\omega(n,E) = e^{f(h,U)}, \qquad h = 2n - 1$$

where

$$f(h,U) = lg(g) + h\ lg(gU) + h\ lg(gU - h^2) - lgh! - lg(h+1)! - lg(2h)!.$$

Using Strirling formula for factorials

$$lgh! = h\ lgh - h + \frac{1}{2}lg2\pi h,$$

the function f(h,U) has a sharp maximum at a value h_0. We now expand f(h,U) at h_0

$$f(h,U) = f(h_0,U) + \frac{1}{2}f''(h_0,U)(h - h_0)^2 + \dots \tag{2.6}$$

In terms of this expression one can immediately carry out the integral over h and obtain the result for the total state density

$$\rho(E) = \sum_n \omega(n,E) = e^{f(h_0,U)} \int e^{\frac{1}{2}f''(h_0,U)(h-h_0)^2}\, dh = \sqrt{\frac{\pi}{|\frac{1}{2}f''(h_0,U)|}}\, e^{f(h_0,U)}. \tag{2.7}$$

The value of h_0 is obtained by the condition

$$f'(h_0,U) = 0. \tag{2.8}$$

In order to obtain an approximate solution, for high value of U, $h_0^2 >> h_0 >> 1$. The resulting equation for h_0 is

$$\frac{2r}{1-r} = lg\frac{1-r}{4r^2}, \tag{2.9}$$

where

$$h_0^2 = rgU.$$

This equation can be solved numerically (r = 0.28431), and the two terms of f and f $''$ have the form

$$f(h_0,U)=\sqrt{gU}[4\sqrt{r}-\sqrt{r}\frac{2r}{1-r}]+lg\frac{g}{4\pi^{\frac{3}{2}}(rgU)^{\frac{5}{4}}}$$

$$f''(h_0,U)=-\frac{4-2r+2r^2}{\sqrt{gU}(1-r)^2\sqrt{r}}.$$

Substituting the value of r and using $g=\frac{6a}{\pi^2}$, we obtain

$$\rho(E)=\frac{1-r}{4\pi Ur\sqrt{2-r-r^2}}e^{2\sqrt{r}\frac{2-r}{1-r}\sqrt{\frac{6}{\pi}}\sqrt{aU}}=C_1\frac{\sqrt{\pi}}{12U}e^{2C_2\sqrt{aU}}. \tag{2.10}$$

$$C_1=1.0118, \qquad C_2=0.9966.$$

It indicates that if summed over all exciton number, the exciton state density of eq.(2.4) reproduce the Fermi gas level density of one type fermion system. In practice for two type fermion system a renormalized factor need to be introduced.

$$\eta(U)=\frac{U}{C_1a^{1/4}(U+T)^{5/4}}. \tag{2.11}$$

The renormalized exciton state density reads

$$\omega^R(n,E)=\eta(U)\omega(n,E). \tag{2.12}$$

Thus, we get the result

$$\sum_n\omega^R(n,E)=\rho_{B-S}(E), \tag{2.13}$$

where $\rho_{B-S}(E)$ is the Back–Shifted Fermi gas level density [5].

The T–factor in H–F model can be rewritten into the form.

$$T_b^{J\pi}(E)=\frac{1}{2\pi h}\sum_{l'l}\int T_{l'l}^b(\varepsilon)\rho(E',j'\pi')d\varepsilon$$

$$=\frac{1}{2\pi h}\sum_{l'l}\int T_{l'l}^b(\varepsilon)\sum_n\omega^R(n,E',J'\pi')d\varepsilon=\sum_n T_b^{J\pi}(n,E') \tag{2.14}$$

where E′ is the residual excitation energy.

This procedure demonstrates that the statistical treatment in H–F model can be decomposed into multi–step processes with a self–consistent fashion.

Eq.(2.13) is correct only for high excitation energy. In the other words, only for high exciton states the independent particle model is held. For lower exciton states, we can introduce some additional correction [6]. If the pick–up mechanism and the two type fermion effects are taken into account, the T factor needs rewriting into

$$T_b^{J\pi}(n,E) = \frac{1}{2\pi h} \sum_{f\ell'} \int T^b_{f\ell'}(\varepsilon) F_b(n,\varepsilon) Q_b(n) \omega^R(n-1, E', j'\pi') d\varepsilon \tag{2.15}$$

The formation factor of composite particles $F_b(n,\varepsilon)$ is given by[3].

$$F(n,\varepsilon) = 1 \quad \textit{for } n, p \textit{ and eq. states}$$

$$F_b(n,\varepsilon) = \frac{1}{(2\pi h)^{A_b - 1}} \int_{\vec{p}_b \text{ fixed}[l,m]} \prod_{i=1}^{A_b-1} d\vec{p}_i d\vec{x}_i . \tag{2.16}$$

where A_b is the mass number of composite particle

$\vec{p}_i, \vec{x}_i$ are the relative momenta and coordinates.

[l,m] stands for the pick–up configuration ($l+m=A_b$),

l: particles above the Fermi surface

m: particles below the Fermi surface

From the calculations we found the fact that when the incident energy is less than 20 MeV, l=1 is the dominant component[7].

Another factor, combination factor, should be introduced to account for memory by the excitation system in the first stages of the reaction for the type of projectile.

If the pick–up configuration is taken into account, the combination factor reads[6].

$$Q_{b\,[l,m]}(n) = \left(\frac{A}{Z}\right)^{Z_b} \left(\frac{A}{N}\right)^{N_b} \begin{bmatrix} p \\ l \end{bmatrix} \begin{bmatrix} A-h \\ m \end{bmatrix} \begin{bmatrix} A_b \\ Z_b \end{bmatrix} \sum_{i=0}^{h} \begin{bmatrix} h \\ i \end{bmatrix}$$

$$\left(\frac{Z}{A}\right)^i \left(\frac{N}{A}\right)^{h-i} \sum_j \begin{bmatrix} i \\ j \end{bmatrix} \begin{bmatrix} h+1-i \\ l-j \end{bmatrix} \begin{bmatrix} Z-i \\ Z_b - j \end{bmatrix} \begin{bmatrix} N-h+i \\ N_b - l + j \end{bmatrix}, \tag{2.17}$$

where A,Z,N (A_b, Z_b, N_b) are the mass number, proton number and neutron number of the target (emitted particle),

As an example, the values of the combination factors are given for $n + {}^{56}Fe$ reaction in Tab. 1.

n	(n,n)	(n,p)	(n,α)	(n,^{3}He)	(n,d)	(n,t)
3	1.4333	0.5000	0.9923	0.8122	0.9758	1.1439
5	1.2889	0.6667	1.0074	0.8800	0.9901	1.1030
7	1.2167	0.7500	1.0151	0.9140	0.9975	1.0827
9	1.1733	0.8000	1.0199	0.9345	1.0021	1.0706

Tab.1 List of Q–Factor of $n+{}^{56}Fe$

In this theory, the exciton state occupation probability $q^{J\pi}(n,t)$ satisfies the

$J\pi$ dependent master equation

$$\frac{dq^{J\pi}(n,t)}{dt} = \lambda_{+}^{J\pi}(n-2)q^{J\pi}(n-2,t) + \lambda_{-}^{J\pi}(n+2)q^{J\pi}(n+2,t)$$
$$- [\lambda_{+}^{J\pi}(n) + \lambda_{-}^{J\pi}(n) + W_{T}^{J\pi}(n)]\ q^{J\pi}(n,t). \tag{2.18}$$

where the transition rates within the exciton states are given by [8]

$$\lambda_{\nu}^{J}(n) = \frac{2\pi}{h}|<M>^{2}|\chi_{\nu}^{J}(n)Y_{\nu}(n) \tag{2.19}$$

$$here\ \nu = +,0,-.$$

The form factor of angular momentum can be obtained in terms of the approach in F.K.K. theory[9], but spin 1 / 2 was used instead of spin 0. The life time of n exciton state in $J\pi$ channel is definded by

$$\tau^{J\pi}(n) = \int_{0}^{\infty} q^{J\pi}(n,t)dt. \tag{2.20}$$

Then the occupation probability of n exciton state in $J\pi$ channel is obtained by

$$P^{J\pi}(n) = \tau^{J\pi}(n)W_{T}^{J\pi}(n), \tag{2.21}$$

where $W_T^{J\pi}$ stands for the total emission rate of n exciton state in $J\pi$ channel. The occupation probabilities satisfy the normalization condition

$$\sum_{n} p^{J\pi}(n) = 1. \tag{2.22}$$

3. Formulation of Double Differential Cross Sections

Based on the leading particle model[10], the generalized master equation is:

$$\frac{dq(n,\Omega,\varepsilon,t)}{dt} = \sum_{m}\int d\Omega'\int d\varepsilon' q(m,\Omega',\varepsilon',t)W_{m\to n}(\Omega'\varepsilon',\Omega\varepsilon)$$
$$- \{\sum_{m}\int d\Omega'\int d\varepsilon' W_{n\to m}(\Omega'\varepsilon',\Omega\varepsilon) + W_{t}(n)\}\ q(n,\Omega,\varepsilon,t) \tag{3.1}$$

where $q(n,\Omega,\varepsilon,t)$ is the occupation probability of the composite system at time t in exciton state n with leading particle energy ε and direction Ω.

$W_{m\to n}(\varepsilon',\Omega',\varepsilon,\Omega)$ is the transition probability of the system from state (m,ε',Ω') to state (n,ε,Ω) in per unit time

$W_T(n)$ is the total emission rate of n exciton state

The solution of eq.(3.1) can be obtained in the form of a partial wave expansion

$$q(n,\Omega,\varepsilon,t) = \sum_{l}\eta_{l}(n,\varepsilon,t)P_{l}(cos\theta), \tag{3.2}$$

where P_l is the Legendre polynomial.
Introducing the partial wave life time

$$\zeta_l(n,\varepsilon) = \int_0^\infty \eta_l(n,\varepsilon,t)dt, \tag{3.3}$$

the double differential cross section of single particle emission is given by the following expression

$$\frac{d^2\sigma}{d\Omega d\varepsilon} = \sum_n \frac{d\sigma(n)}{d\varepsilon} A(n,\Omega), \tag{3.4}$$

where $\frac{d\sigma(n)}{d\varepsilon}$ is the component of n exciton state in the spectrum,

$$A(n,\Omega) = \frac{1}{4\pi} \sum_l \frac{\xi_l(n)}{\xi_o(n)} p_l(cos\theta), \tag{3.5}$$

it satisfies

$$\int A(n,\Omega)d\Omega = 1.$$

The angular distribution of emitted composite particle can be obtained by the following approach. The basic idea is that the leading particle picks up some nucleons in the compound nucleus to form a composite particle and the momentum conservation is taken into account, the normalized angular factor is introduced by[11]

$$A_{lm}(n,\Omega_b,\varepsilon_b) = 1/N_{lm} \int d\vec{p}_1 ... d\vec{p}_{A_b}$$

$$\delta(\vec{p}_b - \sum_{i=1}^{A_b} \vec{p}_i) \prod_{j=2}^{A_b} D_n(\vec{p}_j) \tau(n_1,\Omega_1,\varepsilon_1). \tag{3.6}$$

With $|\vec{p}_1| > p_f$ *and* $|\vec{p}_i| < p_f$ for i=2...A_b. For lower excitation energy [1,m] in the dominant part.

$\tau(n_1,\Omega_1,\varepsilon_1)$in eq.(3.6) is the life time of the leading particle with energy ε_1 in direction Ω_1. The δ function in eq.(3.6) implies the momentum conservation.

$D_n(\vec{p}_j)$in eq.(3.6) is the momentum distribution of the compound nucleus. All of the explicit formulas can be found in Ref.[11].

4. Results and Discussion

Using the semi–classical model mentioned above we performed the calculation of n + ^{56}Fe at E_n = 14.1 MeV with UNF code [12]. The neutron production spectrum in C.M.S. is shown in Fig.1 and the total neutron double differential cross section are shown in Fig.2 and Fig.3 forθ = 30°, and 150°, respectively. the data of

neutron production spectrum is taken from ENDF–B6, while the data of D.D.C.S. are taken from [13].

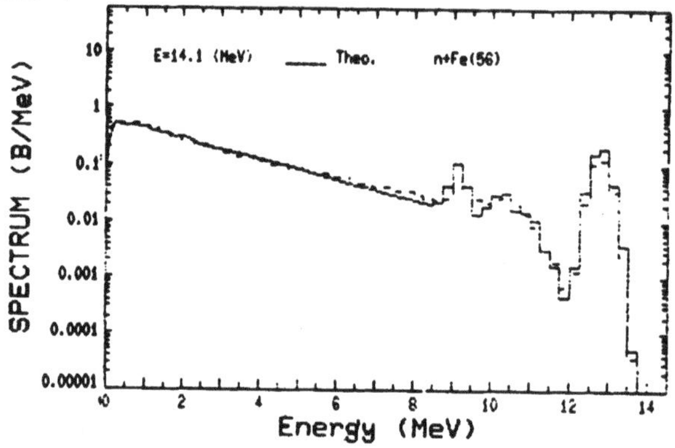

Fig.1 Neutron production spectrum

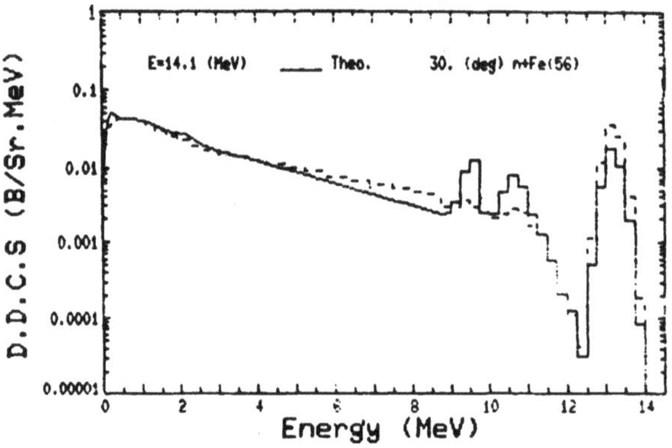

Fig.2 Double differential neutron emission cross section at 30°

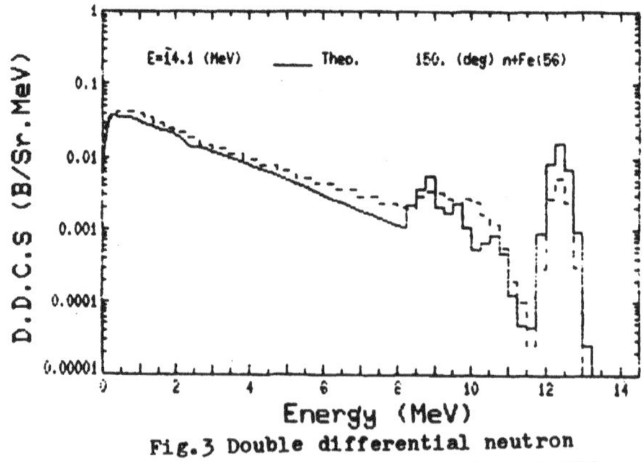

Fig.3 Double differential neutron emission cross section at 150

The proton and α–particle production spectrum are also shown in Fig.4 and Fig.5. The data of proton and α production spectra at $E_n = 14.5$ MeV are taken from Refs [14] and [15].

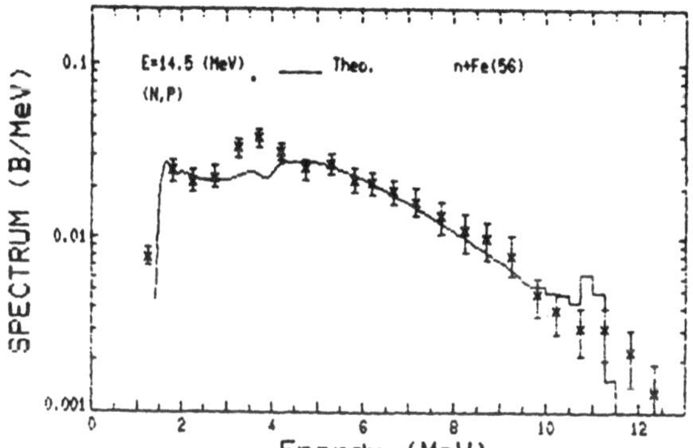

Fig.4 Proton production spectrum in C. M. S. for Fe

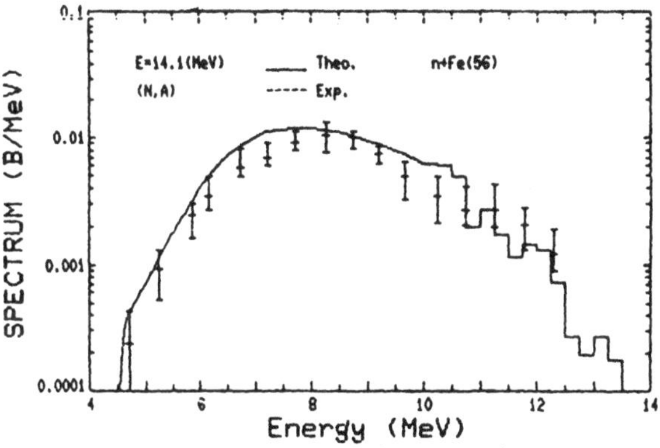

Fig.5 Alpha production spectrum in C. M. S. for Fe

To keep the energy conservation for whole reaction processes, the recoil effect must be taken into account. Because the D.D.C.S. of all kinds of particles have been calculated, thus, one can obtain the kinetic energies released from material (KERMA factor)

The energy released in the whole reaction processes is given by

$$E_{releas} = \sum_i \frac{\sigma_i}{\sigma_T}(E_L + Q_i), \tag{4.1}$$

where σ_i is the cross section of i–th channel
Q_i is the Q value of the i–th channel
and $\sum_i \sigma_i = \sigma_T$ i for all channel

For the case of n + ^{56}Fe at $E_n = 14.1$MeV, the energies released from each reaction channel are listed in Tab.2

Tab.2 List of kerma factors of n+^{56}Fe at $E_n = 14.1$MeV

	(n,γ)	(n,n)	(n,p)	(n,α)	(n,d)	(n,2n)	(n,np)	(n,nα)
γ	0.005	2.182	0.209	0.087	0.0034	0.0849	0.0039	
n		8.038				0.305	0.0284	0.0003
p			0.0235				0.0597	
α				0.131				0.0026
d					0.168			
Recoil	.0001	.320	.0137	.0130	.0013	.0424	.0069	.0003
total	.0052	10.540	0.458	0.231	0.0215	0.4822	0.0988	0.0031

(unit: MeV)

In previous model calculation, only the exciton states in which the particle number p must be equal or great then $A_b$permit to emit the composite particle with mass number A_b. So the initial exciton state for particle A_b emission is given by

$$n_0 = 2A_b - 1. \tag{4.2}$$

When the pick–up mechanism has been employed in the model, the situation has been changed. Since the leading particle can pick up nucleons from the compound nucleus. Therefore the initial exciton state of particle A_b emission is always equal to 3.

The pick–up mechanism plays an important role in the pre–equilibrium reaction processes. To show this fact the percentages of n exciton state in the continuum part of cross sections of (n,α), (n,d) and (n,t) for n+^{56}Fe are given in Tab. 3,4,5, respectively.

Tab.3 Percentage of n exciton state in cont. part of C. S. (n,α)

E_n(MeV)	n=3	n=5	n=7	n=9	eq.
10	8.8	4.7			86.5
14.1	27.1	8.1	8.4		56.5
18	52.6	12.5	6.1	4.0	24.8

Tab.4 Percentage of n exciton state in cont. part of C. S. (n,d)

E_n(MeV)	n=3	n=5	n=7	n=9	eq.
14.1	83.3	0.5			16.2
16.	95.9	2.0	0.1		2.0
18	94.4	3.8	0.3		1.5

Tab.5 Percentage of n exciton state in cont. part of C. S. (n,t)

E_n(MeV)	n=3	n=5	n=7	n=9	eq.
16	94.6	0.2			5.1
18	96.7	1.6			1.7
20	95.0	3.7	0.2		1.1

The results indicate that the percentages of low exciton states increase with the increasing of incident energies while the components of equilibrium state decrease.

In the case of (n,α) reaction, with previous model calculation the exciton states n=3,5 contribute nothing to cross section. But if the pick–up mechanism has been taken into account, they may give important contribution to the cross section. In particular, at high incident energies, they become the dominant parts. This is the reason why the cross section of (n,α) always too small to be calculated by the previous approach. As a matter of fact, the pick–up mechanism stands for the semi–direct processes, which could not be neglected in the composite particle emissions

From Tab. 5 one can see that in the case of triton emission, the dominant part is n=3, which is just omitted in the previous approach which is unable to reproduce the data.

It is easy to draw the conclusion that the semi–direct processes, e.i. the pick–up mechanism must be taken into account in the composite particle emission processes.

References

[1] Mantzouranis,G. Weidenmuller, H. and Agassi, D.*Z.Phys.***A276**,(1976)145
[2] Iwanmoto,A. and Harada,K. *Phys.Rev.* **C26**,(1982)1812
[3] Zhang Jingshang, Wen Yuanqi, Wang Shunuan and Shi Xiangjun *Commun. in Theor. Phys.* **10**,(1980)33
[4] Zhang Jingshang and Yang Xiangjun *Z.Phys.* **A329**,(1988)69
[5] Dilg,W.,Schantl,W.,Vonach,H. and Uhl,M. *Nucl. Phys.* **A217**,(1973)269
[6] Zhang Jingshang and Shi Xiangjun *INDC(CPR)*–**014 / LJ**
[7] Zhang Jingshang *INDC(CRP)*–**017 / LJ**
[8] Wen Yuanqi and Zhang Jingshang to be published

[9] Feshbach,H., Kerman,A. and Koonin,S.
Ann. Phys. (New York) **125**,(1980)429
[10] Sun Ziyang, Wang Shunuan, Zhang Jingshang and Zhuo Yizhong
Z.Phys. **A305**,(1982)61
[11] Zhang Jingshang and Wen Yuanqi
Chin. J. of Nucl. Phys. **Vol.13**, (1991)129
[12] Zhang Jingshang *UNF code* (1991)
[13] Takahashi,A. et al., *MITO 88*,p205
[14] Grimes, S. M., et al., *Phys. Rev.* **C19**, (1979)2127
[15] Haight, R. C., et al., *LLL Report* **UCRL–80235 (19)**

NEUTRON SCATTERING ANALYSIS WITH MICROSCOPIC OPTICAL MODEL POTENTIALS

Luisa F. Hansen
Lawrence Livermore National Laboratory
Livermore, California 94550, USA

ABSTRACT

A review of microscopic optical model potentials used in the analysis of neutron scattering and analyzing power data below 100 Mev (5 $\leq E_n \leq$100 MeV) is presented. The quality of the fits to the data over a wide mass (^{6}Li-^{239}Pu) and energy range is discussed. It is shown that reasonably good agreement with the data is obtained with only three parameters, λ_V, λ_W, and λ_{SO}, which show a smooth mass and energy dependence. These parameters are normalizing constants to the real (V), and imaginary (W) central potentials and the real spin-orbit (V_{SO}) potential.

1. Introduction

Microscopic optical model potentials (MOMP) have been used extensively in the last decade in the analysis of nucleon-nucleus sacattering below 100 MeV. These potentials are based on realistic two-body interactions and contain sufficient physics to carry out rather accurate extrapolations as a function of mass and energy. Consequently, they require a small number of parameters to fit the data. The quality of the fits to neutron and proton elastic and inelastic differential cross sections and analyzing power angular distributions obtained with only three normalizing parameters over a wide range of target masses and incident energies compares well with those obtained with the more widely used phenomenological optical model potentials (POMP). These latter potentials are characterized by the use of a much larger number of parameters. Furthermore, due to interdependences among some of these parameters, there are ambiguities in determining their values.

The MOMP of interest to the energy region of this paper are: the Jeukenne, Lejeune and Mahaux[1] (JLM), the Brieva and Rook[2] (BR), and the Yamaguchi, Nagata and Matsuda[3] (YNM). These three MOMP can be described by a one-step folding model of an effective two-body interaction and the nuclear density. A brief description of these potentials will follow together with comparisons to a large body of experimental data.[4-8] In all the comparisons of the MOM calculations with the data, only three parameters were required: λ_V, λ_W and λ_{SO}, which are normalizing constants to the central real (V) and imaginary (W) potentials and to the spin-orbit real potential (V_{SO}). The values of these constants are a measure of how well the real and imaginary potentials are calculated in these MOMP[1-3].

That is, λ_V is close to unity for the three potentials, which implies that the real central potential is accurately calculated, while the absortion potential needs larger corrections as a function of mass and energy. JLM does a better calculation with a value of λ_W of about 0.9; BR underestimates W by 30 to 50% and YNM's needs to be reduced by as much as 30-40%. Comparisons of MOM calculations with analyzing power data[7] show that V_{SO} is underestimated by 30 to 50%.

2. Microscopic Optical Model Calculations

In the BR, JLM and YNM MOMP, the central potential is calculated by a folding potentials of the type

$$U(r\ E)= \int t(r,r')\rho(r')dr' \qquad (1)$$

where t(r') is a complex effective interaction which is a function of energy and the nuclear density, $\rho(r')$:

$$t(r,r')=t[E,\rho(r'),r,r'] \qquad (2)$$

The effective interaction and the nuclear density account for almost all the physics contain in the MOMP. The differences among these potentials result from the choice of the effective interaction and the approximations carried out to extrapolate from infinite nuclear matter to finite nuclei.

Recently[9], this type of folding potentials (Eq. 1) has been called "standard folding model" potentials to differentiate them from more accurate "full-folding model" optical potentials. In this approach, the off-shell behavior of the nucleon-nucleon t-matrix, its energy dependence, associated knock-on exchange terms and density dependence are treated explicitly in the calculations.

The nuclear densities (Eqs. 1-2) were extracted from electron scattering measurements[10]. Point-proton densities were obtained by deconvoluting the electron charge density, and the point-neutron densities were assumed proportional (N/Z) to the proton ones for nuclei with $A \leq 100$. For heavier nuclei a small increase in the neutron rms radius (up to 2.6%) relative to that of the protons was used, as expected from Hartree-Fock and droplet-model calculations. However, neutron scattering below 30 MeV is rather insensitive[4,6] to this small difference between the proton and neutron radii, due to the weaker neutron-neutron interaction relative to the neutron-proton interaction in the present energy range.

The spin-orbit effective interaction used in the calculations with the JLM and BR potentials was the M3Y interaction with Elliott oscillator G-matrix elements as given by Bertsch et al.[11]. This interaction is energy and density independent.

2a. The Jeukenne, Lejeune and Mahaux (JLM) Potential. The energy and density dependent potential in infinite nuclear matter is calculated[1] from the Reid's hard core nucleon-nucleon interaction using the Brueckner-Hartree-Fock approximation. The potential for a finite nucleus is obtained using the "local density approximation (LDA), where it is assumed that at each point in the nucleus at a distance r from the nuclear center and density $\rho(r)$, the OMP has the same value as in a uniform medium with the same value of the density, neutron excess, and energy. An improved LDA was used to account for surface corrections. In this case, an effective interaction is defined between the projectile and target nucleons, which in the present calculations was calculated[5] at a position given by the average value between the projectile and target particle radii (midpoint LDA). With this approximation, a finite range parameter for the real and imaginary potentials of 1.0 fm was used in all the calculations, compared to the 1.2 fm given in Ref. [1]. The smaller value was required to reproduce the total neutron cross sections in the 20 MeV region.

2b. The Brieva-Rook (BR) Potential. In this potential the effective interaction is calculated from the Hamada-Johnston hard core nucleon-nucleon interaction in infinite nuclear matter. The LDA (midpoint approximation) was also used here to calculate the OMP in the finite nucleus. In the BR potential, the effective interaction is expressed in terms of a direct and exchange components which give rise to the local and non-local parts of the nucleon-nucleus optical potential. In the calculations[4,6] done with the BR MOMP, the non-local exchange term was calculated by making a local approximation using a complex self-consistent wave number,

$$h^2 k(r)^2/2m = E - U(r)$$

where U(r) is the entire complex central potential.

2c. The Yamaguchi, Nagata, Matsuda (YNM) Potential. The OMP in infinite nuclear matter is calculated in the BHF approximatiom using the Hamada-Johnston nucleon-nucleon potential. Only the real part of the spin-orbit potential is calculated. Calculations of the neutron and proton analyzing power done with this effective spin-orbit interaction (real, energy and density dependent), compare very well with the results obtained with the M3Y interaction.

3. Comparisons of MOM Calculations with Measurements

Standard MOMP[1-3] have been remarkably successful in describing a large body of elastic, inelastic and analyzing power data below 100 MeV for incident neutrons and protons on targets spanning from ^{6}Li to ^{239}Pu. In the present paper I will only review those comparisons done for spherical targets. However, calculations for deformed nuclei in the rare earth[12] and actinide[13] regions done with a so called quasi-microscopic optical model potential, which includes the deformation parameters (β_2 and β_4) have given very good agreement with the data.

Calculations of the elastic and analyzing power angular distributions with MOMP are done in two steps: 1) folding model calculations of the V and W central potentials and V_{SO}, 2) these potentials are inserted into a spherical optical-model code to obtain the angular distributions. The comparisons with the experimental data are done by a least-squares adjustment of three normalizing parameters, λ_V, λ_W and λ_{SO}, such that

$$U = \lambda_V V + \lambda_W W + \lambda_{SO} V_{SO} \qquad (3)$$

It must be remembered that these normalizing constants do not have any physical basis, other than to improve the fits to the data. The closer the values of these constants are to unity the better is the representation of the respective potentials by the model.

Comparisons of MOM calculations done with the JLM and BR potential with measurements of the angular distributions for elastic scattered neutrons have shown consistently that the JLM potetial gives overall better fits to the data than BR. In Fig.1 is shown the work of Mellema, et al.[4], for the Fe(n,n) reaction between 8 and 26 MeV. Fig. 2 and 3 show the work of Hansen, et al.[5] at 14.6 MeV and Olson, et al.[6] at 21.6 MeV for targets ranging from Be to Bi. In all these comparisons the BR potential overestimates the differential cross sections at the very forward angles and the total cross section, specially for the lighter mass targets. The main differences between these calculations occur at the larger angles ($\theta > 80^o$). These differences could be the result of the different shape of the real potential at the nuclear surface. The BR potential shows the larger deviation from the Woods-Saxon shape, with an anomalous bump at the surface. For a detailed discussion on this point see Ref. [14]. Independent of the different shapes of V_R, the normalization parameter λ_V is close to unity (1.0±0.1) for both potentials over the whole mass and energy ranges of these measurements[4-6]. The value of λ_W is about 0.90 for JLM and 1.3 for BR with a small decreasing slope toward the higher A values. The value of λ_{SO} was kept equal to 1.0, since no analyzing power data was included in the analysis.

The large body of neutron elastic scattering and analyzing power data measured at TUNL in the energy region of $7 \leq E_n \leq 17$ MeV, motivated a further study of the MOMP, focusing in the spin-orbit potential. This study proceeded to test the M3Y interaction versus the energy and density dependent V_{SO} potential calculated in the YNM potential[3]. Calculations with the YNM and JLM potentials were compared with the TUNL data[7] [^{6}Li, ^{9}Be, ^{10,11}B, ^{27}Al, ^{28}Si, ^{40}Ca, ^{54}Fe, ^{58}Ni, ^{89}Y, ^{93}Nb, ^{116,120}Sn and ^{208}Pb]. The energy range of these comparisons was extended up to 45 Mev by including neutron measurements from Ohio State University (20-26 MeV), and from Michigan State University (30-45 MeV). Figs. 4-6 show the results of these comparisons for ^{6}Li, ^{10}B, ^{40}Ca and ^{208}Pb. The quality of the fits is typical of the results found for all the data in Ref.7. It is interesting to notice that the quality of the fits to the analyzing power data obtained with the M3Y interaction and the YNM spin orbit interaction is very similar. A value of λ_{SO}~1.3 to 1.5 was used in both calculations. As it was found in the previous analysis[4-6], the values of the normalization parameters are well behaved with target mass and neutron energy: λ_V~1.0±0.1 for both, JLM and YNM, potentials. λ_W has an average value of around 0.80±0.10 and 0.6±0.05 for JLM and YNM respectively. However, the deviations from these average values are larger for energies below 12 Mev, especially for the lighter nuclei ($A \leq 12$).

Calculations done with the JLM and YNM potentials for 65 MeV neutrons have been compared with the data of Hjort, et al. from UC Davis [Ref. 8] in Fig. 7. The quality of these fits, which were obtained with the values of the normalizing constants giving above, are comparable or better than those found by Hjort for phenomenological potentials optimized for this enegy region.

Conclusions

The analysis[4-8] of neutron elastic scattering and analyzing power data below 100 MeV with microscopic optical model potentials (JLM[1], BR[2] and YNM[3]] has proven very successful. Reasonably good fits to a large body of data from Li to Bi have been obtained with only three parameters, λ_V, λ_W and λ_{SO} (normalization constants to V, W, and V_{SO}), which show a smooth energy and mass dependency. These parameters have practically constant values (±10%) for E_n>12 MeV and A > 12. λ_V is close to unity in all the calculations. A value of λ_{SO} of 1.4±0.1 fited well the analyzing power and elastic scattering data. The imaginary potentials[1-3] needed the larger corrections, with λ_W about 0.8, 1.4 and 0.6 for the JLM, BR and YNP respectively. The JLM and YNM potentials gave overall better fits to the data than the BR potential, with JLM requiring the smaller corrections to the central and V_{SO} potentials.

Collaborators

Frank S. Dietrich, who implemented the microscopic calculations at Lawrence Livermore National Laboratory (LLNL) has participated in all the work dicussed in Refs. 4-7, and Richard L. Walter, from Duke University and TUNL, collaborator in the work cited in Ref. 7. The present work was performed under the auspices of the U.S DOE by LLNL under Contract No. W-7405-Eng-48. R.L. Walter's work is partially supported by U.S. DOE Grant # DEFG05-91-ER40619.

References

[1] J.P.Jeukenne, A.Lejeune and C.Mahaux, Phys. Rev. C 15, 10 (1977); Phys. Rev. C 16, 80 (1977).

[2] F.A.Brieva and J.R.Rook, Nucl. Phys. A291, 299 (1977); Nucl. Phys. A291, 317 (1977); F.A. Brieva and B.Z. Georgiev, Nucl. Phys. A308, 27 (1978); H.V.Von Geramb, F.A.Brieva and J.R. Rook, Lecture Notes in Physics, Vol. 89, 104 (1979).

[3] N Yamaguchi, S.Nagata and T.Matsuda, Prog. Theor. Phys. 40, 459 (1983).

[4] S.Mellema, R.W.Finlay, F.S.Dietrich, and F.Petrovich, Phys. Rev. C 28, 2267 (1983).

[5] L.F.Hansen, F.S.Dietrich, B.A.Pohl, C.H.Poppe, and C.Wong, Phys. Rev. C 31, 111 (1985)

[6] N.Olson, B.Trostell, E.Ramstrom, B.Holmqvist, and F.S.Dietrich, Nucl. Phys. A472, 237 (1987)

[7] L.F.Hansen, F.S.Dietrich and R.L.Walter, BAPS 32, 1567, (1987); BAPS 33, 1750 (1987); BAPS 34, 1809 (1989); BAPS 35, 1668 (1990).

[8] L.F.Hansen, E.L.Hjort, F.P.Brady, J.Drumond, B.McEachern, G. Osborne, and J.L.Romero, BAPS 35, 1661 (1990).

[9] H.F.Arellano, F.A.Brieva and W.G.Love, Phys. Rev. Lett. 63, 605 (1989); Phys. Rev. C 41,2188 (1990).

[10] C.W.Dejager, H. DeVries and C.DeVries, At. Data Nucl. Data Tables 14, 479 (1974).

[11] G.Bertsch, J.Borysowicz, H.McManus, and W.G.Love, Nucl. Phys. A284, 399 (1977).

[12] Ch.Lagrange and M.Girod, J. Phys. G 9, 197 (1983).

[13] L.F.Hansen, B.A.Pohl, C.Wong, R.C.Haight, and Ch.Lagrange, Phy. Rev. C 34, 2075, (1986).

[14] F.S.Dietrich and F.Petrovich, AIP Conference Proceedings, No 124, 90 (1985).

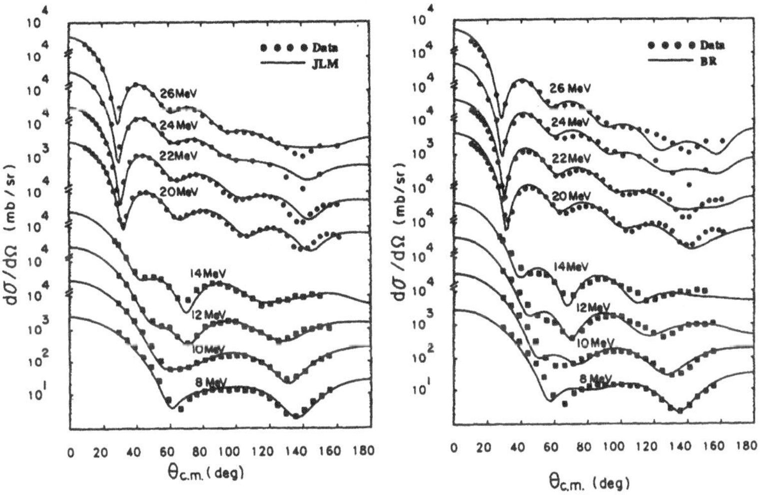

Fig.1 Comparisons of JLM and BR MOM calculations with measurements for neutrons elastic scattered from ^{54}Fe as function of E_n [Ref.4].

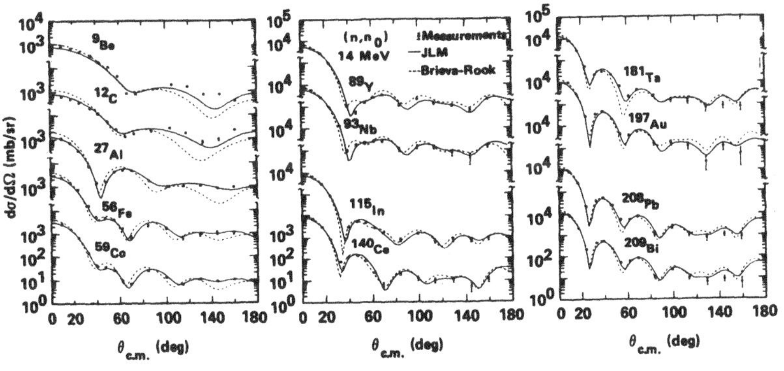

Fig.2 As in Fig.1 for 14.6 MeV neutrons elastic scattered from targets ranging from ^{6}Li to Bi [Ref.5].

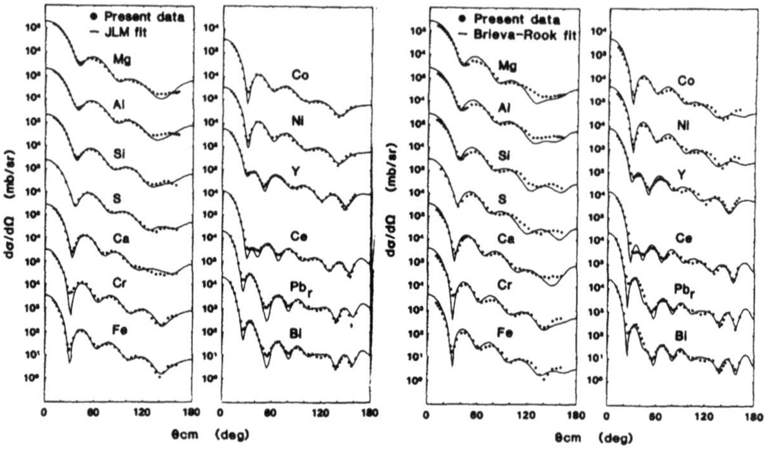

Fig.3 As in Fig.1 for 21.6 Mev neutrons elastic scattered from Mg to Bi targets [Ref.6].

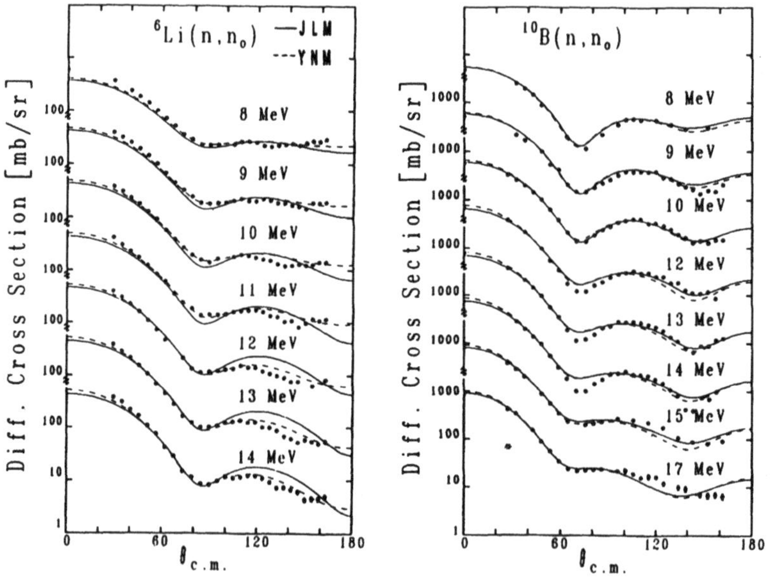

Fig.4 Comparison of JLM and YNM MOM calculations with meaurements for neutrons elastic scattered from ^{6}Li and ^{10}B as function of E_n [Ref 7d].

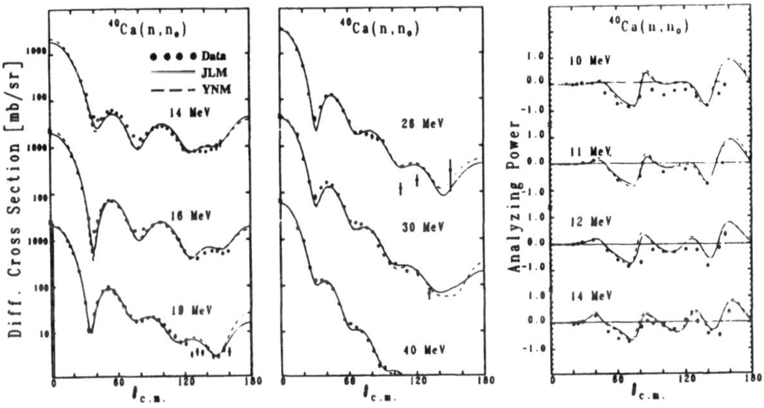

Fig.5 As in Fig.4 for neutron elastic scattering and analyzin power cross sections from ^{40}Ca, as function of E_n [Ref.7c]

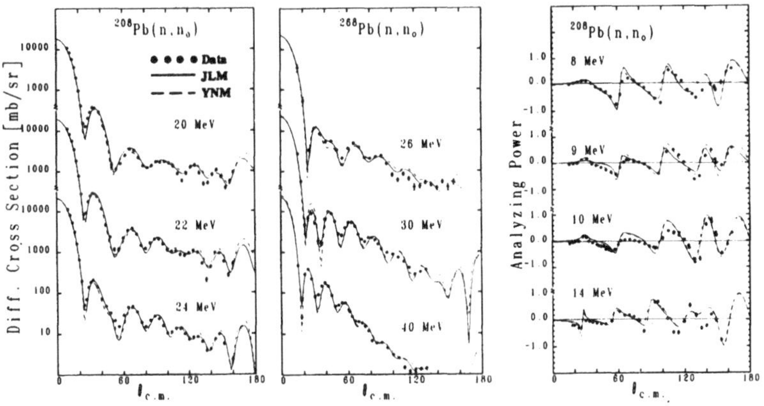

Fig.6 As in Fig.5 for ^{208}Pb [Ref.7c].

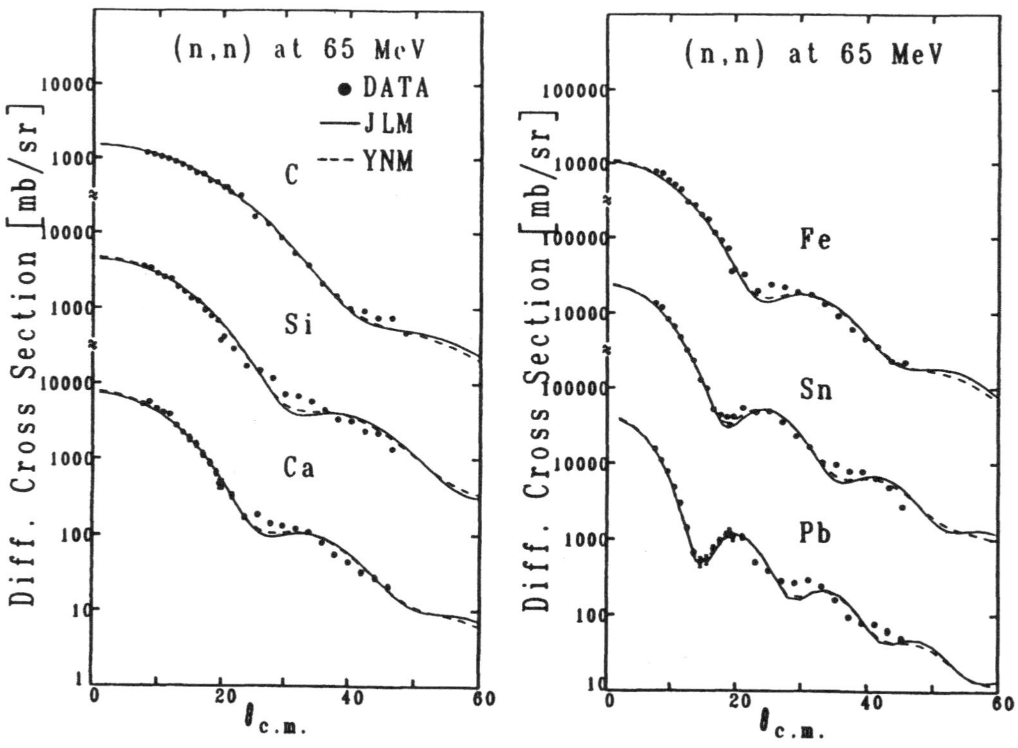

Fig.7 As in Fig.5 for 65 MeV neutrons elastic scattered from C to Pb targets [Ref.8].

ACTIVATION CROSS SECTIONS

REACTION CROSS SECTION MEASUREMENTS IN THE 8 - 13 MeV NEUTRON ENERGY RANGE*

HORST KLEIN
Physikalisch-Technische Bundesanstalt (PTB)
Braunschweig, Fed. Rep. Germany

ABSTRACT

Reliable scattering, activation and reaction cross section data for neutron energies below 15 MeV are required for fusion reactor technology, cancer therapy and various other applications. However, problems already arise in the selection of "monoenergetic" neutron sources appropriate for these kinds of investigations, because the neutron-producing nuclear reactions generally applicable in this energy range are either impracticable, inefficient or unavoidably accompanied by competing reactions producing other neutron groups.

Sophisticated corrections carefully taking into account the influence of the "unwanted" primary neutrons, and special techniques partially developed for use at white neutron sources have been successfully applied to fill the data gap.

This review discusses the various attempts and recent progress made in cross section measurements in this energy range.

---

*The paper was not received before deadline

Study of Fast Neutron Cross Sections at JAERI Tandem Accelerator

Satoshi CHIBA, Masayoshi SUGIMOTO,
Yujiro IKEDA and Yoshimaro YAMANOUTI
Japan Atomic Energy Research Institute

ABSTRACT

Using the JAERI Tandem Accelerator, fast neutron cross sections have been measured. In this paper, activities on these neutron cross section measurements are briefly introduced. Firstly, the neutron sources used in the Tandem facility are explained. Then, methods of the experiments and examples of their results are described, placing emphasis on the results in the energy region between 7 and 13 MeV where the $^{1}H(^{11}B,n)^{11}C$ reaction was used as a neutron source.

1. Introduction

Scheduled operation of the Japan Atomic Energy Research Institute (JAERI) Tandem Accelerator was begun in 1982. This is a general purpose heavy ion accelerator; it has been mainly used in studies of heavy ion nuclear physics, nuclear chemistry, solid state physics and material sciences. Neutron experiment is also one of the major activities being done with this facility/1/.

In the fast neutron cross section measurements, source neutrons are produced either by the $D(d,n)^{3}He$, $H(^{11}B,n)^{11}C$ or $^{7}Li(p,n)$ reaction. By combining these reactions, monoenergetic neutrons having little low energy backgrounds are produced in the energy region between 7 and 30 MeV almost continuously. The ^{11}B-H source is a low background, kinematically collimated monoenergetic neutron source in the "gap" energy region below 14 MeV/2/, and is intensively used now at JAERI Tandem facility. Therefore, this paper will put a focus on the results measured with this neutron source.

In the next section, properties of the neutrons sources used at Tandem are briefly described. Then, some experimental results obtained mainly by the ^{11}B-H neutron source are presented.

2. Some Notes on Properties of the Neutron Sources Used at JAERI Tandem

The source reactions used at the JAERI Tandem and their properties are briefly summarized in **Table 1**.

Table 1 Properties of neutron sources used at JAERI Tandem Accelerator

Reaction	Projectile	Range of En (MeV)	Pulse Intensity (particle/s)	Resolution (ns)
$D(d,n)^3He$	D_2^+	5~11	3~4μA	~1
$D(d,n)^3He$	D^+	10~20	1~2μA	2~3
$H(^{11}B,n)^{11}C$	B^{4+}	7~13	~50nA	~5
$^7Li(p,n)^7Be$	H^+	20~30	~100nA	~1

The difference between the two $D(d,n)^3He$ reactions listed in **Table 1** is that a molecular D_2^+ beam was used in the above reaction, while an atomic beam was used in the below. We found recently that the duo-plasmatron ion source installed in the high voltage (HV) terminal could produce about twice molecular deuterons than the atomic beam. Furthremore the buncher system in the HV terminal is more suited for the molecular beam. Therefore, we gradually shifted to use the molecular beam d-D reaction in the energy range from 5 to 11 MeV.

Typical neutron Time-of-Flight (TOF) spectra obtained with the reactions listed in **Table 1** are compared in **Fig. 1**. The d-D reaction has a large break-up background, while the $H(^{11}B,n)$ and $^7Li(p,n)$ reactions are good monoenergetic neutron sources in 7-13 and 20-30 MeV regions, respectively/2,3/. The FWHM value shown in **Fig. 1(a)** includes a contribution from the timing resolution which in this case is the dominant factor. One big advantage of the $H(^{11}B,n)$ reaction is that produced neutrons are kinematially collimated into a cone of forward angles, as is illustrated in **Fig. 2**. In this example, the neutrons are produced in a cone of 40-degrees, showing a possibility of very-low background experiments. **Figure 3** exhibits an example of background-subtracted energy spectrum of neutrons produced by this reaction, showing essentially no low-energy neutrons. Unfortunately, the intensity of the ^{11}B beam is not high enough for some applications. Thus, we select one of these reactions according to experimental conditions and quantities to be measured.

3. Activation Cross Sections

The activation cross sections were measured in the energy range from 9 to 13 MeV by using the ^{11}B-H reaction/4/. In the typical experiments, a sandwich made of several disk samples was irradiated, then the γ-ray activities were measured by a Ge detector. Energy spectra of the neutrons were monitored by the TOF method with an NE213 scintillator. Corrections were made for the half-lives, branching ratios, γ-ray attenuation, low energy neutrons, and γ-ray backgrounds. Correction factors for the low-energy neutrons are about 0.8 for the $^{115}In(n,n')^{115m}In$ reaction where the cross section has flat maxima in several MeV region. They are much closer to unity for, e.g., $^{64}Zn(n,p)^{64}Cu$ and $^{47}Ti(n,p)^{47}Sc$ reactions.

Examples of the latest results are shown in **Fig. 4**.

4. Gamma-ray production cross sections

The γ-ray production cross sections of several structural materials of D-T fusion reactors have been measured at incident neutron energies from 7.8 to 13.0 MeV/5,6,7/. Most of the previous experiments were carried out mainly with white neutron source. In our study, the $D(d,n)^3He$ reaction was used for 7.8 and 10.0 MeV. However, it was found that corrections for low energy break-up neutrons were enormous in the higher energy region, as can be imagined from **Fig. 1(b)**. Therefore, the $H(^{11}B,n)^{11}C$ reaction was used at 11.5 and 13.0 MeV. These monoenergetic sources generally gave better resolution and hence less corrections compared with the white neutron experiments. In-beam γ-rays were measured with a 7.6 cm diameter by 15.2 cm long NaI(Tl) scintillator. An annular NaI(Tl) scintillator surrounded this detector to give anti-Compton background suppression. The TOF technique was used to further reduce backgrounds. Measured pulse height spectra were converted to the energy spectra by the unfolding method. Corrections, such as angular distributions of source neutrons, neutron multiple scattering, outgoing γ-ray attenuation and Compton scattering in the sample, were done by the Monte-Carlo method. Differential cross sections were given as a function of γ-ray energy over the region from 0.7 to 12 MeV. Discrete and total γ-ray production cross sections were also deduced.

Typical results of the (n,xγ) experiments are shown in **Fig. 5**.

5. Neutron Scattering Cross Sections

The measurement of fast neutron scattering cross sections have been performed in the energy region from 11.0 to 30 MeV by using the D(d,n)^{3}He and ^{7}Li(p,n) source neutrons/8,9,10/. Rusults obtained at 28 MeV for the C(n,n) and (n,n') reactions/1,10/ are illustrated in **Fig. 6**. The optical potential and deformation parameters were deduced, their systematics were discussed and some insights into the structure of the target nucleus were obtained by these experiments.

The H(^{11}B,n)^{11}C reaction has a potential as a neutron source of the (n,n') experiments. However, it has not been fully utilized in actual measurements mainly because of its low intensity. **Figure 7** exhibits an example of such possibility. This is a neutron TOF spectrum scattered from a lead sample at incident neutron energy of 11 MeV, scattering angle of 70-degrees measured with the ^{11}B-H source. Besides the elastic scattering peak, we can see the continuous (n,n') spectrum from which the double-differential cross section, DDX, can be deduced. As is well known, only very few DDX data have been measured in this energy region. Therefore, we feel it worthwhile to exploit the possibility of using this neutron source in the DDX experiments in the 7-13 MeV region.

6. Concluding Remarks

Status of the activities on measurements of fast neutron cross sections at the JAERI Tandem Accelerator was briefly summarized. Properties of the various neutron sources were explained, and some results obtained mainly with the $H(^{11}B,n)^{11}C$ neutron source were presented. This source was proved to be a powerful source in the "gap" energy region between 7 and 13 MeV, and will be continuously used in the future experiments. The $^{1}H(t,n)^{3}He$ reaction available at LANL is an intense neutron source in the same energy region/11/. These two reactions are expected to produce data that supplement each other. Together with the D(d,n) and $^{7}Li(p,n)$ reactions, the JAERI Tandem Accelerator can provide the sources of monoenergetic neutrons in the energy range from 7 to 30 MeV.

Acknowledgement

The authors would like to acknowledge Drs. Y. Kikuchi, Y. Nakajima, T. Nakagawa and T. Fukahori for valuable comments.

References

1. e.g., JAERI TANDEM, LINAC & V.D.G. Annual Report 1989, JAERI-M 90-139 (1990).
2. S. Chiba et al., Nucl. Instr. Methods, A281(1989)581.
3. Y. Yamanouti, IAEA-TECDOC-410, p.251, IAEA (1987).
4. Y. Ikeda et al., Presented at the Int. Conf. on Nuclear Data for Science and Technology, May 1991, Juelich, Germany.
5. M. Mizumoto et al., Proc. Int. Conf. on Nuclear Data for Science and Technology, May/June 1988, Mito, Japan (Edited by S. Igarasi), p.197, Saikon, Tokyo(1988).
6. K. Hasegawa et al., Presented at the Int. Conf. on Nuclear Data for Science and Technology, May 1991, Juelich, Germany.
7. K. Hasegawa, Ph. D. Thesis, Tohoku University (1990).
8. S. Chiba et al., Jour. Nucl. Sci. Technol., 25(1988)210.
9. S. Chiba et al., Jour. Nucl. Sci. Technol., 25(1988)511.
10. Y. Yamanouti et al., Presented at the Int. Conf. on Nuclear Data for Science and Technology, May 1991, Juelich, Germany.
11. R. C. Haight, NEANDC-259 'U', p.95, OECD NEA (1989).

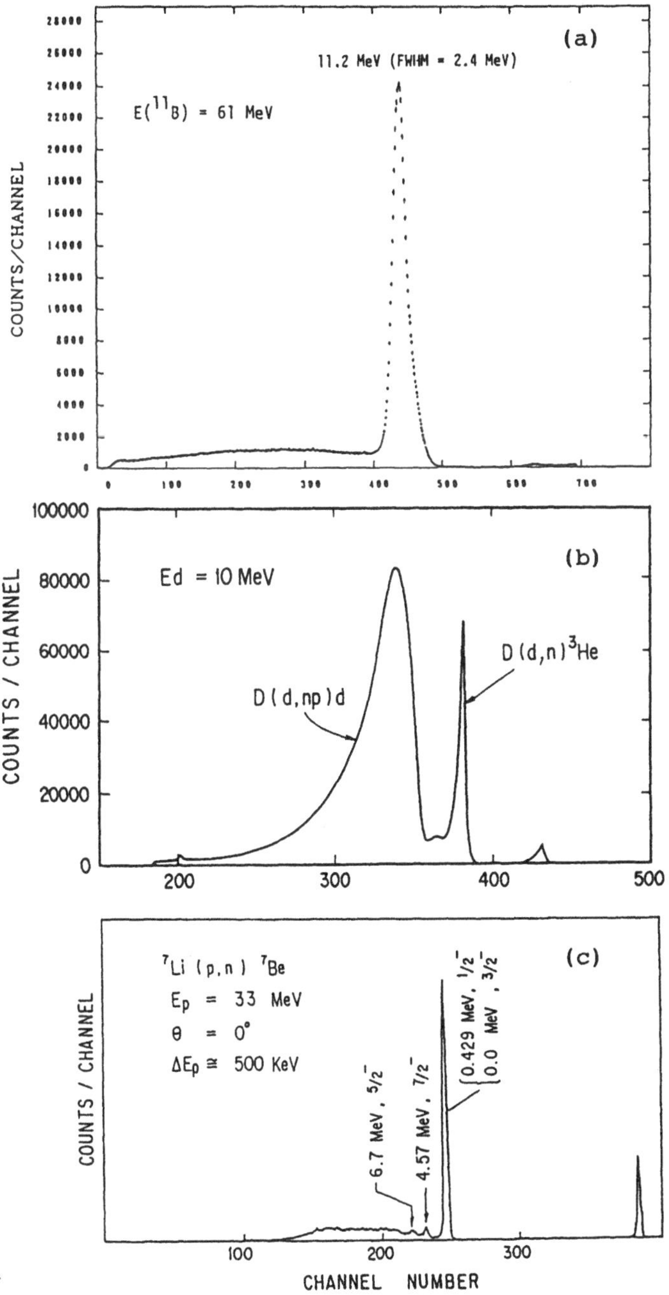

Fig. 1 Typical neutron TOF spectra obtained with (a) $H(^{11}B,n)^{11}C$/2/, (b) $D(d,n)^{3}He$, and (c) $^{7}Li(p,n)^{7}Be$/3/ reactions.

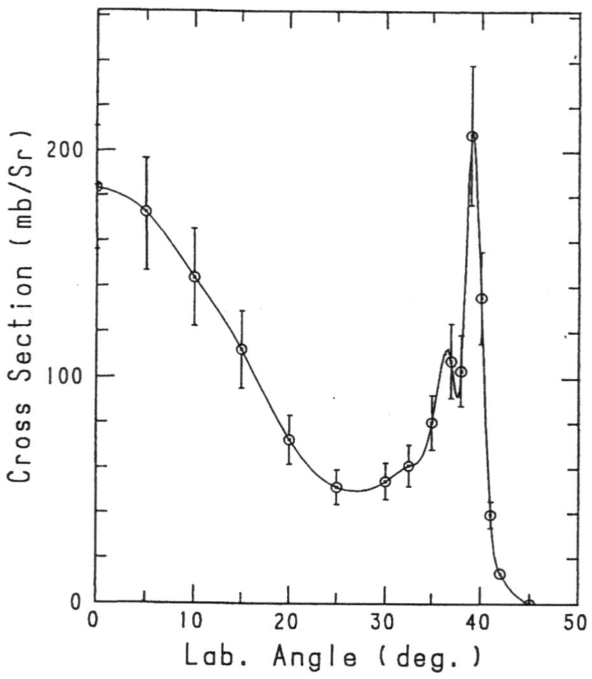

Fig. 2 Angular distribution of the neutrons produced from the $H(^{11}B,n)^{11}C$ reaction at the incident B energy of 61 MeV/2/.

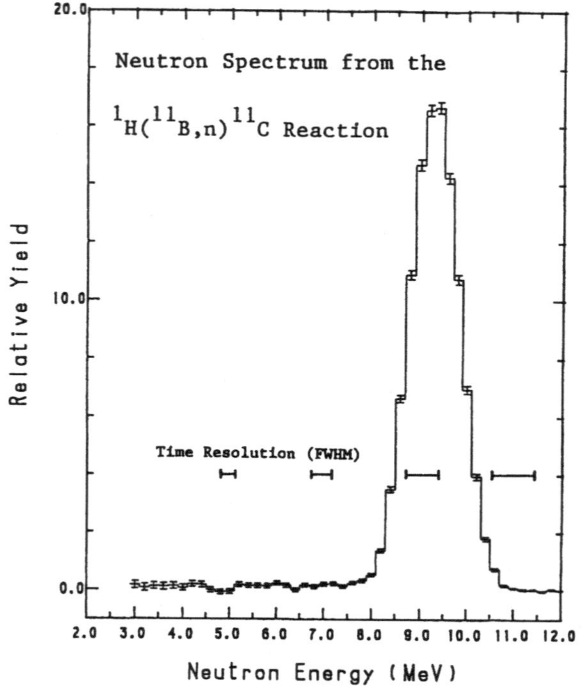

Fig. 3 Background-subtracted energy spectrum of neutrons produced from the $H(^{11}B,n)^{11}C$ reaction.

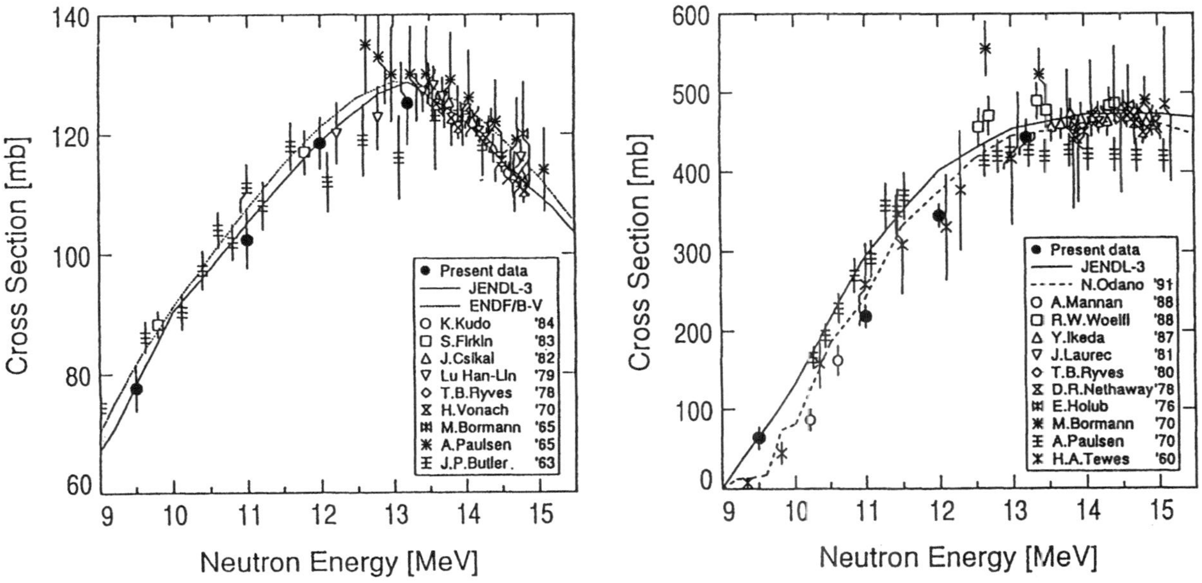

Fig. 4 Cross sections of $^{27}Al(n,\alpha)^{24}Na$ (left) and $^{93}Nb(n,2n)^{92m}Nb$ (right) reactions/4/.

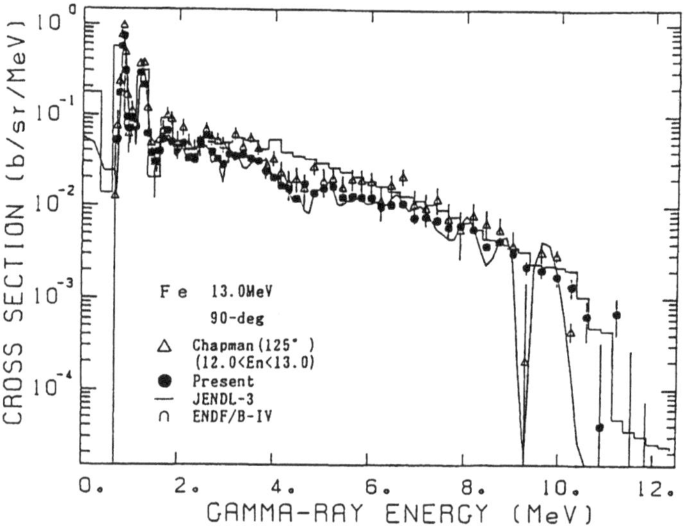

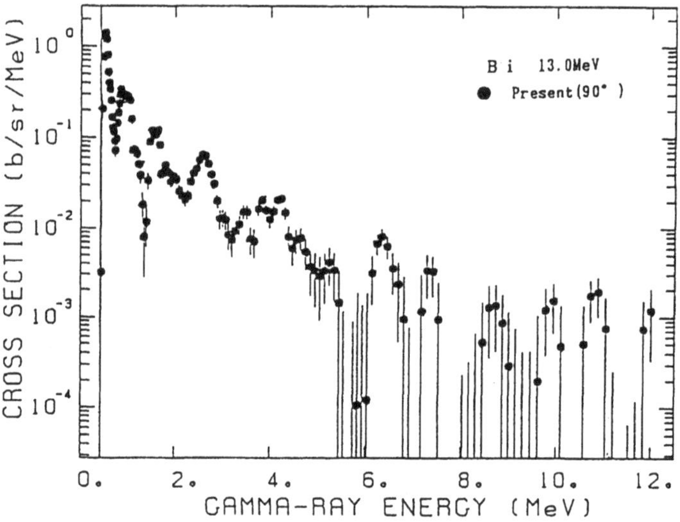

Fig. 5 Gamma-ray production cross sections measured with the ^{11}B-H neutron source/7/.

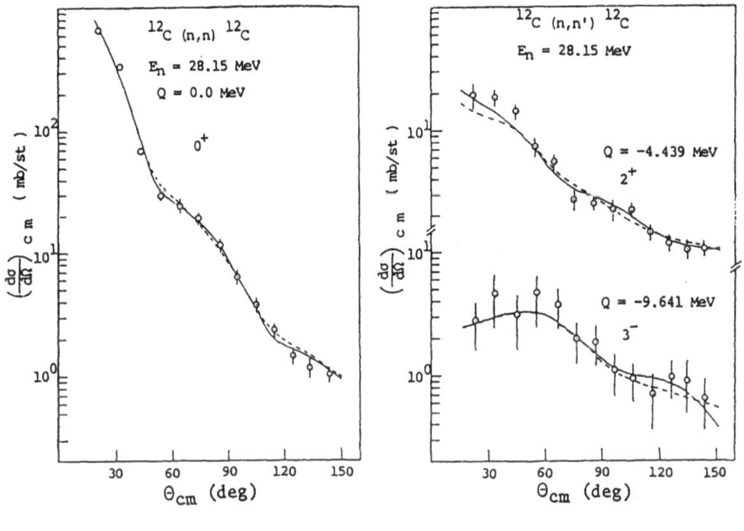

Fig. 6 Neutron elastic and inelastic scattering cross sections of carbon at 28.2 MeV/1/.

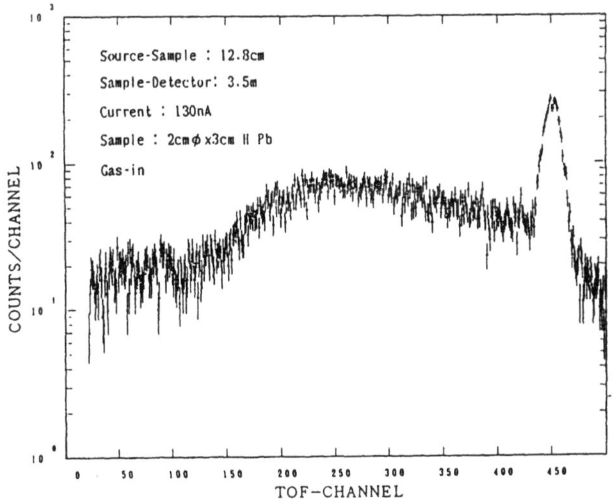

Fig. 7 Neutron TOF spectrum scattered from a lead sample at En = 11.0 MeV and scattering angle of 70 degrees measured with the ^{11}B-H source.

ABSOLUTE MEASUREMENT OF CROSS SECTIONS OF $^{27}Al(n, \alpha)^{24}Na$ AND $^{56}Fe(n, p)^{56}Mn$ AT 14.6 MeV

BAO ZONGYU, RONG CHAOFAN, YANG XIAOYUN,
YE ZONGYUAN and DING SHENGYAO
Radiometrology Division,
China Institute of Atomic Energy
P. O. Box 275, Beijing 102413, P. R. China

ABSTRACT

The cross sections for the reactions $^{27}Al(n, \alpha)^{24}Na$ and $^{56}Fe(n, p)^{56}Mn$ were absolutely measured by the activation method at E_n = 14.6 MeV. The neutron fluence was determined by the associated α-particle method. The absolute activities of the reaction products were measured by a 4π β-γ coincidence device. Taken accout of various corrections, the total uncertainty was 1.3%, and the two cross sections were obtained as 113.2 $\pm$ 1.5 mb and 111.4 $\pm$ 1.5mb respectively.

1. Introduction

The cross sections of $^{27}Al(n, \alpha)^{24}Na$ and $^{56}Fe(n, p)^{56}Mn$ are often used in determining fast neutron fluence. The $^{27}Al(n, \alpha)$ cross section adopted as a standard by IAEA in 1983, and $^{56}Fe(n, p)$ cross section has been used as a secondary standard by several laboratories. Up to now, more than 90 measurements for $^{27}Al(n, \alpha)$ and 55 for $^{56}Fe(n, p)$ have been published. However, the considerable deviations have been found between the data reported before and after 1975, such as between ENDF/B-V and recent measurements.

One of the tasks of our division is to measure precisely the standard cross sections. In recent years, we have developed the methods for fast neutron fluence determination, which have been successfully examined through the international comparisons organized by BIPM. As for measuring the activities of the reaction products, a 4π $\beta-\gamma$ coincidence counter with computer has been setup, and adopted as a standard facility through several domestic comparisons. To varify if our facilities are reliable for precise measurement, we chose these two reaction cross sections as the first step for measuring a variety of the standard cross sections.

2. Experiment

2.1 Measurement of the neutron fluence

The neutrons were produced by the $T(d, n)^4He$ reaction at the 600 KV Cockcroft Walton with an experimental room $20 \times 10 \times 7\ m^3$. A $\phi 8$ mm Ti-T target with thickness of $0.5 mg/cm^2$ was so placed that its plane was 45° to the incident deuteron beam. For E_d = 200KeV, the neutron energy at 45° with respect to the deuteron beam was 14.57 ± 0.02 MeV, measured by the silicon detector method. The neutron fluence was determined by the associated α-particle method. The length of α-particle tube is 1004.2 mm and there are 3 diaphragms in it to minimize the background. An aluminum foil in thickness of 1 μm covered the diaphragm just in front of the Au-Si detector.

2.2 Sample and irradiation

The aluminium and iron samples with purity of 99.99% and size of $\phi 20 \times 0.2$ mm were put at the place 80-130 mm from the neutron target. The sample was held by a ring made of a thin copper alloy wire with a diameter of 0.5 mm. The Al and Fe foils were irradiated separately or sandwiched. Total of 7 experimental runs were carried out.

2.3 Measurement of activities

The absolute activities of the samples after irradiation were measured by 4π $\beta-\gamma$ coincidence device. A methane gas flow proportional β-counter positioned between two $\phi 7.6 \times 7.6$ cm NaI(Tl) scintillator γ-counters was used. The signals were recorded and processed by an on-line micro-computer.

3. Correction, uncertainty and results

In addition to the routine decay corrections where the half - lives were taken as 14.959 hr. for ^{24}Na and 2.5785 hr. for ^{56}Mn, respectively, the other main corrections were as follows:

3.1 The room background

The room background was determined by $1/Re^2$ law where Re was the effective mean distance, deduced according to the real distance R between the target spot and the sample, and their diameters. This correction varied 1.3%-0.3% for R = 80 - 130 mm.

3.2 The absorption and the scattering of the neutron target assembly

The target assembly consisted of 1 mm thickness Ti-T target, 0.6 mm thick Mo backing and about 0.6 mm thick cooling water. The correction was 1.0%. If cooled by air, however, the correction will be 0.5%.

3.3 The dead time correction

The dead time correction of the associated α-particle detection system was 0.4 -0.7 %, depending on the deuteron beam intensity.

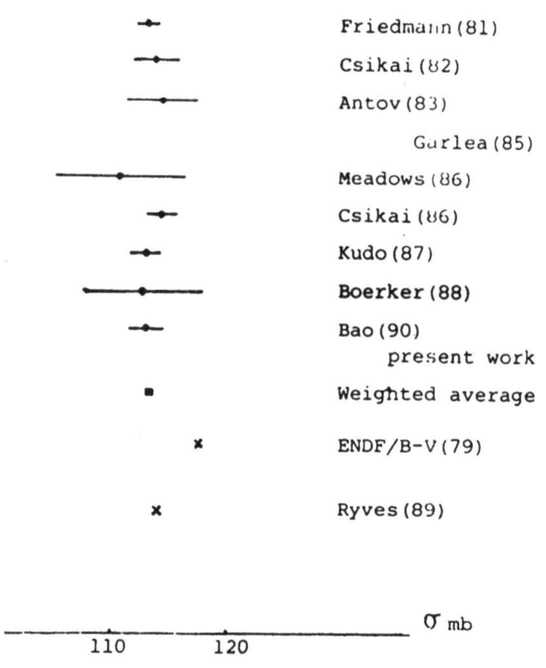

Fig. 1. The cross section of $^{27}Al(n, \alpha)^{24}Na$ at E_n = 14.6 MeV

3.4 The effect of charged particles

In order to reduce the effect of charged particles produced by D(d, n)^{3}He and D(d, p)^{3}H reactions, a fast discriminator and fresh Ti-T target were used in every experiment run. The α-background was less than 0.2% within 15 hr. irradiation period.

3.5 The influence of Al foil in front of semiconductor detector

To prevent the scattering deuteron beam into the α-detector, it was covered by a foil with 1 μm thickness. In such case, 0.5% correction for the loss of α-particles was needed.

Total of 7 experimental runs were carried out, and their deviations are less than 3 σ(σ is statistical error). The weighted average cross sections were obtained as 113.2 ± 1.5 mb for $^{27}Al(n, \alpha)^{24}Na$, and 111.4 ± 1.5 mb for $^{56}Fe(n, p)^{56}Mn$

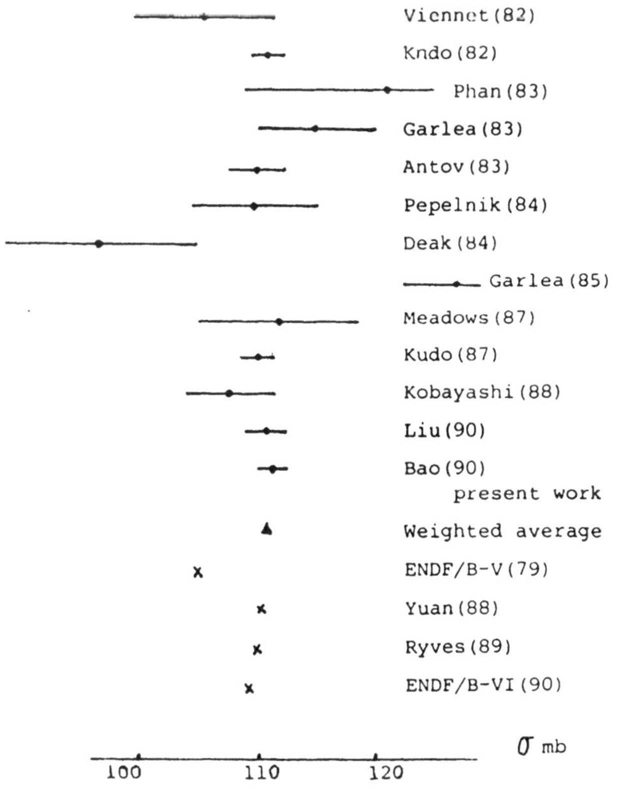

Fig. 2. The cross section of $^{56}Fe(n, \alpha)^{56}Mn$ at E_n = 14.6 MeV

respectively. The experimental and some evaluated cross sections for the two reactions published in the last ten years are shown in Fig. 1 and Fig. 2. For comparison, all the data have been normalized to the point of E_n = 14.6 MeV. Except a few data with big uncertainties, the weighted average cross sections are 113.5 ± 0.5 mb and 110.7 ± 0.7 mb, respectively, in a good agreement with our results. But according to ENDF/B-VI, the results are 118.1 mb and 109.3 mb. The divations with our results are larger than 1.0%. It seems that a new comprehensive evaluation for the two reaction cross sections is needed.

RECENT RADIOCHEMICAL STUDIES OF FAST NEUTRON INDUCED REACTIONS

S.M. Qaim
Institut für Chemie 1 (Nuklearchemie)
Forschungszentrum Jülich GmbH
5170 Jülich, Germany

ABSTRACT

The radiochemical technique used in the study of fast neutron induced reactions is briefly discussed. Some recent results on low-yield trinucleon emission reactions are reviewed. A short description of experimental and theoretical studies on isomeric cross section ratios is given. The data of significance to fusion reactor technology are discussed.

1. Introduction

Studies of fast neutron induced reactions are of fundamental significance and also provide nuclear data useful for fusion reactor technology (FRT). Radiochemical investigations constitute a small but meaningful fraction of the total effort in the field. This paper briefly reviews some of the recent studies in this area.

2. Radiochemical Technique

Radiochemical investigations involve off-line identification of the activation products. The technique is not applicable if the reaction product is stable. In special cases where the emitted particle is radioactive (e.g. tritium or ^{7}Be), measurement of the emitted particle rather than the product can also be applied.

Activation cross section measurements generally involve three salient features (for a detailed review cf. [1]):

- Determination of absolute activity of the product
- Measurement of neutron flux density
- Characterization of the neutron field

High resolution counting methods are essential for an unambiguous identification of the activation product. In recent years conventional Ge detector γ-ray spectroscopy has superseded almost all the other methods of counting. There is, however, considerable need and scope of soft γ-ray and X-ray counting, especially in investigations on low-lying isomeric states. In all these spectroscopic methods care is demanded in the determination of peak area and the efficiency of the detector.

The fast neutron flux density can be determined via one of the several available techniques, viz. associated particle method, long counter using the H(n,n') process, fission chamber, or more simply via monitor reactions. For a better check, in general more than one technique is applied. If only monitor reactions are used, it is imperative that several reactions with different thresholds are employed.

The characterization of the neutron field is of considerable importance. Since the activation measurement does not distinguish between various nuclear processes contributing to the formation of a product nucleus, it is likely that low energy neutrons in a predominantly fast neutron field may give rise to erroneous results for low threshold reactions. It is therefore urged that neutron fields be well characterized prior to activation cross section measurements.

Specific radiochemical separations facilitate isolation and identification of the transmutation products. A variety of radiochemical methods like precipitation, coprecipitation, solvent extraction, distillation, thermochromatography, high performance ion-exchange chromatography, vacuum extraction of soft β^- emitting gaseous products, etc. has been employed. Separations, if carried out at low carrier concentrations, also facilitate the preparation of thin sources for β^- counting or soft X-ray spectroscopy.

The radiochemical technique has high sensitivity and is thus a powerful tool in the study of low-yield reactions. It is ideally suited for the study of closely-spaced low-lying isomeric states. In contrast, the emitted neutron and charged particle spectral distributions cannot be resolved to allow distinction between transitions to different levels of the product nucleus. Furthermore, (n,x) and (n,n'x) type reactions can be more easily distinguished using the radiochemical technique as compared to the spectral method. The major limitation of the radiochemical method is that it yields only integral cross sections. Such data are of considerable practical significance but are only of limited mechanistic value. Only through detailed nuclear model calculations on the excitation functions can some light be shed on the reaction mechanisms.

3. Trinucleon Emission Reactions

The probability of trinucleon emission in fast neutron induced reactions is very low, especially in the medium and heavy mass regions. Most of the studies to date have been performed radiochemically. In

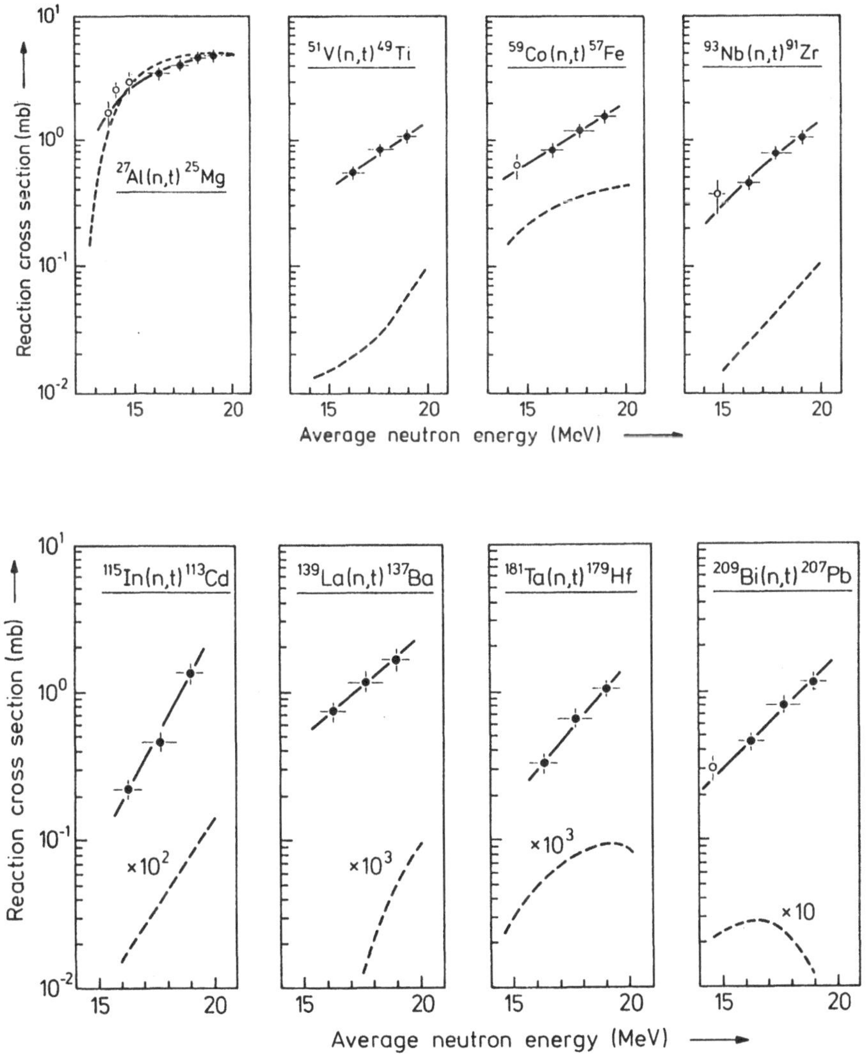

Fig. 1 Excitation functions of (n,t) reactions on medium and heavy mass nuclei. The points give experimental data and the solid curve is an eye-guide through those data. The dashed curve describes results of Hauser-Feshbach calculations [3].

recent years the emphasis has been on investigation of excitation functions [cf. 2-5].

In (n,t) reactions on light mass nuclei the excitation function may show resonance type structures [cf. 2]. However, for medium and heavy mass nuclei smooth curves were found. The results for nuclei from ^{27}Al to ^{209}Bi are summarized in Fig. 1 [3]. It is evident that for a given target nucleus the (n,t) cross section increases with E_n. For a given energy, however, the cross section decreases rapidly from Al to V, but remains practically constant between V and Bi. The results of Hauser-Feshbach calculations given in Fig. 1 depict that the statistical model reproduces the excitation function of (n,t) reaction on ^{27}Al very well. For heavier nuclei, however, the contribution of statistical processes is small and decreases with the increasing mass of the target nucleus.

As far as the (n,^{3}He) reaction is concerned, so far the excitation function on only two target nuclei, namely ^{93}Nb and ^{139}La, has been

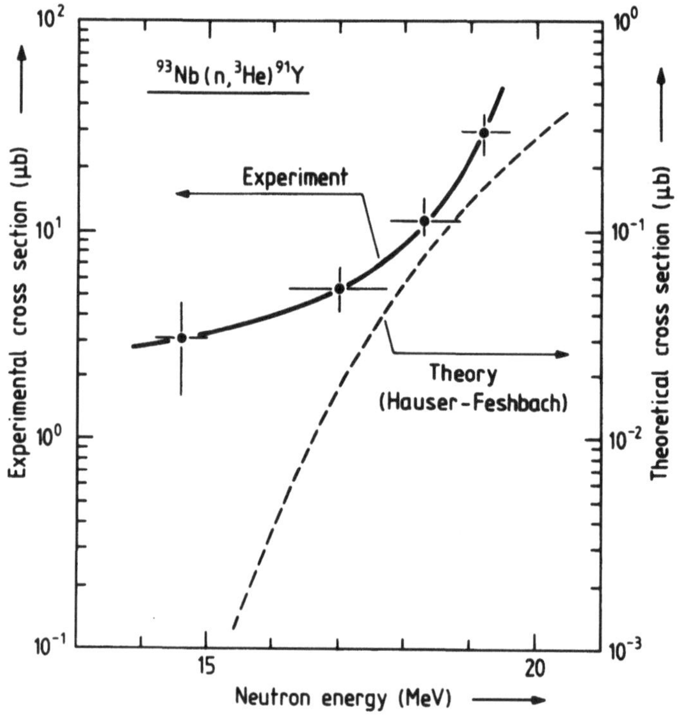

Fig. 2 Excitation function of ^{93}Nb(n,^{3}He)^{91}Y reaction [4].

measured [4,5]. The results for the $^{93}Nb(n,^{3}He)^{91}Y$ process are given in Fig. 2 [4]. Evidently the cross section increases rapidly with increasing neutron energy. The results of model calculations differ from the experimental data by a factor of > 10^2, showing that the contribution of the statistical processes to ^3He emission is small.

4. Isomeric Cross Section Ratios

Studies of isomeric cross section ratios are of considerable fundamental significance. They shed some light on the role of nuclear spin in the population of ground and excited states of a reaction product as well as on the effect of kinetic energy and angular momentum brought by the incident particle.

At 14 MeV rather extensive data exist on the formation of metastable states in major nuclear reactions like (n,p), (n,α) and (n,2n) [cf. 6-8]. The data on low-yield reactions are scarce. The available information as a function of neutron energy is also rather small. Presently at Jülich a programme is underway to perform such studies in the neutron energy range of 4 to 11 MeV. The results for $^{91}Zr(n,p)^{91m,g}Y$ are shown in Fig. 3 [9]. At low energies only about 30 % of the transitions occur to the high spin ($9/2^+$) state; around 14 MeV, however, the contribution increases to about 55 %.

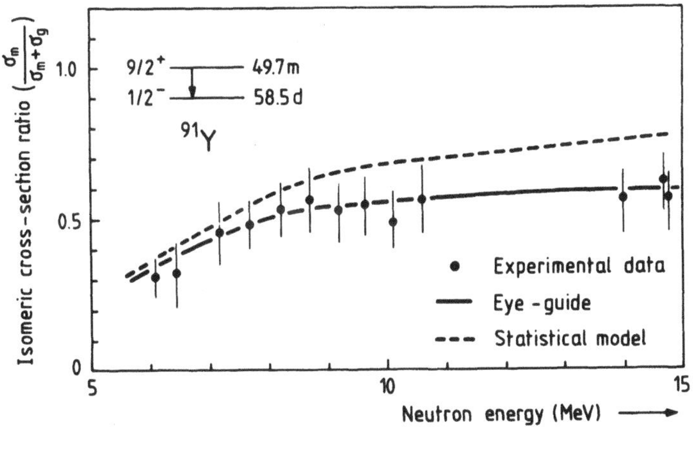

Fig. 3 Isomeric cross section ratio for the isomeric pair $^{91m,g}Y$ (formed via (n,p) reaction on ^{91}Zr) plotted as a function of incident neutron energy [9].

The nuclear model calculations done so far involve Hauser-Feshbach formalism and incorporate precompound effects. The results given in Fig. 3 show that the isomeric cross section ratio is described well up to a neutron energy of about 10 MeV; at higher energies some deviations occur. It was found that whereas the summed cross section can be described rather well by these calculations [cf. 9, 10], the calculated cross section for the formation of the metastable state is rather strongly dependent on the input level scheme of the product nucleus [10]. Calculation of very high spin isomers is also rather uncertain [cf. 11]. Evidently, a considerable amount of further experimental and theoretical work is essential for a better understanding of isomeric cross section ratios.

5. Nuclear Data for FRT

Radiochemical measurements of cross section data relevant to fusion reactor technology (FRT) generally deal with:

- Activation and transmutation products
- Tritium breeding
- Gas production

The activation of reactor components leading to long-lived products is one of the major concerns in design studies [cf. 12]. Nuclear model calculations can yield reliable results if ground states formed via common reactions like (n,p), (n,α), (n,2n), etc. are involved. For other reactions, especially if metastable states are concerned, experimental investigations are mandatory. In recent years efforts have been increased [cf. 1, 13-15] and measurements on several potential first wall materials and blanket constituents have been done.

For tritium breeding the (n,t) reactions on the two isotopes of lithium, namely ^{6}Li and ^{7}Li, are important. For ^{6}Li the data were well known. In the case of ^{7}Li, however, there was considerable discrepancy. A careful analysis of the tritium emission data and some new measurements revealed [cf. 16] that the discrepancy had arisen due to a combination of neutron emission and tritium emission data. From the viewpoint of tritium breeding the radiochemical measurement is thus more reliable.

Hydrogen and helium gas production cross sections are important in the context of radiation damage to first wall materials. These involve studies of (n,xp) and (n,xα) reactions. The radiochemical method has been successfully applied to the investigation of (n,p), (n,n'p), (n,α) and (n,n'α) reactions at 14 MeV, so that total gas production can be estimated. The results are generally in agreement with the charged particle detection and mass spectrometric data. At lower neutron

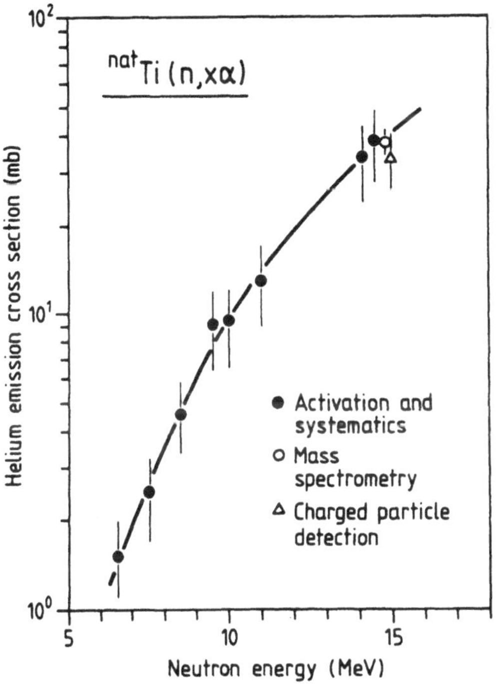

Fig. 4 Helium emission cross sections for elemental titanium as a function of neutron energy [17].

energies the available information is insufficient, especially if the element consists of several isotopes. The results on helium production from titanium reported recently [17] are reproduced in Fig. 4. The cross sections for (n,α) reactions on the major isotopes ^{48}Ti and ^{50}Ti were determined experimentally and those for ^{46,47,49}Ti via model calculations. The accuracy of the data thus deduced is not very high but is sufficient to meet the requirements for radiation damage calculations [cf. 18].

6. Conclusion

The radiochemical method is of great significance in the study of low-yield reactions and low-lying isomeric states. Considerable progress has been achieved in the study of trinucleon emission reactions in medium and heavy mass regions. Some of the radiochemically measured integral data are unique and find application in fusion reactor technology.

References

[1] Proc. Specialists' Meeting on Neutron Activation Cross Sections for Fission and Fusion Energy Applications, September 1989, Argonne, USA (Edited by M. Wagner and H. Vonach), NEANDC-259 "U", OECD, Paris (1990)
[2] A. Suhaimi, R. Wölfle, S.M. Qaim, P. Warwick and G. Stöcklin, Radiochimica Acta 43 (1988) 133
[3] R. Wölfle, S.M. Qaim, H. Liskien and R. Widera, Radiochimica Acta 50 (1990) 5
[4] S.M. Qaim, R. Wölfle and H. Liskien, Phys. Rev. C34 (1986) 489
[5] S.M. Qaim and R. Wölfle, Acta Phys. Hungarica 69 (1991) 139
[6] V. McLane, C.L. Dunford and P.F. Rose, Neutron Cross Sections, Vol. 2, Academic, San Diego, 1988
[7] R.A. Forrest, AERE R 12419 (1986)
[8] J. Kopecky and H. Gruppelaar, ECN-200 (1987)
[9] S.M. Qaim, M. Ibn Majah, R. Wölfle and B. Strohmaier, Phys. Rev. C42 (1990) 363
[10] S.M. Qaim, A. Mushtaq and M. Uhl, Phys. Rev. C38 (1988) 645
[11] M.B. Chadwick and P.G. Young, in ref. [1], p. 241
[12] R.A. Forrest and G.J. Butterworth, Proc. Int. Conf. on Nuclear Data for Science and Technology, May 1991, Jülich, Germany, in press
[13] L.R. Greenwood, D.G. Doran and H.L. Heinisch, Phys. Rev. C35 (1987) 76
[14] Proc. IAEA Consultants' Meeting on Activation Cross Sections for the Generation of Long-Lived Radionuclides of Importance in Fusion Reactor Technology, September 1989, Argonne, USA (Edited by Wang DaHai), INDC(NDS)-232/L, IAEA, Vienna (1990)
[15] Y. Ikeda, A. Kumar and C. Konno, Proc. Int. Conf. on Nuclear Data for Science and Technology, May 1991, Jülich, Germany, in press
[16] S.M. Qaim and R. Wölfle, Nucl. Sci. Eng. 96 (1987) 52
[17] S.M. Qaim, N.I. Molla, R. Wölfle and G. Stöcklin, Proc. Int. Conf. on Nuclear Data for Science and Technology, May 1991, Jülich, Germany, in press
[18] S. Cierjacks and K. Ehrlich, Proc. Int. Conf. on Nuclear Data for Science and Technology, May 1991, Jülich, Germany, in press

MEASUREMENT AND EVALUATION OF ACTIVATION CROSS SECTION

LU HANLIN, ZHAO WENRONG and YU WEIXIANG
China Institute of Atomic Energy,
P. O. Box275-3, Beijing 102413, China

ABSTRACT

The cross sections of 79 reactions (including 21 reactions which were done in recent 3 years) of interest for nuclear technology applications were measured by the activation method from 1975 - 1991. For some reactions compilation and evaluation have been made. Table 1 lists all of them.

A brief description of the experimental method main corrections and some results of activation cross sections is given. The description is focused on the results of $Pt(n, 2n)^{195m}Pt$, $^{23}Na(n, 2n)^{22}Na$, $^{60}Ni(n, p)^{60}Co$ and $^{62}Ni(n, \alpha)^{59}Fe$ reactions and the measurements of reaction cross sections induced by 8.5 MeV neutrons.

1. Introduction

The neutron activation cross sections are very useful in nuclear technology applications and nuclear theory test. As development of fusion reactor technology, more and more accurate data are required.

One of pressing needs for nuclear activation data is from dosimetry safety and environmental assessments on accidental hazard potential, decay heat, shutdown doserate, waste disposal, maintenance, and so on need activation cross sections with higher accuracy. In addition, there are some conventional applications of activation data, for example, measurement of neutron fluence and spectra, neutron activation analysis, etc.

Many kinds of cross sections have been measured and published, and several sets of evaluated results have been given by some labs. But among them there are large discrepancies. In this case, a decision was made to measure and evaluate some reactions in our lab so as to check and improve the existing data.

Up to now, we have measured seventy-nine reactions and evaluated part of them, which are listed in table 1.

2. Experimental method

2.1 Irradiation

The investigation is performed by means of activation method. The neutron source used in our measurement is as the follows: the Li(p, n)^{7}Be, T(p, n)^{3}He, D(d, n)^{3}He and T(d, n)^{4}He reactions were used as neutron sources in 0.1 - 0.5, 0.5 - 2.0, 3 - 12 and 12, 5 - 18 MeV neutron energy ranges respectively at Cockcroft-Walton, Van-de-graaff and Cyclotron accelerators.

2.2 Neutron fluence rate

Neutron flux are determined using various methods in the whole measured energy range: proportional counters filled with hydrogen or methane were used for neutrons below 2 MeV; the method of counting protons in the associated reaction for about 3 MeV neutrons; proton recoil semiconductor and emiconductor telescope for about 1.1 - 5.5 MeV neutrons; proton recoil scintillation telescope for about 8 - 18 MeV neutrons; α associated particle method for about 14 MeV neutrons; the reference standard cross sections were used in a very wide energy region.

In order to check and improve the accuracy or neutron flux determination, we had participated in the second international fluence-rate inter comparison for 0.144, 0.561 and 14 MeV neutrons and got quite satisfactory results.

2.3 Determination of activities

The activities of the residual nuclides investigated were detected by a Ge(Li) (136cm^3) or a NaI(Tl) (ϕ8 $\times$ 8cm) detector which is connected with a computer system. Their efficiencies were calibrated using a set of standard γ - ray sources. The orrections were made for the effects of γ - ray cascade in the detector and γ - ray self-absorpt in the samples, as well as for the efficiency change owing to the size different of the sample.

3. Main corrections

A main correction is the contamination of the neutron field. The neutron field of the T(d, n)^{4}He reaction is always contaminated by the neutrons of d - D reaction, and evaporation and scattered neutrons produced in the vicinity of the target assembly and samples. The break-up neutron is another ontamination source in the d - D neutron field.

3.1 Evaporation and scattering neutron from the target assembly and in the sample

One important source of contamination for investigated neutron field comes from the interaction of the incident neutron with the target assembly and samples.

The low energy neutrons are produced by (n, n') and (n, 2n) reactions. So less mass of the target and sample holder must be used. As an example, Fig. 1 shows the correction factor of scattered neutrons induced by target assembly. And Fig. 2 shows the correction factor of scattered and attenuated neutrons induced by target assembly for the ^{140}Ce(n, 2n)^{139}Ce reaction.

The relation between correction factor of scattered neutrons and reaction Q value at 0° is plotted in Fig. 3, which was calculated for each reaction using a Monte-Carlo program.

3.2 Contamination of neutron source

The d - D neutrons are produced in the d - T neutron field by the incident deuteron beam on deuterium nuclei accumulated on the target. Generally, deuterons accumulate on the analyzing magnet, beam collimator and target due to the bombardment of deuteron beam. Special experimental arrangement was used to determine the relation of the ratio $\Phi_{d\text{-}D}/\Phi_{d\text{-}T}$ vs. the irradiation time, which is shown in Fig. 4.

In the d - D neutron field, if 8 - 12 MeV neutrons are obtained, the d - D break-up neutron will be appeared from the D(d, np)D reaction. Its fraction and spectrum must be determined for the low threshold or no threshold reaction studies.

3.3 Room background

The relation of correction factor (%) of room background for ^{56}Fe(n, p)^{56}Mn reaction at 14 MeV, vs. distance square from target is given in Fig. 5. For no threshold reaction, for example, for the ^{115}In(n, γ)^{116m}In reaction at 144 and 565 KeV, the ratio of activities of In foil due to room background and direct neutrons at distance of 10 cm from the target is 49%.

3.4 Selection of reference cross section

If the reference cross section is used as a neutron fluence monitor, how to choose a reasonable reference cross section is an important problem. Besides the suitable and accurate cross section, and convenient measurement for activities, the shapes of reference and investigated excitation functions are also important for determination of the cross section uncertainties. There are three cases for the shapes of excitation functions of the reference and investigated reactions as shown in Fig. 6. The best one is that the trend of the shapes for two excitation functions is about the same.

4. Results

4.1 Activation cross sections for 8.50 MeV neutrons

The cross sections of (n, p), (n, α), (n, n'), (n, 2n) and (n, x) reactions for ^{24}Mg, ^{46,47,48}Ti, ^{51}V, ^{54}Fe, ^{58}Ni, ^{93}Nb, ^{113,115}In, ^{181}Ta and Pt were measured by activation method at 8.50 $\pm$ 0.27 MeV neutron energy. The cross section of ^{27}Al(n, α)^{24}Na reaction was used as a neutron fluence monitor.

The main corrections were made for neutron scattering at target assembly, monitors and samples, and for break-up neutrons (E_n ~ 1.0 MeV).

The results are summarized in Table 3. Some of them agree with existing data (such as ^{46,48}Ti(n, p), ^{54}Fe(n, α), ^{113,115}In(n, n')) and evaluated data (^{51}V(n, α), ^{181}Ta(n, 2n)). But for the reactions of ^{47}Ti(n, p), ^{54}Fe(n, p) and ^{58}Ni(n, p) our values are different from the evaluated data of the ENDF/B - VI dosimetry file, as shown in Figs. 7, 8, 9.

4.2 Cross section for Pt(n, x)^{195m}Pt reaction

The activity of ^{195m}Pt is produced through three reactions of ^{194}Pt(n, γ), ^{195}Pt(n, n') and ^{196}Pt(n, 2n). The contribution of the ^{194}Pt(n, γ) reaction is in low energy region, ^{195}Pt(n, n') in 1 - 6 MeV, while ^{196}Pt(n, 2n) in the region above 8 MeV. Since the energy region covered by Pt(n, x)^{195m}Pt is quite wide, in principle, no threshold exists for ^{195m}Pt production. Therefore, Pt(n, x)^{195m}Pt is favourable for dosimetry applications. Unfortunately, besides our measurements, only one experimental point at 14.4 MeV was found.

4.3 Activation cross sections for ^{23}Na(n, 2n)^{22}Na, ^{60}Ni(n, p)^{60}Co and ^{62}Ni(n, α)^{59}Fe reactions

^{23}Na(n, 2n)^{22}Na; The experimental results for the ^{23}Na(n, 2n)^{22}Na cross sections, evaluations of ENDF/B-VI and IRDF-85 are summarized in Fig. 11. There is a large discrepancy in existing experimental data and between two famous evaluations. The present data are situated between the results of two evaluations. Newly published data of Y. Ikeda (88) and M,Wagner (90) are much closer to the present values, and consistent with the data of G, N. Maslov and Shani at 14 MeV. Almost all of the data sets by different authors are used to get the relative shape of excitation function from threshold to 21.5 MeV (Fig. 12). The present evaluation lies between two evaluations of ENDF/B - VI and IRDF - 85 with a slight trend to IRDF - 85.

^{60}Ni(n, P)^{60}Co and ^{62}Ni(n, α)^{59}Fe: Figs. 13, 14 show a comparison of our results with other measurements, calculations and evaluations. As Fig. 13 shows, there is a good agreement among the values of the $^{60}Ni(n, p)^{60}Co$ cross section of present work, H. Vonach (89) at 12·MeV and Wang Yongchang (89) at 14-15 MeV, and calculations of M. Uhi (89) at 12.8 - 15 MeV and N. Yamamuro (89) over 14 MeV. But there is some disagreement with the evaluated data of the ENDF/B - V dosimetry file, which is based on the data of A. Paulsen(67).

For the $^{62}Ni(n, \alpha)^{59}Fe$ reaction our results are consistent with the other experimental data at 14 MeV, which are about 30 - 40% lower than the value of ENDF/B - V dosimetry file.

Table.1 List of measured reactions and associated parameters

No.	Reaction	En (MeV)	Exp. Point	Isotope Abundance (%)	Product Nucleus	E_γ (keV)	I_γ (%)	$T_{1/2}$
1	$^{24}Mg(n,p)^{24}Na$	8.50–18.3	37	78.99	^{24}Na	1368	99.993	15.02 h
2	$^{27}Al(n,\alpha)^{24}Na$	12.2–18	11	100	^{24}Na			
3	$^{45}Sc(n,2n)^{44m}Sc$	13.2–18.3	12	100	^{44m}Sc	271	86.6	2.44 d
4	$^{45}Sc(n,2n)^{44g}Sc$	14.6–18.3	5*		^{44g}Sc	1157	99.9	3.93 h
5	$^{45}Sc(n,2n)^{44m+g}Sc$	14.6–18.3	5					
6	$^{46}Ti(n,p)^{46}Sc$	4.5–18	10	8.0	^{46}Sc	889 1121	100	83.83 d
7	$^{47}Ti(n,p)^{47}Sc$	8.50	1	7.3	^{47}Sc	159	68	3.341 d
8	$^{48}Ti(n,p)^{48}Sc$	5–18	15	73.8	^{48}Sc	984	100	43.7 h
9	$^{51}V(n,\alpha)^{48}Sc$	8.50–18	12	99.75				
10	$^{55}Mn(n,2n)^{54}Mn$	12.4–18.3	15	100	^{54}Mn	835	99.975	312.2 d
11	$^{54}Fe(n,p)^{54}Mn$	8.50–17.2	25	5.8				
12	$^{54}Fe(n,\alpha)^{51}Cr$	8.50–18.3	22		^{51}Cr	320	9.85	27.70 d
13	$^{56}Fe(n,p)^{56}Mn$	12.8–18.3	25	91.72	^{56}Mn	847	98.9	2.578 h
14	$^{59}Co(n,2n)^{58}Co$	12.5–18.3	20/9	100	^{58}Co	810	99.44	70.91 d
15	$^{59}Co(n,p)^{59}Fe$	13.6–17.8	8		^{59}Fe	1099	56.5	44.496 d
16	$^{59}Co(n,\alpha)^{56}Mn$	12.5–18.3	20/6					
17	$^{58}Ni(n,2n)^{57}Ni$	12.8–18.3	13	68.27	^{57}Ni	1377	77.9	1.503 d
18	$^{58}Ni(n,p)^{58}Co$	8.50–17.8	10					
19	$^{58}Ni(n,x)^{57}Co$	12.8–18.3	13		^{57}Co	122	85.5	271.2 d
20	$^{60}Ni(n,p)^{60}Co$	14	2	26.10	^{60}Co	1173	99.89	5.271 y
21	$^{62}Ni(n,\alpha)^{59}Co$	14	2	3.59				
22	$^{63}Cu(n,\alpha)^{60}Co$	12.8–17.7	7	69.17				
23	$^{66}Zn(n,2n)^{65}Zn$	12.8–17.7	9	27.9	^{65}Zn	1116	50.57	244.1 d
24	$^{67}Zn(n,p)^{67}Cu$	12.8–17.7	6	4.1	^{67}Cu	185	47	61.9 h
25	$^{70}Zn(n,2n)^{69m}Zn$	12.8–17.7	5	0.6	^{69m}Zn	439	94.8	13.76 h
26	$^{85}Rb(n,2n)^{84m}Rb$	12.5–17.5	11	72.165	^{84m}Rb	248	63	20.5 m
27	$^{85}Rb(n,2n)^{84m+g}Rb$	12.5–17.5	11		^{84g}Rb	881	67.9	32.87 d
28	$^{85}Rb(n,p)^{85m}Kr$	13.5–14.8	6		^{85m}Kr	151	76.1	4.48 h
29	$^{85}Rb(n,\alpha)^{82}Br$	13.5–14.8	6		^{82}Br	776	83.3	35.3 h
30	$^{87}Rb(n,2n)^{86}Rb$	13.5–14.8	6	27.835	^{86}Rb	1079	8.8	18.82 d
31	$^{87}Rb(n,p)^{87}Kr$	13.5–14.8	6		^{87}Kr	403	48.3	76 m
32	$^{89}Y(n,2n)^{88}Y$	12.8–18	13	100	^{88}Y	898	92.7	106.6 d
33	$^{90}Zr(n,2n)^{89}Zr$	12.4–18.3	25	51.45	^{89}Zr	909	99.01	78.43 h
34	$^{96}Zr(n,2n)^{95}Zr$	12.8–17.7	7	2.80	^{95}Zr	724 765	44.2 54.5	64.02 d
35	$^{93}Nb(n,2n)^{92m}Nb$	12.8–18.3	14	100	^{92m}Nb	934	99.0	10.15 d
36	$^{93}Nb(n,\alpha)^{90m}Y$	8.50–18	8		^{90m}Y	480	91.7	3.19 h
37	$^{113}In(n,2n)^{112m}In$	14	1	4.3	^{112m}In	156	12.8	20.9 m
38	$^{113}In(n,n')^{113m}In$	8.50–14.8	9		^{113m}In	391	64.89	1.65 h

Table.1 (Continued)

39	^{115}In(n,2n)^{114m}In	11.4–18.0	11 / 18	95.7	^{114m}In	191	15.7	49.51 d
40	^{115}In(n,n')^{115m}In	0.88–14.8	53		^{115m}In	336	45.9	4.486 h
41	^{115}In(n,γ)^{116m}In	0.144,0.565	2		^{116m}In	$4\pi\beta$		54.2 m
42	^{115}In(n,p)^{115}Cd	14	3		^{115}Cd	527	27.5	53.38 h
43	^{115}In(n,α)^{112}Ag	14	2		^{112}Ag	617	42.5	3.14 h
44	^{127}I(n,2n)^{126}I	11.4–18.0	10	100	^{126}I	389	34.1	13.0 d
45	^{136}Ce(n,2n)^{135}Ce	13.5–14.8	8	0.19	^{135}Ce	265	42.1	17.76 h
46	^{138}Ce(n,2n)^{137m}Ce	14	1	0.25	^{137m}Ce	254	10.96	34.4 d
47	^{140}Ce(n,2n)^{139}Ce	12.8–18.3	23	88.48	^{139}Ce	166	79.9	137.2 d
48	^{140}Ce(n,p)^{140}La	13.5–14.8	8		^{140}La	487	45.9	40.27 h
49	^{142}Ce(n,2n)^{141}Ce	12.4–18.3	23	11.08	^{141}Ce	145	48.4	32.55 d
50	^{169}Tm(n,2n)^{168}Tm	12.4–18.3	16	100	^{168}Tm	198	50.3	93.1 d
51	^{169}Tm(n,3n)^{167}Tm	15–18.2	9		^{167}Tm	208	41.1	9.25 d
52	^{181}Ta(n,2n)^{180m}Ta	8.50–18.3	34	99.988	^{180m}Ta	93.3 103	4.27 0.79	8.152 h
53	^{181}Ta(n,p)^{181}Hf	12.8–18.3	18		^{181}Hf	482	80.6	42.2 d
54	Pt(n,x)^{195m}Pt	0.144–17.76	39	100	^{195m}Pt	99	11.11	4.02 d
55	^{198}Pt(n,2n)^{197}Pt	8.50	1	7.2	^{197}Pt	191	3.5	18.3 h
56	^{197}Au(n,2n)^{196}Au	9.41–18.2	17 / 7	100	^{196}Au	356	87.6	6.18 h
57	^{197}Au(n,3n)^{195}Au	16–18.2	4		^{195}Au	100	9.347	183 d
58	^{204}Pb(n,2n)^{203}Pb	9.4–15	12	1.4	^{203}Pb	279	80.1	52.05 d
59	^{23}Na(n,2n)^{22}Na	13.5–18.0	6	100	^{22}Na	1274	99.95	2.602 y
60	^{60}Ni(n,p)^{60}Co	12.8–17.6	3	26.1	^{60}Co	1173	99.89	5.271 y
61	^{62}Ni(n,α)^{59}Fe	12.8–17.6	3	3.59	^{59}Fe	1099	56.5	44.496 d
62	^{132}Ba(n,2n)^{131}Ba	14.2–18.0	4	0.101	^{131}Ba	496	44	11.8 d
63	^{134}Ba(n,2n)^{133m}Ba	14.2–18.0	4	2.417	^{133m}Ba	276	17.5	1.621 d
64	^{134}Ba(n,2n)$^{133m+g}$Ba	13.5–18.0	6	2.417	^{133}Ba	355	62.2	10.54 y
65	^{134}Ba(n,p)$^{134m+g}$Cs	13.5–18.0	6	2.417	^{134}Cs	795	85.4	2.062 y
66	^{134}Ba(n,α)^{131m}Xe	17–18	2	2.417	^{131m}Xe	164	1.96	11.9 d
67	^{136}Ba(n,2n)^{135m}Ba	14.2–18.0	4	7.854	^{135m}Ba	268	15.6	1.196 d
68	^{136}Ba(n,p)^{136}Cs	14.2–18.0	4	7.854	^{136}Cs	341	48.6	13.16 d
69	^{138}Ba(n,α)^{135}Xe	14.2–18.0	4	71.70	^{135}Xe	250	90	9.104 h
70	^{109}Ag(n,2n)^{108m}Ag	14.19–14.83	5	48.161	^{108m}Ag	434	90.5	127 y
71	^{151}Eu(n,2n)^{150m}Eu	14.19–14.77	3	47.8	^{150m}Eu	334	94	35.8 y
72	^{151}Eu(n,γ)^{152m}Eu	0.03–1.2	3	47.8	^{152m}Eu	842	14.6	9.32 h
73	^{151}Eu(n,γ)^{152g}Eu	0.03–1.2	5	47.8	^{152g}Eu	344	26.6	13.33 y
74	^{153}Eu(n,2n)^{152g}Eu	14.77	1	52.2	^{152g}Eu	344	26.58	13.33 y
75	^{153}Eu(n,γ)^{154}Eu	0.03–1.2	5	52.2	^{154}Eu	1274	35.5	8.8 y
76	^{159}Tb(n,2n)^{158}Tb	14.19–14.77	2	100	^{158}Tb	944	43	180 y
77	^{179}Hf(n,2n)^{178m2}Hf	14.19–14.77	2	13.629	^{178m2}Hf	495 574	68.92 83.80	31 y
78	^{180}Hf(n,2n)^{179m2}Hf	14.77	1	35.1	^{179m2}Hf	454	66.0	25.1 d
79	^{187}Re(n,2n)^{186m}Re	14–15	4	62.6	^{186m}Re	137	9.54	3.777 d

Table 2 Reference standard cross section used in our lab.

En (MeV)	Cross section (mb)		
	$^{27}Al(n,\alpha)^{24}Na$	$^{56}Fe(n,p)^{56}Mn$	$^{93}Nb(n,2n)^{92m}Nb$
13.2	126.3±4.2		444.0±10.0
13.25		116.6±1.3	
13.4	125.9±4.1		447.4±8.9
13.5		116.9±1.2	
13.6	125.1±2.9		450.7±7.4
13.75		116.4±1.2	
13.8	123.9±2.9		453.5±6.0
14.0	122.3±2.2	115.2±1.1	455.8±5.0
14.2	120.3±1.9	113.8±1.0	457.6±4.5
14.4	118.2±1.9	111.8±1.0	458.7±4.5
14.6	115.8±1.6	109.5±1.0	459.1±4.9
14.8	113.2±1.1	106.9±1.0	458.6±5.6
15.0	110.3±1.5	103.9±1.0	457.0±7.0

Table 3 Cross sections at 8.50 MeV

Reaction	$K=\frac{\sigma_X}{\sigma_{Al}}$	Correction (%)		Cross section	References			
		Scattering	D(d,np)D	σ (mb)	σ (mb)	En (MeV)		
$^{24}Mg(n,p)^{24}Na$	2.33	0.24		131±9	126±4	8.6	J.P.Butler	(63)
					122±6	8.2	H.Liskien	(66)
$^{46}Ti(n,p)^{46}Sc$	4.421	3.23		241±16	246.9±24.7	8.491	D.L.Smith	(75)
$^{47}Ti(n,p)^{47}Sc$	1.828	3.15		99.7±7.0	123.8±6.31	8.491	D.L.Smith	(75)
$^{48}Ti(n,p)^{48}Sc$	0.308	2.93		16.8±1.1	15.87±0.79	8.491	D.L.Smith	(75)
$^{51}V(n,a)^{48}Sc$	0.0566	2.14		1.61±0.11				
$^{54}Fe(n,p)^{54}Mn$	10.47	3.08		571.3±38.0	462.5±25.9	8.491	D.L.Smith	(75)
					511±27	8.40	A.Paulsen	(79)
$^{58}Ni(n,p)^{58m+g}Co$	12.52	2.22		689±47	657±74	8.33	J.F.Barry	(62)
					630±50	8.40	K.Debertin	(65)
$^{54}Fe(n,a)^{51}Cr$	0.758	2.14		41.8±3.7	37.5±2.5	8.40	A.Paulsen	(79)
					41.2±2.7	8.90	A.Paulsen	(79)
$^{93}Nb(n,a)^{90m}Y$	0.0141	0.6		0.80±0.16				
$^{115}In(n,n')^{115m}In$	5.52	1.83	2.6	297±21	315	8.61	D.C.Santry	(75)
					302	8.533	D.L.Smith	(75)
$^{113}In(n.n')^{113m}In$	4.23	1.78	2.0	229±17	244	8.533	D.L.Smith	(75)
$^{181}Ta(n,2n)^{180m}Ta$	4.323	0.6		243±18				
$Pt(n,X)^{195m}Pt$	4.561	2.32	1.7	246±18				
$^{198}Pt(n,2n)^{197}Pt$	0.7113	0.19	0.5	39.9±2.8				

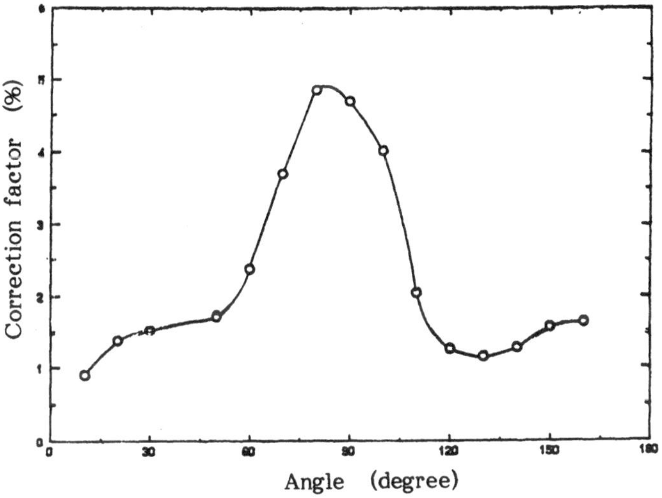

Fig.1 Correction factor of scattered neutrons induced by target assomble for the Ce(n,2n)Ce reaction

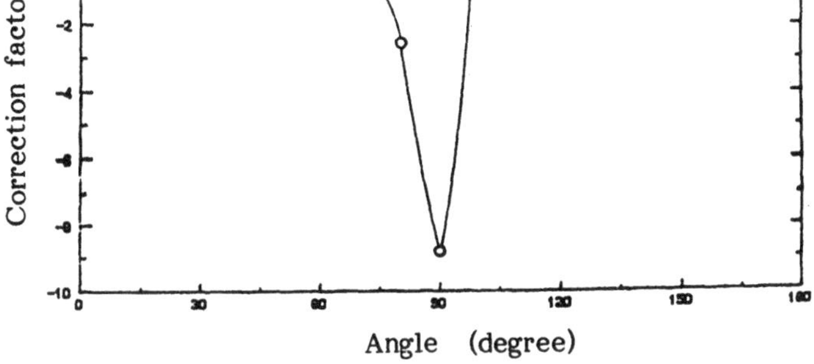

Fig.2 Correction factor of scattered and attenuated neutrons induced by target assomble for the Ce(n,2n)Ce reaction

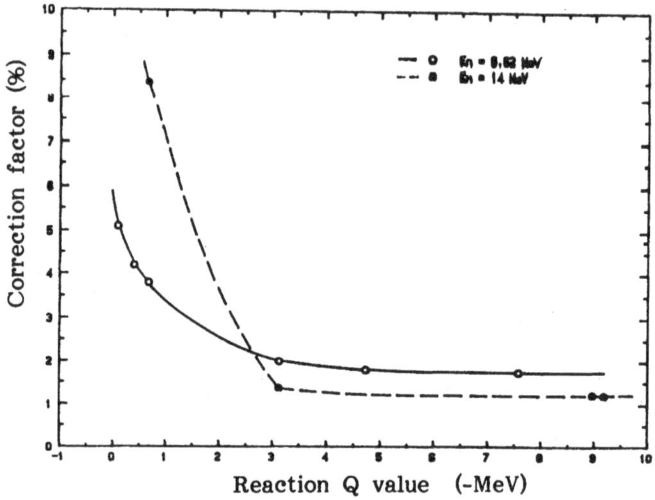

Fig.3 Relation between correction of scattered neutrons and reaction Q value

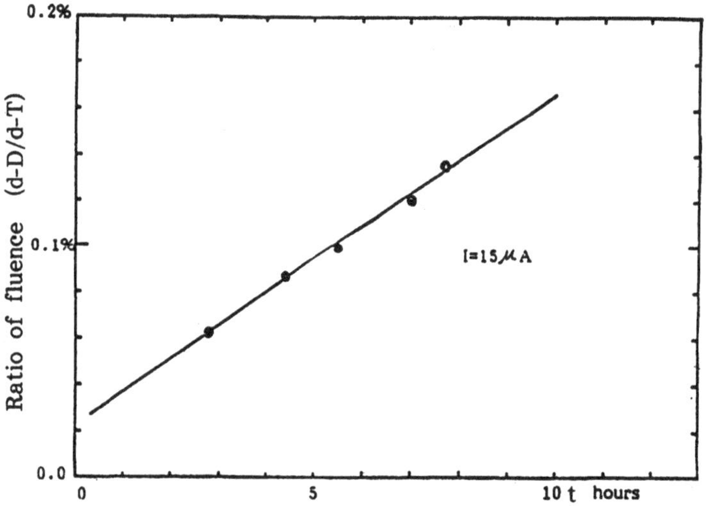

Fig.4 Ratio of neutron fluence of d-D to d-T reaction with time irradiation

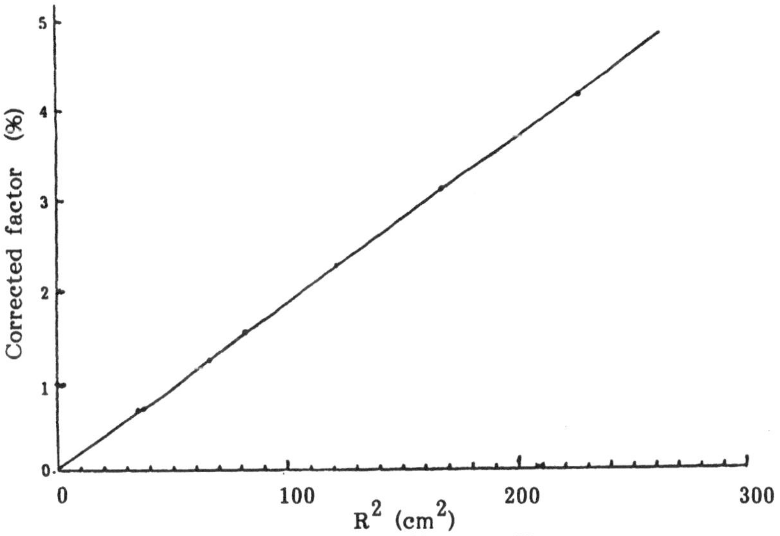

Fig.5 Correction factor of the $^{56}Fe(n,p)^{56}Mn$ reaction vs. distance square from the target

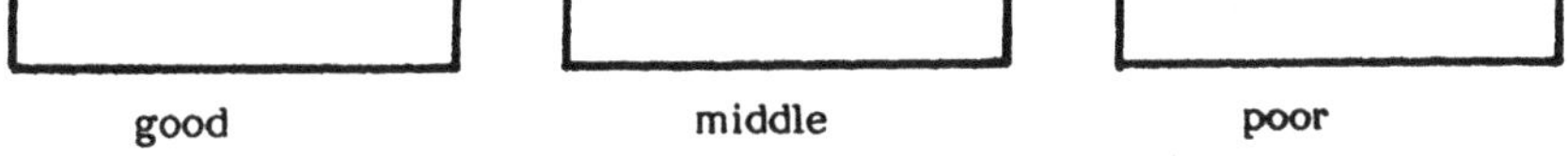

Fig.6 Three cases of excitation function shapes for the investigated and reference reactions

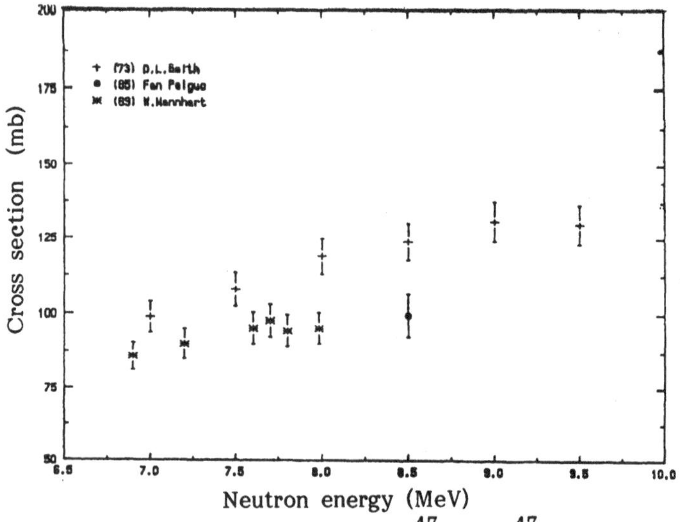

Fig.7 Cross section of the $^{47}Ti(n,p)^{47}Sc$ reaction

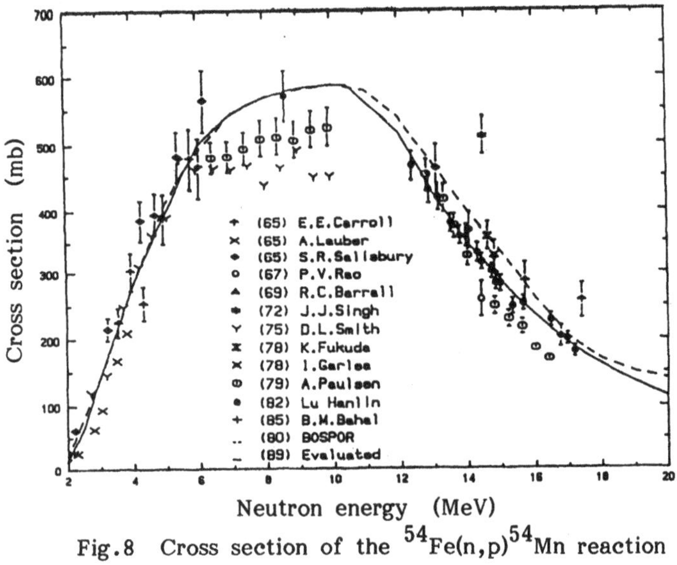

Fig.8 Cross section of the $^{54}Fe(n,p)^{54}Mn$ reaction

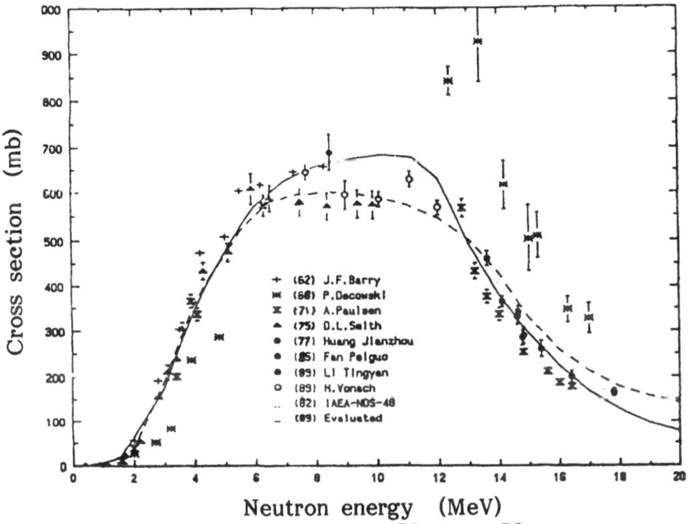

Fig.9 Cross section of the $^{58}Ni(n,p)^{58}Co$ reaction

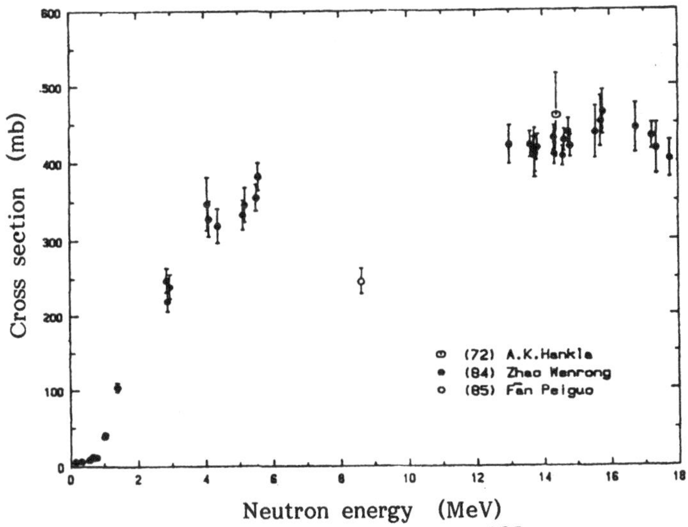

Fig.10 Cross section of the $Pt(n,x)^{195m}Pt$ reaction

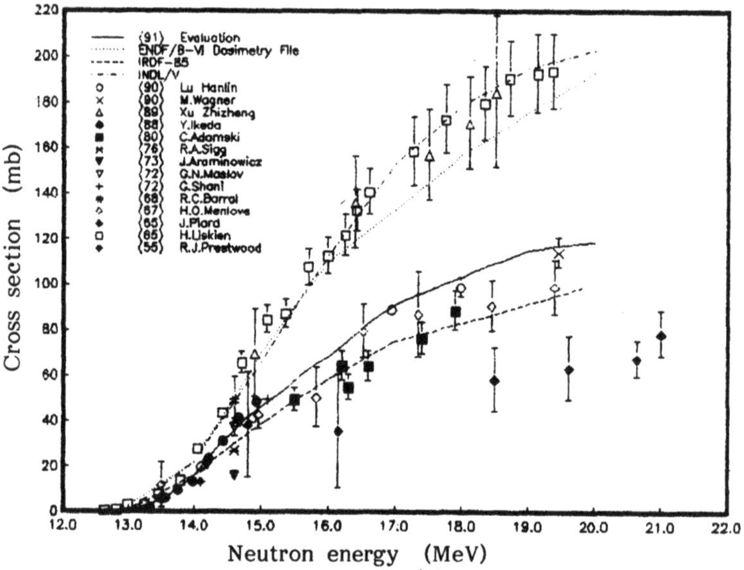

Fig.11 Cross section of the $^{23}Na(n,2n)^{22}Na$ reaction

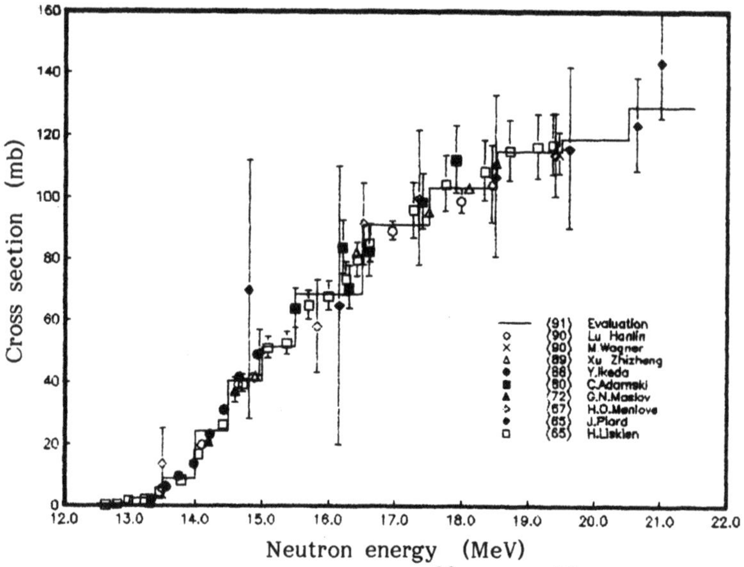

Fig.12 Cross section of the $^{23}Na(n,2n)^{22}Na$ reaction

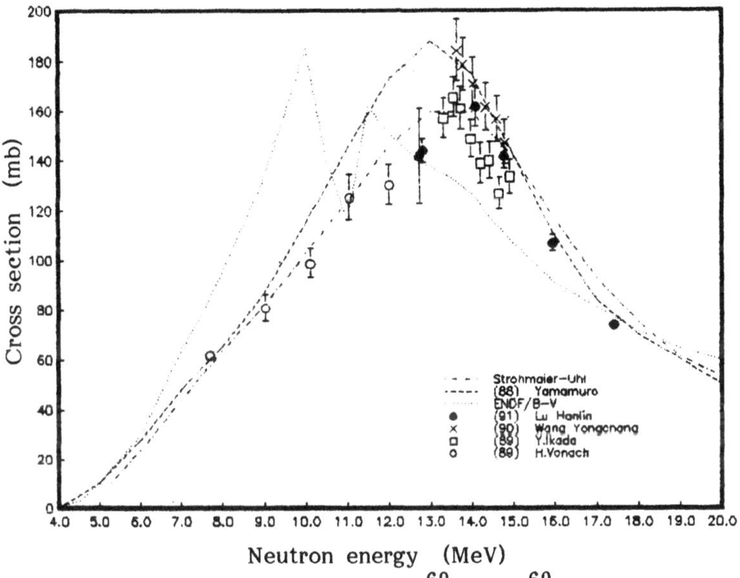

Fig.13 Cross section of the $^{60}Ni(n,p)^{60}Co$ reaction

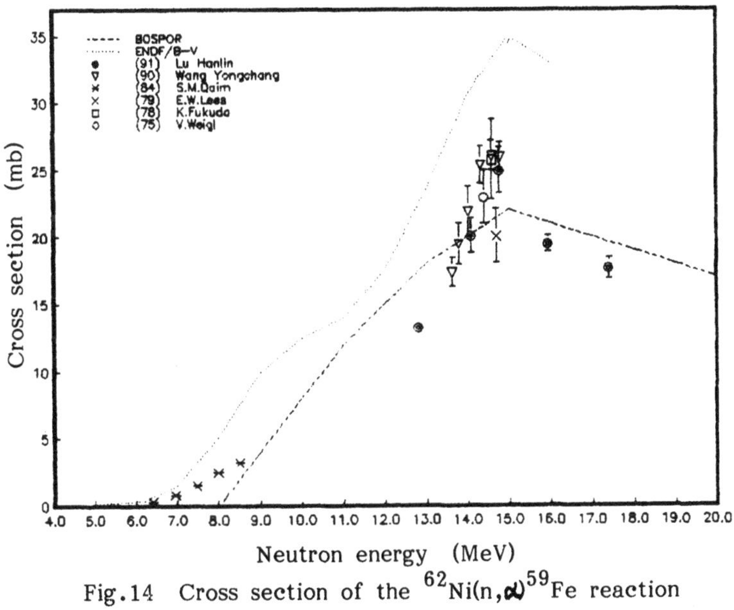

Fig.14 Cross section of the $^{62}Ni(n,\alpha)^{59}Fe$ reaction

NUCLEAR REACTIONS

NUCLEAR STUDIES AT TUNL USING POLARIZED NEUTRON BEAMS*

Richard L. Walter, Calvin R. Howell and Werner Tornow
Triangle Universities Nuclear Laboratory, Duke Station, Durham, NC 27706, USA

ABSTRACT

Experimental data obtained using polarized neutron beams has proven to be essential for determining the nucleon-nucleon and the nucleon-nucleus interaction. The present paper reviews the experimental methods and some results of the Triangle Universities Nuclear Laboratory for a variety of polarization experiments involving neutron elastic scattering. A brief introduction to the nucleon-nucleon problem and its relation to the three-nucleon problem is presented; data for n-p and n-d analyzing powers are highlighted. Measurements involving heavier targets (^{93}Nb and ^{208}Pb) and their connection to the development of conventional and dispersive optical models are shown. The importance of the dispersive model for ^{27}Al in relation to conclusions about the nucleon-nucleus spin-spin potential is presented. Comparisons of microscopic models to data for ^{10}B and ^{28}Si are described.

1. Introduction

The ultimate goal of nuclear physics is to describe nuclear reactions in terms of fundamental interactions. Conceptually, one hopes to do this in a QCD description involving quarks and gluons; however, this fundamental approach is far from being solvable quantitatively. In the meantime, theorists continue to develop models based on phenomenological nucleon-nucleon (NN) interactions that are described in terms of meson-exchange or merely in terms of NN phase shifts. Considerable understanding of meson-based NN potentials has been achieved in the last ten years, as evidenced by the successes of predictions based on the potential originally developed by the Bonn group[1] and then successively by Machleidt[2] who was a member of this group. The theoretical

*Work partially supported by the USDOE Grant # DEFG05-91-ER40619

group at Bochum[3] has taken advantage of the speed and size of newer computer systems and used these meson-based models to calculate rigorously for the first time scattering observables for three-nucleon systems. Since two- and three-nucleon systems are closely connected, the first section of the paper combines common details of these topics. Comparisons are made between analyzing power $A_y(\theta)$ data and theoretical predictions.

The next section of the paper introduces a typical analysis based on a phenomenological spherical optical model (SOM) of the conventional type. The recent analysis for ^{93}Nb(n,n) is chosen for this example. Next the development intensely pursued by Mahaux and co-workers[4] over the last eight years and picked up by others over the last few years is touched upon. This breakthrough is the insertion of the dispersion relation into the SOM to give the so-called dispersive optical model (DOM). Three cases are considered here, ^{93}Nb and ^{208}Pb, as well as ^{27}Al, where the focus is on the interpretation of the nucleon-nucleus spin-spin interaction. Lastly, the results of some of our microscopic calculations that use effective NN interactions obtained by Jeukenne, Lejeune and Mahaux[5] (JLM) and Yamaguchi and co-workers[6] (YNM) are compared to our data. Although this latter project, which is being carried out in collaboration with L. Hansen and F. Dietrich of LLNL, involves about 20 nuclei ranging from ^{6}Li to ^{208}Pb, we will only present results from two fairly light nuclei, ^{10}B and ^{28}Si.

2. Experimental Methods

At TUNL the source of polarized neutrons is the ^{2}H(d,n)^{3}He reaction induced by vector polarized deuterons. For n-p and n-d scattering, a direct current beam can be used because the hydrogenous scatterer also serves as a recoil particle detector, i.e., it is a scintillator. For experiments involving heavier targets or for ^{2}H(n,np)n deuteron breakup experiments, a pulsed beam with time-of-flight neutron detection is employed. All of the data presented in this paper were obtained with a Lamb-shift polarized-ion source which was constructed at TUNL in 1970. During the last two years we have been installing an intense atomic-beam polarized-ion source capable of delivering 5 μA of negative ions for injection into our tandem; this beam is about ten times more intense than that from our previous source. This intensity will allow us the luxury for shielding the neutron source reaction to obtain a collimated beam of neutrons, thereby producing a cleaner environment for an array of neutron detectors.

Figures 1 and 2 show two scattering arrangements. Figure 1 is a simple open-geometry plan that can be used with d.c. beams and scintillating scatterers. Much of the n-p and n-d elastic scattering data were obtained with this type of setup. The flight time

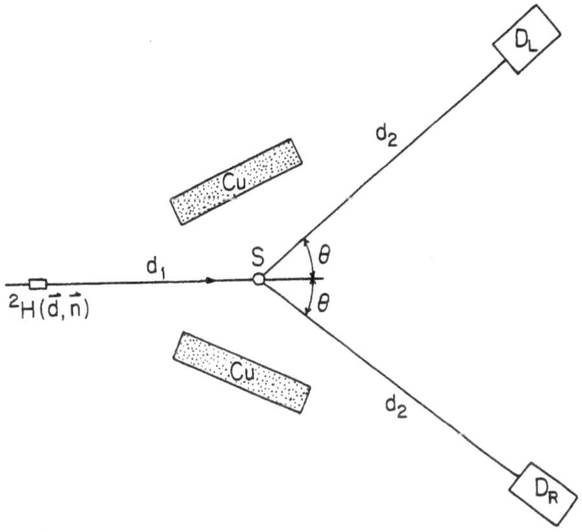

Fig. 1
Open-geometry setup
for $A_y(\theta)$ measurements.

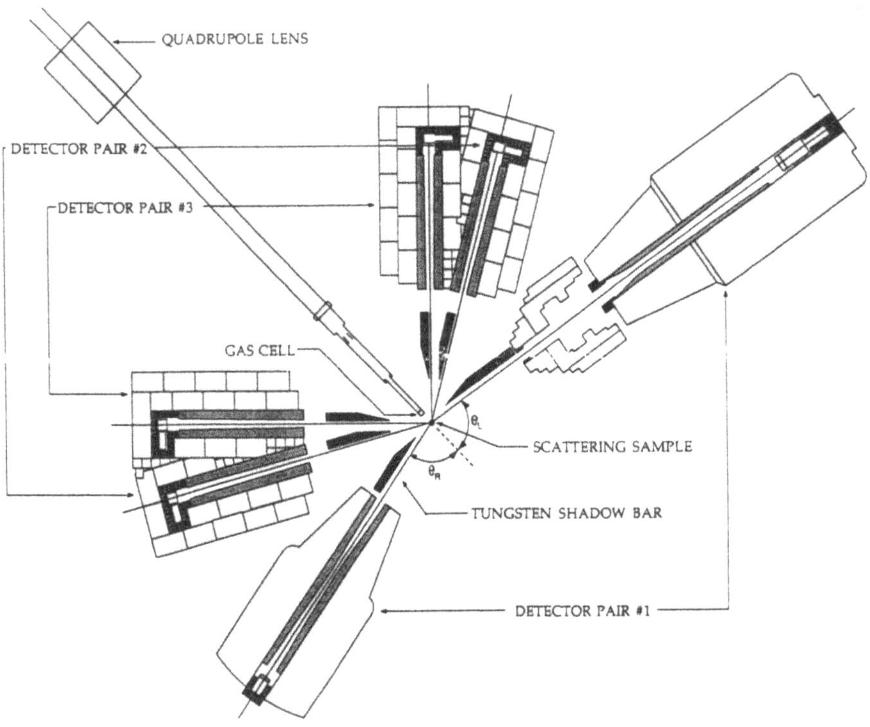

Fig. 2. Multi-detector array for $A_y(\theta)$ measurements using time-of-flight.

between events in the scatterer and each side detector allows one to determine the scattered neutron energy. The pulse height for the event in the center detector gives the recoil energy. This overdetermination leads to very clean spectrometry. Figure 2 is the normal shielded detector arrangement for time-of-flight spectrometry of the scattered neutrons; analyzing powers can be determined for both elastic and inelastic scattering. Although the neutron-beam polarization direction can be flipped by rotating the deuteron beam polarization by 180° at the ion source, usually a pair of detectors are employed at equal scattering angles on opposite sides of the incident neutron beam axis in order to further diminish instrumental asymmetries. In the arrangement we used for the ^{208}Pb experiment, the three pairs of shielded detectors in Fig. 2 were employed. To give an impression of scale, the flight paths ranged from 4 to 6 m. We have constructed a third neutron source, a shielded source, in which the neutron production target (the $^2H(d,n)$ reaction) is buried inside a 10 m^3 block of iron, copper, concrete, paraffin and lead. A narrow aperture allows a collimated beam of neutrons emitted from the reaction at 0° to pass through to a scatterer located 1.5 m from the "monoenergetic" source. The neutron detectors are 0.5-2.0 m from the scatterer. The advantage of the shielded source is that the complication of inscattering from the shadow bars is eliminated, thereby allowing for a large array of detectors for simultaneous measurements; for our next generation of $A_y(\theta)$ measurements for n-p and n-d scattering, we will use six pairs of detectors.

3. The Nucleon-Nucleon Interaction: n-p and n-d Elastic Scattering

Although the solution has seemed to be just around the corner for nearly two decades, the nucleon-nucleon (NN) interaction at low energies continues to be a puzzle. As one tries to explain NN and nucleon-deuteron (Nd) scattering data, and the ^3H and ^3He binding energies using various NN potentials, new questions need to be addressed. One recent concern is charge independence breaking (CIB) in the NN P-waves; i.e., how large is the difference between the n-p and p-p force for the various P-wave components? Another is the size of the $^3S_1 - {}^3D_1$ mixing parameter ε_1, which is a direct measure of the NN tensor force. Tied tightly to full descriptions of three-nucleon observables in terms of the NN interaction are the size and nature of the three-nucleon force.

In general polarization measurements have a large part to play in solving these fundamental questions. In particular, *neutron* polarization data has a special role. First, only through n-p measurements can one fully verify the size of CIB. Indeed, only in the last few years have adequate n-p data been available to even attempt to determine the n-p interaction without reliance on p-p data and the assumption of CIB in the isospin T = 1

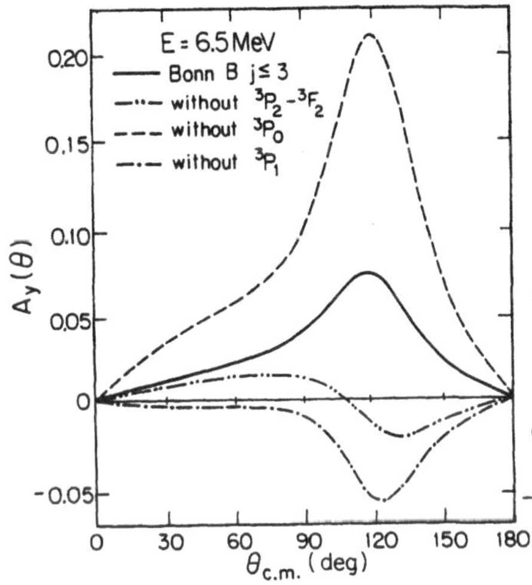

Fig. 3. Sensitivity of $A_y(\theta)$ for $^2H(n,n)$ to nucleon-nucleon P-waves.

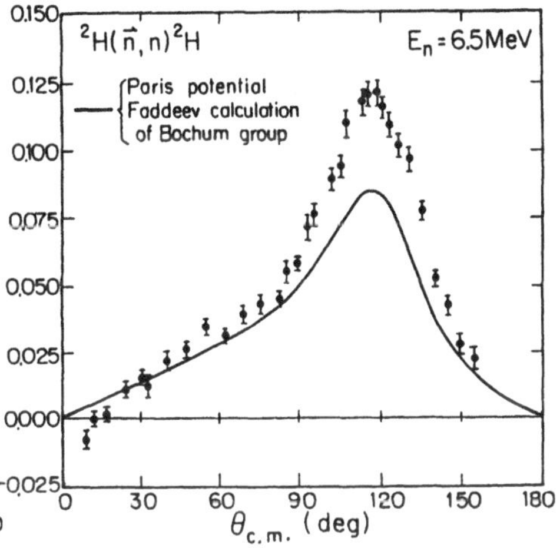

Fig. 4. $^2H(n,n)$ analyzing power and rigorous calculation.

phase shifts for $l > 0$. In fact, a recent phase-shift analysis[7] that analyzed the 0-30 MeV n-p and p-p data separately found surprisingly large CIB in the triplet P-waves. From a meson interaction viewpoint, their findings indicate a large difference between the π^0NN and $\pi^{\pm}NN$ coupling constants G_0^2 and G_c^2. We have shown in ref. [8] that $A_y(\theta)$ for n-d scattering at low energies is particularly sensitive to the size of the NN triplet P-waves. This is illustrated in Fig. 3 where the effect of turning off each ^3P wave in the Bonn B potential of Machleidt[2] is shown for $E_n = 6.5$ MeV. The calculation with the full potential is indicated by the solid curve.

A second notable feature in regard to neutron experiments is that, although 3-nucleon calculations have progressed a long way in recent years with the advent of powerful computers, the long-range Coulomb force acting in p-d scattering has inhibited theoreticians from conducting exact calculations of scattering observables above the deuteron breakup threshold for the p-d system, a system for which much data exists. Thus, one finds that the newly available rigorous calculations[3] that utilize meson-based potentials, e.g., the Paris[9] or Bonn B[2] potential, have been completed only for n-d scattering.

How well do the n-d $A_y(\theta)$ data compare to these rigorous calculations? First, we note that the measurements have advanced to a fairly high level of *absolute* accuracy, e.g., to better than ±0.003. This is sufficient to provide stringent tests of calculations, as

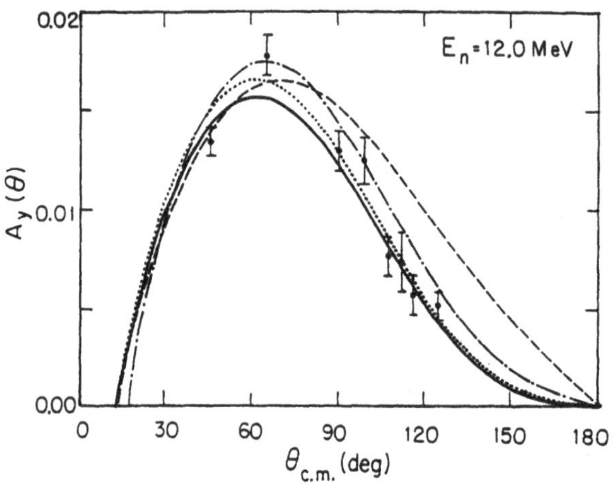

Fig. 5.
Analyzing power for $^1H(n,n)$ compared to available potential or phase-shift models.

can be seen in Fig. 4. The calculations shown here, which were made by our Bochum collaborators, clearly indicate that the familiar Paris NN potential[9] is insufficient to describe our data. Similar systematic differences were also seen in our recent measurements at 5.0 and 8.5 MeV. The difference between the prediction and the data might be attributed to CIB, but also could be evidence for 3-nucleon forces, which have not been included yet in the calculations. But even before one can consider such conclusions, it is important to test our current knowledge of the triplet P-wave NN phase shifts.

Thus, we return to n-p scattering. In this case the $A_y(\theta)$ has been shown[10] to be the most sensitive NN observable to P-wave variations. This is true, even though the magnitude of $A_y(\theta)$ is about an order of magnitude smaller for n-p than for n-d. We have recently completed measurements at five energies between 7.6 and 18.5 MeV. Typical overall uncertainties are less than ±0.0015 (absolute). The data for 12 MeV are compared in Fig. 5 to calculations based on the Bonn B (solid curve) and Paris (dotted) potentials and recent phase shift analyses by the Nijmegen group[7] (dot-dashed) and Arndt (0-350 MeV, VL35)[11] (dashed). The data favor the Paris and Bonn B predictions and definitely rule out VL35. As can be seen, in the case of n-p scattering the Bonn B potential (that was used in the n-d calculations for Fig. 5 above) appears to provide an accurate description of the P-waves in this energy range, as most of the magnitude of $A_y(\theta)$ is produced by P-waves. However, some other solution, one which gives a better prediction of the n-d data above, might produce similar agreement with the n-p data. As indicated in section 2, the use of our shielded source will allow for higher accuracy experiments. Our plan is to use that setup to repeat the n-p and n-d measurements in order to impose even more stringent tests on the NN interaction and 3-nucleon forces.

Lastly, experiments[12] underway at TUNL are determining the ε_1 mixing parameter. Precision measurements, which entail transmission of polarized neutrons through cryogenically cooled, polarized titanium-hydride, will be conducted from about 2 to 10 MeV.

4. Conventional Spherical Optical Model Studies

We have completed $A_y(\theta)$ measurements and usually also $\sigma(\theta)$ measurements for 20 isotopes ranging from ^{6}Li to ^{209}Bi. Our energy range of interest has been 8-17 MeV, the region above which compound-nucleus contributions to the elastic scattering typically are negligible. In all cases we have made model analyses of our data in conjunction with published data up to about 40 MeV for $\sigma(\theta)$ and total cross section σ_T. In many cases we studied the isovector dependence by including (p,p) data. Although the analyses in many cases were made with deformed potentials (i.e., the coupled-channels approach) and involved $A_y(\theta)$ data for inelastic scattering as well, the focus of the present paper is elastic scattering. Therefore, in this section we will concern ourselves with the spherical OM. We will show results for ^{93}Nb (n,n), one of our most recent studies.

Our potentials have the standard form: Woods-Saxon (WS) and derivative WS. The strengths usually depend linearly upon energy and the geometry parameters are held constant. Since our focus here is on polarization data, i.e., $A_y(\theta)$ data, special mention is made about the spin-orbit (s.o.) interaction. We find that V_{SO} = 5.5-6.5 MeV, r_{SO} =1.0-1.1 fm and a_{SO} = 0.5-0.6 fm. Although the $\sigma(\theta)$ distributions are almost perfectly described the $A_y(\theta)$ fits are not as successful. Thus, some fluidity in the s.o. parameters remains. In many cases chi-square searches on the complete data set for a specific isotope result in the desire for an imaginary s.o. potential with the same sign as the real s.o. term and a magnitude of about 13% of V_{SO}. Typically, in the current searches on nuclei with A>40, we will fix the energy dependence of both V_{SO} and W_{SO} to have the slope of that in the SOM global model of Walter and Guss[13]: $\alpha_{V_{so}}$ = -0.015 and $\alpha_{W_{so}}$ = -0.018. The introduction of W_{SO} has the effect of lowering the absolute value of $A_y(\theta)$ over almost the entire distribution.[14] We find that this effect reduces chi-square and cannot be simulated by changing any other parameters or combination thereof; it seems to be a unique feature of the W_{SO} term.

In Fig. 6 comparisons are shown for the SOM fits for ^{93}Nb(n,n) to the TUNL $\sigma(\theta)$ and $A_y(\theta)$ data. The $\sigma_T(E)$ data from 8-20 MeV was also included in the SOM search. In fact, the model describes σ_T to within 2% from 2 to 20 MeV. The parameters of the models (see ref. [14]) are in the standard range for SOM's. The accuracy of the $A_y(\theta)$ data are typical of our work prior to our recent studies on ^{208}Pb. One calculation in Fig.

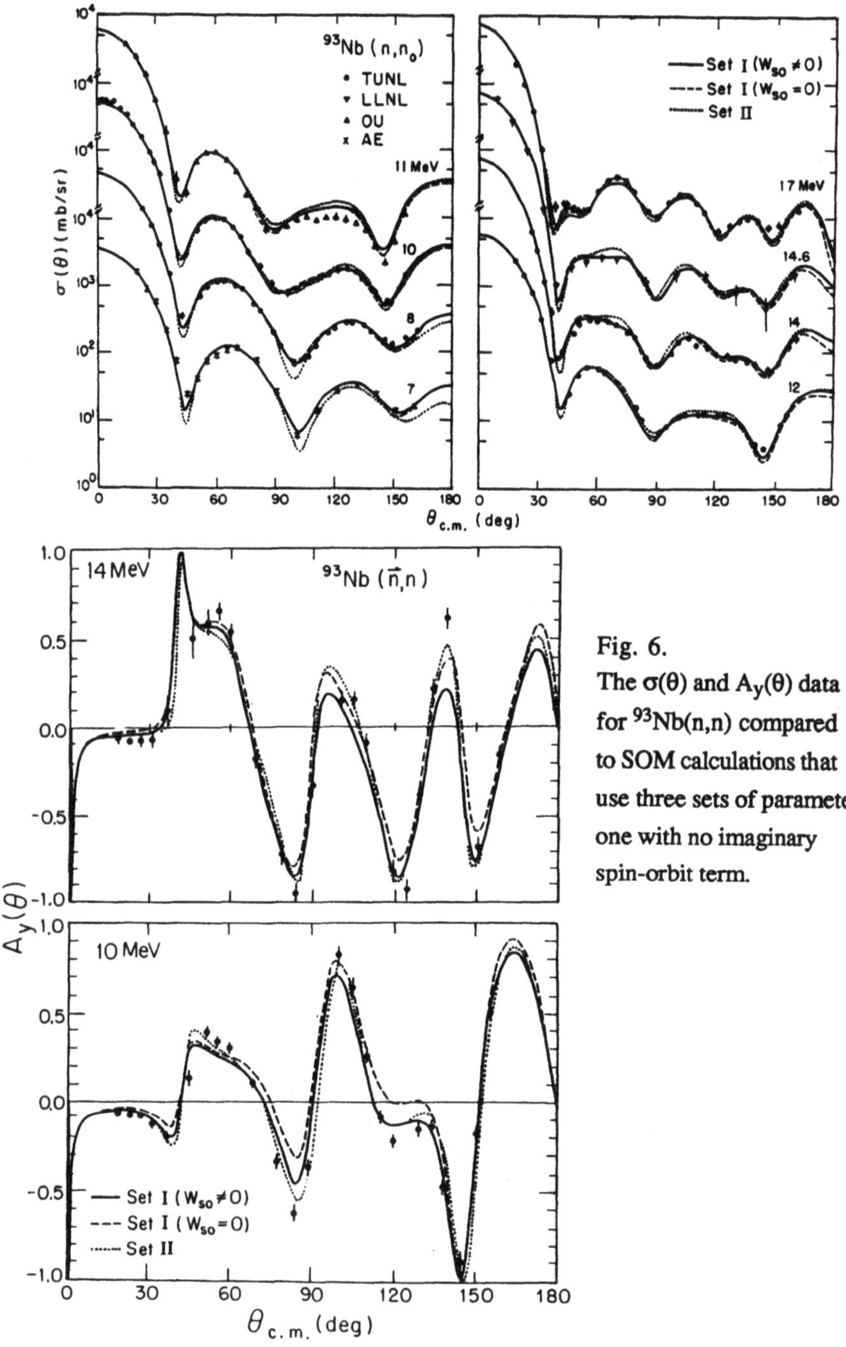

Fig. 6.
The $\sigma(\theta)$ and $A_y(\theta)$ data for ^{93}Nb(n,n) compared to SOM calculations that use three sets of parameters, one with no imaginary spin-orbit term.

6 (long-dashed curves) shows the effect of turning off W_{SO}; note that setting $W_{SO} = 0$ raises $A_y(\theta)$ everywhere. Our SOM fails to give a good description to $A_y(\theta)$, particularly at backward angles. We have not begun to explore other form factors for $V_{SO}(r)$ in order to attempt to address this problem. In general, we have found that coupled-channels models also have difficulties with $A_y(\theta)$ at backward angles, so the omission of explicitly handling inelastic processes probably is not the source of the problem.

5. The Dispersive Optical Model

The dispersion relation (DR), which is a result of causality, had been ignored in low energy nuclear physics modeling for nucleon-nucleus scattering until recent years when it was recognized that the DR provided a way of obtaining the systematic properties of the nuclear mean field, both at positive energies (scattering) and negative energies (bound states). We write the central part of the mean field as M(r,E) = V(r,E) + iW(r,E). The DR connects the imaginary part of M to the real part. The real part is written as $V = V_{HF} + \Delta V$ where V_{HF} is the Hartree-Fock type of contribution and ΔV is the dispersive contribution. Mahaux has shown that the ΔV is obtained from the integral relation:

$$\Delta V(r,E) = \frac{1}{\pi} \int_{-\infty}^{\infty} W(r,E') \left[\frac{1}{E'-E} - \frac{1}{E'-E_F} \right] dE'.$$

Thus, the imaginary potential gives an E-dependent contributions to the real potentials. Here, W is comprised of a volume contribution W_V and a surface contribution W_D; thus, one breaks up the integral and obtains separate contributions: $\Delta V = \Delta V_V + \Delta V_D$, which have a volume and suface radial form, respectively.

In order to evaluate the integral, W must be defined over all energies, both positive and negative. Until recently, the W are taken to be symmetric about the Fermi energy E_F. Both V_{HF} and W_V are usually of the Woods-Saxon (WS) radial shape and W_D is a derivative WS shape. To simplify the dispersive optical model (DOM) treatment, it is convenient to assume V_{HF} and W_V have the same radius and diffuseness parameters, r_V and a_V. The W_D usually has a larger radius and this means that V no longer has a WS shape nor an energy-independent half radius, since W_D has a sizeable energy dependence, particularly near $E = E_F$.

Since the DOM is mathematically rigorous, technically it should be introduced into all SOM (and coupled-channels) analyses. (The complexity of this and the problem of defining W(r,E) for all energies makes the DOM of questionable value if one is just constructing a phenomenological model over a limited energy range for applied purposes.) The

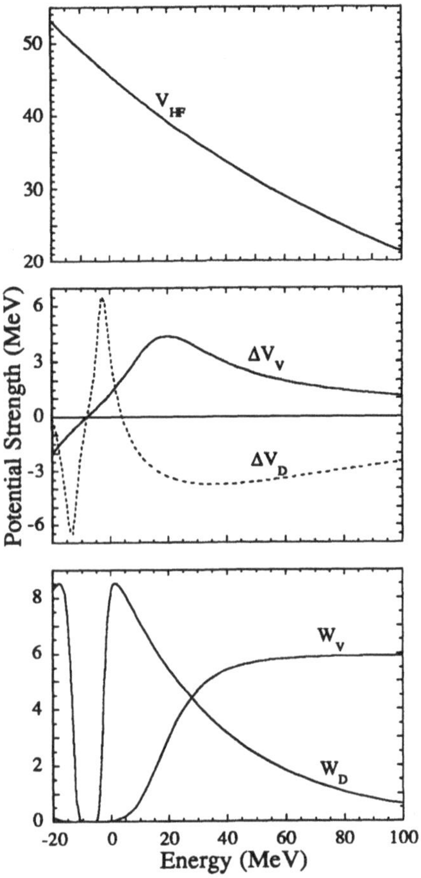

Fig. 7. Potentials of DOM for ^{93}Nb(n,n).

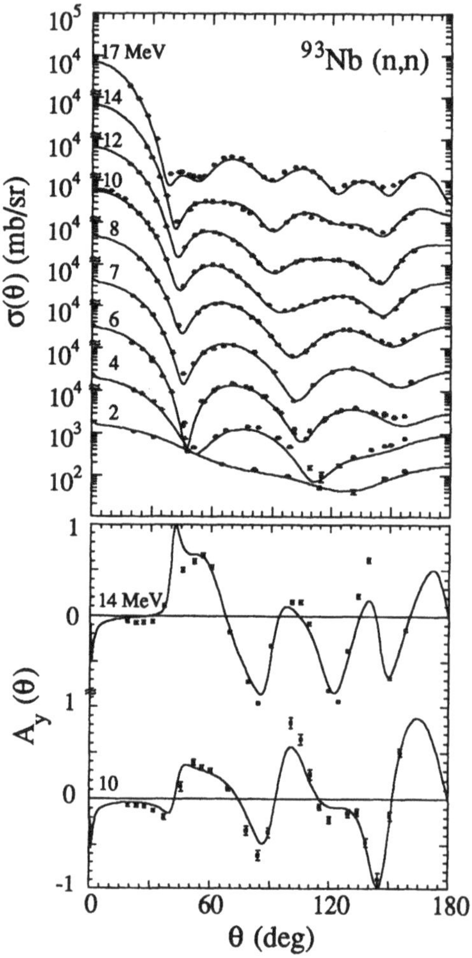

Fig. 8. Data and DOM calculations.

significance of the DOM, as it becomes more understood, is that it will be useful for connecting the wealth of scattering information at positive energies to the potential at negative energies from which bound state properties can be investigated: single-particle binding energies, spectral functions, occupation probabilities, and spectroscopic factors.

We have recently developed a global spherical-optical-model search code that has the DR explicitly introduced using the above formula. We take W_V and W_D to be symmetric functions and make them both go to zero near E_F. At present, one of the reactions that we are studying is n + ^{93}Nb; until now our model is based only on data for E>0, i.e., no

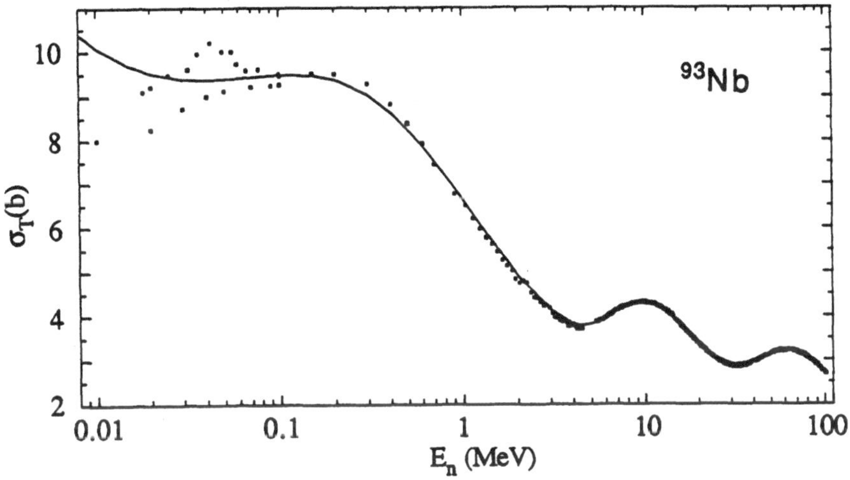

Fig. 9. Total cross-section data compared to DOM calculations.

bound state information was used to constrain the parameters. Data from TUNL and several other sources including ANL $\sigma(\theta)$ and σ_T measurements and evaluations[15,16] and the new σ_T data obtained at LANL by Finlay *et al.*[17] from 5-100 MeV were crucial in establishing the model parameters. In Fig. 7 we show the energy dependences of the various parts of the mean field potential. Nothing there is startling. The fits to the data are shown in Figs. 8 and 9. Here, the agreement with the $\sigma(\theta)$ and σ_T (over four orders of magnitude in energy) are truly beautiful! The σ_T data above 0.5 MeV are described to better than ±2%; below 0.5 MeV the data scatter quite a bit. On the other hand the $A_y(\theta)$ data again are not as well described. In this model we fixed the s.o. energy dependence according to ref. [13] and then obtained V_{SO} =(6.9 – 0.015 E) MeV, r_{SO} = 1.12 fm and a_{SO} = 0.59 fm from the global search. At this point we have not calculated the single particle properties yet. For ^{93}Nb this spectroscopic information may have to be derived from information for nearby nuclei.[16] In summary, the DOM works well for describing cross-section data for ^{93}Nb, but the $A_y(\theta)$ data indicate some weakness in the spin-dependent part of the model.

Earlier[18] we had performed several DOM analyses for ^{208}Pb, in which case there is a much larger $\sigma(\theta)$ and $A_y(\theta)$ data base. The first model was based on scattering and σ_T data from 2 to 80 MeV, while the second included bound-state information as well. At the time of this work the search code did not contain the DR explicitly, so the fitting procedure was relatively crude and slow. (Presently we are reanalyzing the ^{208}Pb data with the DOM automatic search code.) In the second model, a slight change in the magnitude and E-dependence of V_{HF} of the first model was all that was needed to give a quite good prediction for the location of the single-particle and single-hole states (see middle column in Fig. 10). About 30% of the $\sigma(\theta)$ and 50% of the $A_y(\theta)$ data bases are compared in Fig. 11 to the DOM fits. (Incidentally, some of our $A_y(\theta)$ data were obtained with the three detector pairs shown in Fig. 2; the data have the highest accuracy of any data that we have obtained for medium- to heavy-weight scatterers.) There is almost no perceptible difference in the fits for the two models. The σ_T data agrees to within about 2% between 2-40 MeV. Clearly, the DOM model works well for this nucleus. Recently, Mahaux and Sartor[19] compared additional single-particle features obtained with their DOM to observations from pickup and stripping reactions; their comparisons opened new questions about such nuclear spectroscopy, but also about the simplified assumption in DOM models of using $W_D(E)$ and $W_V(E)$ that are symmetric about the Fermi energy.

6. <u>Microscopic Optical Model Studies</u>

In collaboration with L. Hansen and F. Dietrich[20] we have investigated the successes and failures of a microscopic model that folds in the JLM[5] and YNM[6] effective interactions to see how well $\sigma(\theta)$ and, in particular, $A_y(\theta)$ are described for a wide range of nuclei. The energy range of concentration was typically 8-17 MeV. These models had not been well tested for $A_y(\theta)$ data for neutron scattering. Since a detailed description of the project is presented at this conference,[20] we will only show some of the calculations for two fairly light nuclei, ^{10}B and ^{28}Si. These are selected here to show the capability of the model at this end of the periodic table. (Heavier nuclei have been studied by others as well as us, with great success in most cases.) The results are compared to the TUNL data in Fig. 12. The JLM calculations use an effective interaction based on the Reid-hard-core NN interaction and use the M3Y spin-orbit potential, whereas the YNM uses the Hamada-Johnson NN potential, which contains a s.o. interaction. The two models predict quite similar values, although YNM appears to be more successful for ^{28}Si and JLM for ^{10}B. For each model there are three adjustable parameters to be optimized, the scale factors for the strengths; λ_V, λ_W, and λ_{SO}. For the light nuclei λ_W is quite energy dependent. The λ_{SO} for the ^{10}B calculations shown here are 1.4 and 1.6, respectively, for JLM and YNM, and for ^{28}Si are 1.3 for both models. The ^{28}Si findings compare well with our results for heavier nuclei where we obtained relatively good agreement with measurements when the s.o. strength also was multiplied by 1.3. In summary, we find

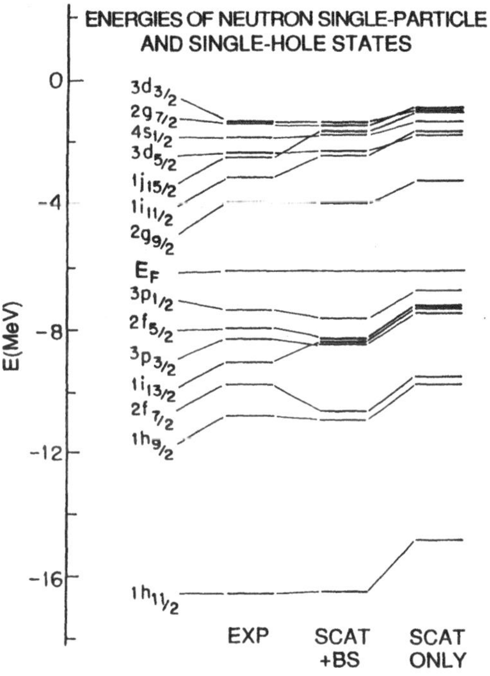

Fig. 10.
Energies of single-particle and single-hole states based on states in ^{209}Pb and ^{207}Pb, respectively.

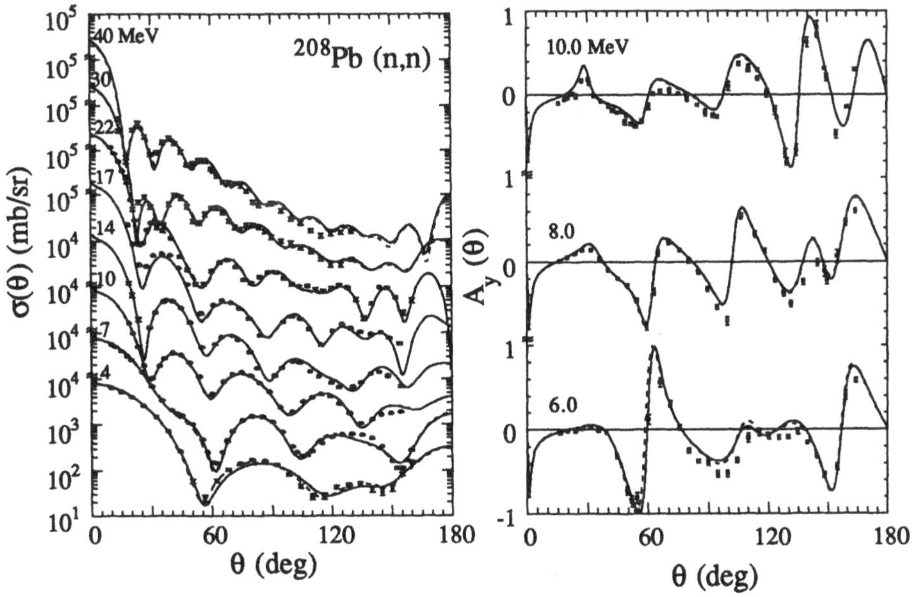

Fig. 11. DOM calculations compared to data for ^{208}Pb(n,n).

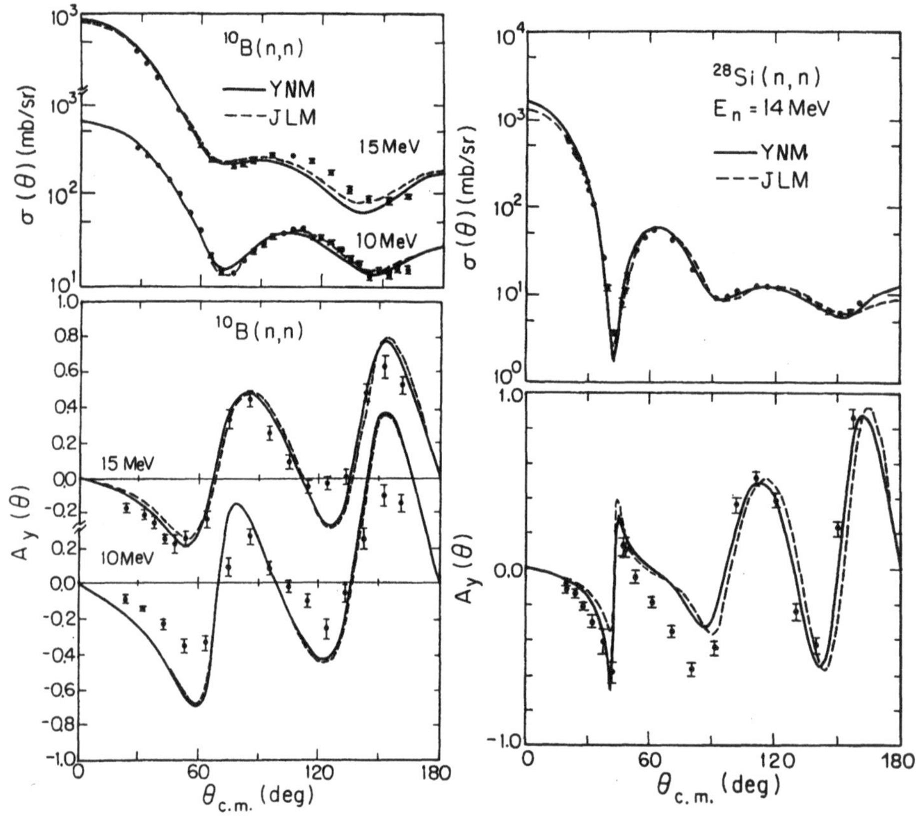

Fig. 12. Microscopic model calculations compared to data for ^{10}B(n,n) and ^{28}Si(n,n).

that the microscopic model gives a good description to both $\sigma(\theta)$ and $A_y(\theta)$ for heavy nuclei for the 8-17 MeV energy range and that it is reasonably successful for light nuclei for E $\gtrsim$ 12 MeV. However, one must allow for a smooth E-dependent scaling for the potential strengths.

7. The Phenomenological Spin-Spin Interaction

The final item is a brief mention of a project involving M. Nagadi and J. P. Delaroche.[21] We are developing phenomenological DOM's that describe $\sigma(\theta)$, $A_y(\theta)$, and $\sigma_T(E)$ for ^{27}Al and ^{59}Co, in order to provide OM's for energies around 2-30 MeV. We add to these spherical models a spin-spin interaction and compare the resulting calculations to so-called "spin-spin cross-section" data obtained in transmission

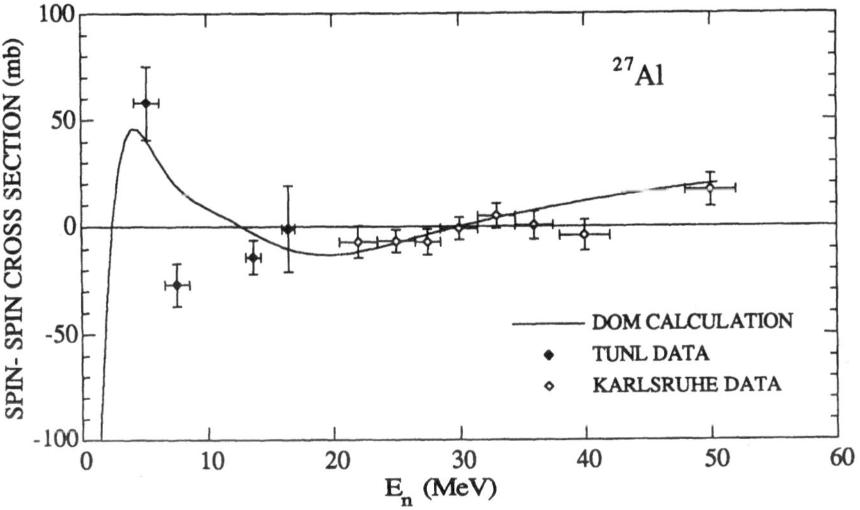

Fig. 13. The spin-spin cross-section data compared to DOM calculations.

measurements of polarized neutrons through polarized targets. (See references in [21] for details of this type of OM calculation and the data.) Contrary to the findings of Gould et al. who introduced a complex spin-spin potential, we obtain a qualitatively good description of the data with only a real spin-spin term. It has a strength $V_{SS} = 0.8$ MeV. The results of the calculation are compared to the data in Fig. 13 for ^{27}Al. Clearly, more data below 10 MeV would be informative.

8. Conclusion and Acknowledgement

We conclude that neutron polarization experiments are contributing in an important way to developing detailed but essential information about the nucleon-nucleon and the nucleon-nucleus interaction. All the investigations mentioned here are still being pursued actively, and probably will continue to be so into the long-term future. We acknowledge the assistance of numerous collaborators, graduate students, and fellows. We appreciate the many stimulating and enlightening discussions that we had with many colleagues.

References

[1] R.Machleidt, K.Holinde and C.Elster, Phys. Rep. 149 (1987) 1.

[2] R.Machleidt, Adv. Nucl. Phys. 19 (1989) 189.

[3] H.Witała, W.Glöckle and T.Cornelius, Few Body Systems Suppl. 2 (1987) 555.

[4] C.Mahaux and R.Sartor, Nucl. Phys. A528 (1991) 253 and references therein.

[5] P.Jeukenne, A.Lejeune and C.Mahaux, Phys. Rev. C16 (1977) 80.

[6] N.Yamaguchi, S.Nagata and J.Matsuda, Prog. Theor. Phys. 70 (1983) 459, 76 (1986) 1289.

[7] V.G.J.Stoks, P.C.van Campen, T.A.Rijken and J.J.de Swart, Phys. Rev. Lett. 61 (1988) 1702, see also T.A.Rijken et al., Nucl. Phys. A508 (1990) 173c and J.R.Bergervoet et al., Phys. Rev. C41 (1990) 1435.

[8] W.Tornow et al., Phys. Lett. B257 (1991) 273.

[9] M.Lacombe et al., Phys. Rev. C21 (1980) 861.

[10] W.Tornow and R.L.Walter, Few Body Systems 8 (1990) 11.

[11] R.A.Arndt (private communication).

[12] W.Tornow et al., in Physics with Polarized Beams on Polarized Targets, Indiana, 1989, (J. Sowinski and S.E. Vigdor, ed., World Scientific Publ. Co. 1990, p. 75)

[13] R.L.Walter and P.P.Guss, Nuclear Data for Basic and Applied Science, Proc. of the Int. Conf., Santa Fe, NM, 1985, P.G. Young et al., ed., (Gordon and Breach, NY, 1986) Vol. 2, p. 1079.

[14] R.S.Pedroni et al., Phys. Rev. C43 (1991) 2336.

[15] A.B.Smith and D.L.Smith, ANL/NDM-88 (1985) and references therein.

[16] A.B.Smith, P.T.Geunther, and R.D.Lawson, Nucl. Phys. A455 (1986) 344.

[17] R.W.Finlay et al., Nuclear Data for Science and Technology, Jülich, Germany (1991) (to be published).

[18] M.L.Roberts et al., (Phys. Rev. C, in press).

[19] C.Mahaux and R.Sartor, Nucl. Phys. A493 (1989) 157; A528 (1991) 253.

[20] L.F.Hansen, F.S.Dietrich and R.L.Walter (these proceedings).

[21] M. Nagadi et al., Bull. Am. Phys. Soc. 34 (1989) 1831.

DOUBLE DIFFERENTIAL NEUTRON EMISSION CROSS SECTIONS OF (p, n) REACTIONS ON ^{59}Co AND Mo ISOTOPES

ZUYING ZHOU, HONGQING TANG, QINGCHANG SUI,
BUJIA QI, CHENWEI ZHOU, ZUNJIAN KE,
GUANREN SHEN and HAIHONG XIA
China Institute of Atomic Energy (CIAE)
P. O. Box 275 (46), Beijing 102413, P. R. China

ABSTRACT

Double differential neutron emission cross sections for the (p, n) reactions on ^{59}Co and Mo isotopes in the proton energy of 9 to 15 MeV have been measured. The level density parameters were derived.

1. Introduction

At low excitation energy the nuclear level densities can be determined by direct counting of the number of levels and at about neutron binding energy from resonance data, but level missing could be caused due to the limitation of experimental energy resolution. Alternatively, nuclear level densities can be derived from energy differential particle emission cross sections in compound nucleus reactions. For this purpose (p, n) and (α, n) reactions are recommended since they have less non-compound contributions at excitation energy below neutron binding energy.

Double differential neutron emission cross sections for (p, n) reactions on ^{59}Co and Mo isotopes (^{96}Mo, ^{97}Mo ^{98}Mo and ^{100}Mo at proton energy of 9 MeV, ^{95}Mo at 9, 13 and 15 MeV and ^{59}Co at 9, 11, 13, 15 MeV) have been measured. The presented level density parameters were derived from comparing the calculations using the conventional evaporation model with the experimental data.

2. Experiment and Results

The experiments were carried out at the HI-13 tandem Van de Graaff accelerator in CIAE. A schematic layout of the experimental arrangement including the neutron time-of-flight setup, target chamber and the Faraday cup is shown in Fig. 1.

The neutron time-of-flight setup consists of three neutron detectors and pulsed beam system. Each detector is composed of a liquid scintillator (10.5 cm

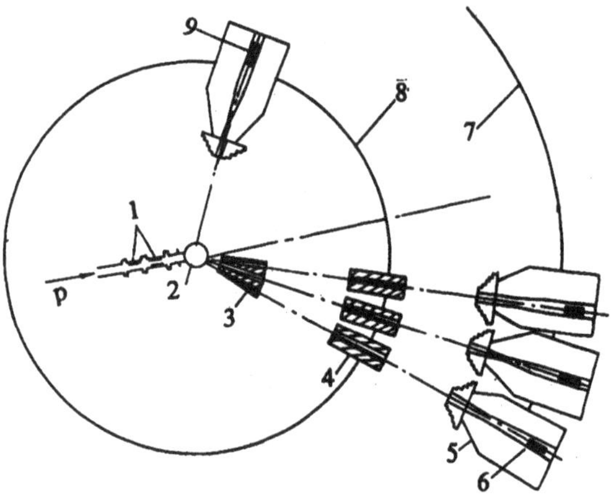

Fig. 1 Three-detector fast neutron TOF setup
1. Beam pick off 2. Target chamber 3. Collimator-1
4. Collimator-2 5. Shielding 6. Neutron detector
7. R = 6m rail 8. R=3m rail 9. Neutron monitor

diameter × 5 cm, filled with liquid scintillator ST-451) and a photomultiplier tube XP-2041. The detectors are massively shielded and well collimated to the target. The flight path is 5-7 m. The target chamber is an aluminium cylinder with a diameter of 44 cm, a height of 45 cm and a thin wall of 3 mm. The pulsed proton beam impinged on a self supporting target foil. A beam burst of about 1 ns width (FWHM) with repeatation of 2 MHz was used. The beam pick-off signal was taken out right in the upstream of the target chamber. The beam average current is about 100 nA. The detector bias was set at about 0.85 MeV neutron energy. To reduce γ-ray background an n-γ pulse shape discriminator modular (CANBERRA 2160A) was used. Detector efficiency was calibrated with n-p scattering experiments[1] and Monte-Carlo calculation. Three sets of standard electronics and a data acquisition system were used in the TOF spectrometers. Another TOF spectrometer with a shorter flight path of 3 m was used as a monitor.

For each incident energy and each target both neutron spectra with and without target were measured, the latter was used for background subtraction. The correction has been done considering the neutron attenuation through 3 mm aluminium of chamber wall for each angle (< 2%) and through target and target frame for 90 deg. (5%-11%).

According to the position of the target γ peak and time calibration of the system, after background subtraction, detector efficiency and neutron attenuation correction the TOF spectra were converted into the energy spectra with bin size 200 keV in the centre of mass system. The neutron energy spectra are shown in Fig. 2 and some angular distributions are given in Fig. 3. The angle integrated neutron energy spectra were obtained using Legendre polynomial fitting to the data.

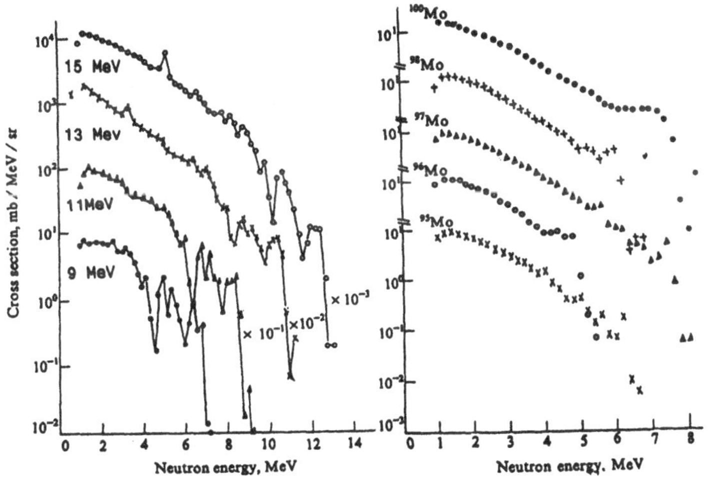

Fig. 2 Neutron emission spectra at 50 deg.
Left: $^{59}Co(p, n)^{59}Ni$
Right: Mo isotopes (p, n) reactions, E_p=9 MeV

Total uncertainties of 7 - 8.5 % for double differential cross sections come from the error sources of statistics, detector efficiency, beam current integration and target thickness, etc..

Neutrons with energies less than 1.4 MeV were not included in the spectra since the detector efficiency changed rapidly with energy near the bias.

3. Discussion

The isotropic angular distributions for compound nucleus reactions predominate in that interesting region. Neutron emission spectra from (p, n) reaction were calculated using the conventional evaporation model. The optical potential parameters of Perey[2] for proton and of Wilmore and Hodgson[3] for neutron were applied. The level density formula used in the calculation is in

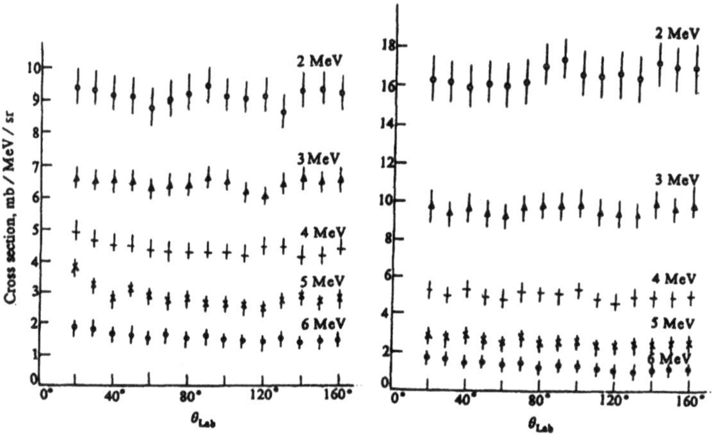

Fig. 3 Angular distribution
Left: $^{59}Co(p, n)^{59}Ni$, E_p = 15 MeV
Right: $^{95}Mo(p, n)^{95}Tc$, E_p = 15 MeV

Table 1 Level density parameter a and δ

nucleus	^{59}Ni	^{95}Tc	^{96}Tc	^{97}Tc	^{98}Tc	^{100}Tc
a $(MeV)^{-1}$	6.3	10.7	11.0	11.2	11.2	11.8
δ (MeV)	1.00	0.60	-0.45	0.97	-0.15	-0.15

Gilbert-Cameron form with the following new statistics of Su Zongdi et al.[4], their pairing and shell correction values as well

$$U_x = 1.4 + 263A$$

$$a = [0.00880(S(z) + S(n)) + QB]A$$

where QB is taken as:

$$QB = \begin{cases} 0.142, & \text{for } 29 \le z \le 62 \text{ (and } n \le 89); \ 79 \le z \le 85 . \\ 0.120, & \text{for } z < 28; \ 62 \text{ (and } n > 89) \le z \le 78; \ z \ge 86. \end{cases}$$

The comparison of the calculations with the experimental data is shown in Fig. 4.

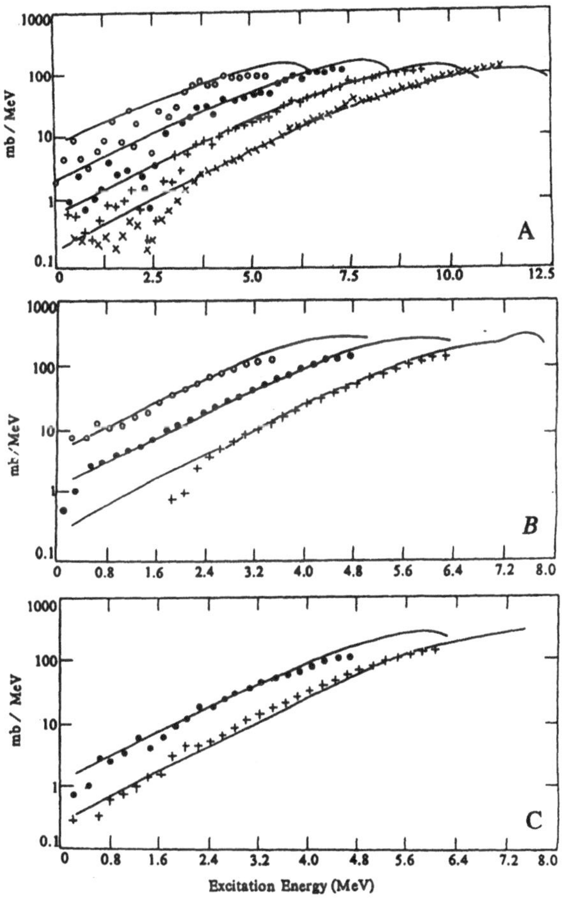

Fig. 4 Comparison of experimental neutron spectra and calculation
(solid lines represent calculation results)

A: For $^{59}Co(p, n)^{59}Ni$,
∘: $E_p = 9$ MeV •: $E_p = 11$ MeV
+: $E_p = 13$ MeV ×: $E_p = 15$ Mev

B: ∘ for $^{96}Mo(p, n)^{96}Tc$, • for $^{98}Mo(p, n)^{98}Tc$ and + for $^{100}Mo(p, n)^{100}Tc$

C: • for $^{95}Mo(p, n)^{95}Tc$ and + for $^{97}Mo(p,n)^{97}Tc$

The level density parameters "a" which were determined from the optimal description for experimental spectra and are given in table 1. To make the experimental spectra contain as low non-compound component as possible the data shown in the Fig. 4 are angle integrated based on those in backward hemisphere. The cross sections for the competition channels from proton induced reaction on Mo isotopes are much lower as compared with those of (p, n) reactions. If changes of transmission coefficients can be ignored, the cross sections of (p, n) reactions on Mo isotopes will be determined by level densities of the residual nuclei. However in the case of ^{59}Co, the cross sections of (p, p') process are comparable with those of (p, n) reaction and there may be some ambiguity in the analysis of level density of ^{59}Ni.

Acknowledgement

The authors are grateful to Professor W. Scoble for his help with Mo isotopic targets.

References

1. Shen Guanren, Sa Jun, Huang Tangzi, Yu Chunying, Tang Hongqing and Sui Qingchang, *Chinese Atomic Energy Science and Technology*, **211** (1987) 9
2. F. G. Perey and C. M. Perey, *Phys. Rev.* **1311** (1963) 745
3. D. Wilmore and P. E. Hodgson, *Nucl. Phys.* **551** (1964) 673
4. Su Zongdi, Wang Cuilan, Zhuang Youxiang and Zhou Chunmei, *INDC(CPR)-2* (1985)

NEUTRON SHELL EFFECT IN THE (n, 2n) CROSS SECTIONS AT 14.6 MeV

K. Z. XIE, S. M. ZHOU, Q. Q. YAN and X. S. CHEN
Shanghai Institute of Nuclear Research,
Academia Sinica,
P. O. Box 800-204, Shanghai 201800, China

and

M. Y. ZHOU and C. S. WANG
Shanghai University of Science and Technology,
Shanghai 201800, China

ABSTRACT

Neutron activation technique was used to measure the cross sections of 14.8 MeV (n, 2n) reactions of even-even target nuclei. The absolute values were calculated by using an integrated pre-equilibrium statistical model. The result gives a good fit to the experimental data and shows the existence of shell effect on the cross section. Problems related to this effect is discussed

1. Introduction

A large number of publications have appeared which deal with the measurement of cross sections of 14 MeV neutron reactions. Most of them involved investigations of (n, 2n) cross sections. In the study of fission and fusion reactions, (n, 2n) reactions are of great importance. Their reaction cross sections are essential for research of reactor physics.

For many years there has been a controversy as to whether there exist shell effects on (n, 2n) cross sections at 14 to 15 MeV energy. The effect is manifested in maxima or minima in cross section values primarily in closed shell nuclides. Some authors claimed the existence of shell effects[3-6] while others demonstrated that there were no shell effects in measured cross sections[7-15]. The experimental cross sections deviate systematically from the statistical theory which implies an incomplete understanding of the mechanism of these reactions. The absolute values were calculated by using an integrated pre-equilibrium-statistical model. The result gives a good fit to the experimental data, and provides an evidence of shell effect.

2. Experiment

The measurement of reaction cross sections by use of neutron activation method involves the measurement of radioactivity of the product radionuclide by pre-determined fluence of neutrons of specific energy. The reaction cross section is then obtained from these values and the related reaction parameters. The equation for the cross section is the following:

$$\sigma = \frac{N_\gamma \cdot R^2}{M \cdot \varepsilon_\gamma \cdot f_s \cdot f_\alpha \cdot \frac{1}{1+\alpha} \cdot \sum_{i=1}^{l} \phi_i \, [1-\exp(-\lambda T_i)] \exp(-\lambda t_i)},$$

where N_γ is the characteristic gamma-ray total energy peak counting rate of the product radionuclide, R is the distance between the neutron source and the sample, M is the number of isotopic nuclei, ε_r is the total energy peak efficiency, f_s is the self-absorption correction factor for characteristic gamma-rays in the sample, f_α is the branching ratio of characteristic gamma-ray, α is the internal conversion factor, ϕ_i is the neutron fluence in the time interval i of the irradiation period (neutrons/s·sr), λ radioactivity decay constant, l is the number of time intervals (depending on the condition of the steadiness of neutron fluence), t_i is the time interval of irradiation, T_i is the time interval from the end time of t_i to the starting of measurement or the cooling time.

The sample was irradiated under the 14 MeV neutrons by T (d, n) ^{4}He reaction using 400 KV high voltage multiplier generator. The sample was placed at the position 5 cm from the tritium target and at an angle of 45° from the deuteron direction (parallel to the target plane). The incident deuteron energy was 200 KeV and the tritium titanium target thickness was 0.5mg/cm^2, the incident neutron energy was 14.6 ± 0.20 MeV, neutron source strength was 5×10^9 n/s.

For the measurement of gamma-ray radioactivity, a coaxial 100 cm^3 Ge(Li) detector and a ϕ7.6 cm × 7.6 cm NaI (Tl) crystal were used. The detection efficiency for total energy peak of the spectrometer was calibrated by $4\pi\ \beta-\gamma$ coincindence method, in the whole energy region, an accuracy of 3% was abtained. The stability of the detector efficiency was monitored with a standard ^{137}Cs source. The results indicated long period stability of the efficiency is within 1%. Corrections were made for the geometrical dependence of the detector efficiency and variation of gamma-ray self-absorption due to sample.

The associated particle method was used to measure neutron fluence. The detector for recording alpha particle is a gold surface barrier silicon detector placed at 90° with respect to the incident deuteron beam. Table 1 shows the sources of the errors in determination of the neutron fluence.

Table 1 The error in determination of the neutron fluence

Source of error	α statistic deviation	Dead time correction	Back-ground	Target drift	Solid angle error	Competing reaction	Anisotropy correction	Coulomb scat.	Total deviation
%	0.1	0.3	0.2	0.6	0.1	0.4	0.3	0.5	1

3. Experimental Rsults and Error Analysis

Activation method has been used to determine the reaction cross sections of even nuclei with N in the neighbourhood of 28 to 50. The cross sections measured and the source of errors are shown in the table 2 and table 3.

Table 2 Source of errors of absolute measurement

Source of error	Neutron fluence rate	γ detection efficiency and area of total energy peak	Sample self absorption correction	Sample geometry correction of γ-ray efficiency	Sample weight	Distance between target and sample	Count statistics	Target scattering and absorption	Total deviation
%	1.0	3.5	0.6	0.3	<0.5	1.0	1.3	0.5	4.2

Table 3 (n, 2n) reaction cross section of even proton target nuclide ($\times 10^{-31}$ m^2)

	^{46}Ti	^{50}Cr	^{54}Fe	^{58}Ni	^{64}Zn	^{70}Ge	^{76}Se	^{80}Se	^{86}Sr	^{88}Sr	^{96}Ru	^{100}Mo	^{102}Pd	^{110}Pd
Measurement	51	31	27	30	153	550	890	750	650	389	610	1229	1141	2510
	±2	±1	±1	±1	±6	±16	±27	±35	±21	±12	±20	±42	±50	±110
Calculation	26	6	0.2	190	248	605	743	833	851	442	965	1078	1081	1191

4. Theoretical Calculation

Since the introduction of compound nucleus model by N. Bohr et al. in 1936, equilibrium state statistical theory found successes in describing the physics of nuclear reactions. However, as the energy of the incident particles rises, statistic theory was found to be unable to explain the hard tails of the high energy regions and angular distribution showing forward peaking. In 1966 pre-equilibrium model was proposed by Griffin and some others. Their theories depicted that in addition to emission of particles from the compound nucleus after reaching equilibrium, there existed probability of particle emission from the nucleus not reaching the equilibrium state. The reaction mechanism then lies in between the direct interaction mechanism and the compound nucleus mechanism. Since the pre-equilibrium theory and the equilibrium state statistic theory are both capable to explain the nuclear reactions within their applicable regions, by interlacing the above two theories, one should acquire theoretically calculated results consistent with the experimental measurements. Calculation of (n, 2n) cross sections of 14 nuclides by using the work of A. Chattejee as reference was undertaken.

1. Cross Section Calculation

According to the work of Chattejee, the (n, 2n) cross section is expressed as a sum of a preequilibrium (PE) component and a statistical evaporation or quilibrium (EQ) component. The two parts are:

$$\sigma\,(n,\, 2n) = \sigma_\gamma\,[f_{PE}\,(n,\, 2n) + f_{EQ}\,(n,\, 2n)\,(1\text{-}\delta)] \cdot F_n,$$

where σ_γ is the cross section of formation of compound nucleus, f_{PE} (n, 2n) and f_{EQ} (n, 2n) are the pre-equilibrium and equilibrium fraction of emission of first neutron, δ sum of all pre-equilibrium fractions of probable exit channels induced by incident neutrons, F_n the probability of emission of second neutrons in (n, 2n) reactions. Here the only proton competing reaction is (n, np).

At the second stage of the reactions the decay is assumed to be purely statistical. For a pure equilibrium model f_{PE} as well as δ are assumed to be zero.

2. Preequilibrium Decay

The equation for pre-equilibrium fraction was given by Gadioli originated from the excition model of Griffin-William.

$$\frac{df_{PE}}{d\varepsilon} = \frac{(2s+1)m\varepsilon\sigma_{inv}(\varepsilon)}{4\pi^3\hbar^3|M|^2g^4E^3} \sum_{\substack{t-1 \\ (\Delta t = 2)}}^{\bar{t}} \left(\frac{U}{E}\right)^{t-2}(t+1)^2(t-1)\left(\frac{t+k}{t}\right),$$

where m is the reduced mass of neutron, E excitation energy of compound nucleus, k constant which is 1 (or -1) for neutron (or proton) emission, t number of excitons for equilibrium state, g average single particle level density of the Fermi gas model. $|M|^2$ is expressed in simple equation

$$|M|^2g^4 = \alpha A,$$

where α is an adjustable parameter. In our work $\alpha = 1.889 \times 10^{-4}\ \mathrm{MeV}^{-2}$

3. Equilibrium Decay

The nucleus after reaching exciton equilibrium state will undergo equilibrium decay. This process can be represented by the following evaporation model:

$$f_{EQ} = \frac{\int_0^{E_n - S_{n2}} \varepsilon \sigma_{inv}^{(n)}(\varepsilon) \rho^{(n)}(E_n - \varepsilon) d\varepsilon}{\int_0^{E_n} \varepsilon \sigma_{inv}^{(n)}(\varepsilon) \rho^{(n)}(E_n - \varepsilon) d\varepsilon + \int_0^{E_n + Q(np)} \varepsilon \sigma_{inv}^{(p)}(\varepsilon) \rho^{(p)}(E_n + Q(n,p) - \varepsilon) d\varepsilon},$$

where $\sigma_{inv}^{(n)}(\varepsilon)$ and $\sigma_{inv}^{(p)}(\varepsilon)$ are (n, n') and (n, p) inversion reaction cross section respectively. $\rho(n)$ and $\rho(p)$ are the corresponding residual nucleus level densities.

After emission of first neutron, the nucleus might remain at the excited state, deexcitation by second emission can occur. Because the excitation energy is rather low, therefore in our calculation, only equilibrium state emission is considered. F_n is given as

$$\frac{1-F_n}{F_n} = \frac{\mu_p}{\mu_n} \cdot \frac{1-(\mu_p+1)e^{-\mu_p}}{1-\gamma_n-(\mu_n+\gamma_n+1)e^{-\mu_p}} \cdot \exp(\Delta\delta+\Delta e_0-\gamma_p),$$

where $T=(0.0165A)-1$, $\rho(E)=\frac{1}{T}\times\exp[(E-E_o)/T]$, $\gamma_n=\nu_n/T\mu_n$, $\mu_n=(E_n-S_{n\gamma}-E_{on})/T$, $\mu_p = (E_n - S_{p\gamma} - E_{op})/T$, $\Delta S = (S_{n\gamma} - S_{p\gamma})/T$, $\Delta e_0 = (E_{on} - E_{op})/T$, $S_{n\gamma}$ and $S_{p\gamma}$ are the binding energies of the last neutron and the last proton in the target nucleus respectively.

The values of E_{vi} (i = n or p) have been approximated as $E_v = 6\Delta/\sqrt{A}$ MeV with $\Delta = \pm 1, 0$ for even-even, odd-odd, and odd mass residual nuclei.

In our calculation, α value used is the same for different nuclide. In the papers published, variable values of α or functional relations were used for cross section calculation. In addition for nuclide of $A < 60$, the emission of (n, α) must not be overlooked. In our calculation, on account of emission of (n, α) competition, results showed greater discrepancy for some nuclides. If KRK equation for level density is used for correction of shell effect in the calculation, the results can be improved.

The calculated reaction cross section of ^{58}Ni needs further explanation. It is 6 times larger than the experimental result. It is believed that because the structure of this nuclide is very special, although it belongs to light nuclei, its shape is distorted by use of distorted wave Born approximation. It was discovered that the direct reaction fraction of the (n, 2n) cross section is quite big. William[17] used statistic model with correction of direct reaction and pre-equilibrium and got a result of 3.0×20^{-30} m^2 for σ(n, 2n)which is in consistence with experiment.

The cross sections do exhibit a characteristic systematic behaviour. On the average, the values increase with increasing mass number A, but in more detail the cross sections vary with maximum and minima which are closely connected with the magic neutron numbers. For even-proton nuclei with neutron numbers around N=28 (A ≈ 54), there appears to be a minimum. A second minimum appears for even-proton nuclei around neutron number N = 50 (A≈90).

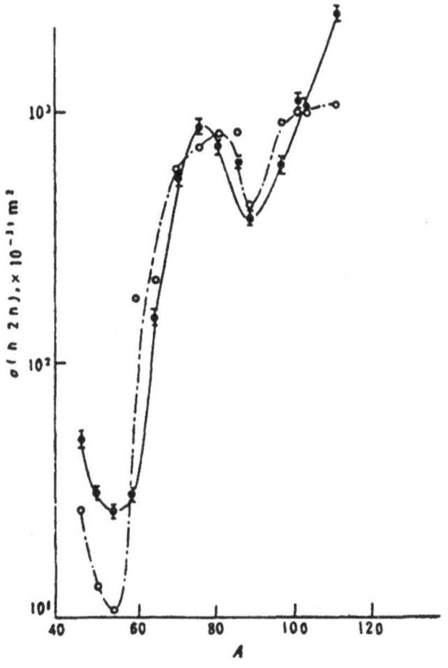

Fig. 1 (n, 2n) cross sections at 14 MeV versus mass number A for even-proton nuclei
∘ Calculated
• Experimental

In the region of the magic neutron number N = 28 (A ≈ 54) the thresholds for the (n, 2n) reaction lie only slightly below 14 MeV, so that the minimun of the (n, 2n) cross sections in this region can be understood as a Q value effect.

Around N=50(A≈90) the Q values go through a minimum and this is probably the reason for the observed striking minimum in the (n, 2n) cross section for even nuclei. The present results indicate the existence of shell effects in the cross section values for even-even closed neutron shell nuclei at 14.6 MeV neutron energy. Although the present measurements are in good agreement with values in recent literatures, more measurements are needed. The lack of experimental data required to normalize the cross section values to the same excitation energy hinders an accurate systematic study of (n, 2n) reactions.

References

1. Sothras S. L. and Inorg J., *Nucl.Chem.* **40** (1978) 587
2. Chattejee S., *Nucl.Phys.* **A121** (1969) 593
3. Borman M., *ibid*, **65** (1965) 257
4. Borman M. et al., *ibid*, **A130** (1965) 195

5. Moretto L.G., *ibid,* **A180** (1972) 337
6. Moretto L.G., *ibid,* **A182** (1972) 641
7. Wenden L.N. et al., *Phys. Rev.* **C11** (1970) 350
8. Csikai J. et al., *Phys. Lett.* **20** (1966) 52
9. Risder R., *Sitibet Osterr Akad Wiss.* **175** (1966) 53
10. Hille P., *Nucl. phys.* **A107** (1968) 49
11. Qaim S.M., *ibid,* **A185** (1972) 614
12. Pearlstein S., *Nucl. Data* **A3** (1967) 327
13. Kondaish E., *J. Phys. A. Math. Nucl. Gen.* **7** (1974) 1457
14. Chatterjee A., *Phys, Rev.* (1978) 2118
15. Gilbert A. et al., *Canadian J. Phys.* **43** (1965) 1446
16. Braga-Mariazzan G.M., *Phys. Rev.* **C6** (1972) 1398
17. William L., *J. Phys. G. Nucl. Phys.* **8** (1982) 153

INTERMEDIATE ENERGY AND OTHERS

PRECISION TOTAL CROSS SECTIONS and THE OPTICAL MODEL AT INTERMEDIATE ENERGY

Roger W. Finlay
Institute of Nuclear and Particle Physics
Ohio University
Athens, OH 45701 USA

ABSTRACT

New measurements of neutron total cross sections between 5 MeV and 600 MeV are presented. These data do not provide an adequate basis for the development of an optical model potential, but they certainly can be used to test existing models. This presentation describes preliminary comparisons of theory with experiment for both phenomenological and microscopic models using both the Schroedinger equation and the Dirac equation. These comparisons indicate the need for additional information about the neutron–nucleus interaction at intermediate energy.

1. Introduction

The nucleon–nucleus optical potential has been a topic of enduring interest to nuclear physicists. Indeed, nearly all of the presentations to this conference are strongly influenced by the optical potential and its complicated parametrizations. The global neutron–nucleus interaction at low energy has been clarified considerably in recent years [1,2], and, with the incorporation of the long–neglected dispersion correction [3], a reasonably satisfactory parametrization of the basic phenomenology has begun to emerge. However, most of this work is based on measurements below 40 MeV. Above that energy we must either make new measurements or else learn what we can from measurements with proton beams, since there is very little neutron data above that energy. Shen and co–workers [4] reviewed the situation very recently: The data set for differential cross section measurements consisted of five forward–angle data points for five targets at 155 MeV and seven points for ^{12}C at 96 MeV. This data library has not changed in thirty years!

Yet this is potentially a very interesting region for nuclear physics. In the intermediate energy region, the pion production channel opens, pion–nucleon resonance states (Δ) are formed inside the nuclear medium. Above about 200 MeV, an impulse approximation might be expected to give a good description of the data, while at lower energy Pauli blocking and medium corrections must be important. Phenomenologically, the imaginary term in the optical potential increases rapidly above the meson production threshold, while the real term is thought to pass through zero around 500 MeV. Our long–cherished Woods–Saxon form factors are known to fail for intermediate energy proton scattering. It would not be an easy task to develop a neutron–nucleus optical potential at this time given the paucity of the data. Yet, such potentials are needed not only for the insight they provide for reaction theory but also for such practical matters as the calculation of distorted waves for (n,p), (p,n) and (e,e′n) reactions, transport and shielding calculations at the very large accelerators, and for radiation dosimetry.

The present contribution is an attempt to reopen interest in the role of neutron data in the intermediate energy region. New, high–precision measurements of total cross sections are presented and compared with various optical model calculations. The basic attitudes behind this presentation are a) energy–dependent total cross section data alone are not sufficient to develop an optical model (differential cross sections and spin observables would be needed for that task) and b) precision total cross section measurements do provide a valid test of existing models and theories of the nucleon–nucleus interaction.

2. The Experiment

New, precise measurements of neutron total cross sections for 5 MeV $\lesssim E_n <$ 600 MeV are presented for eighteen target nuclei with $9 \leq A \leq 209$ including isotopically enriched samples for ^{40}Ca, ^{90}Zr and ^{208}Pb. A detailed account of experimental conditions and procedures has been prepared for publication elsewhere [5] so the present description will be brief. A traditional good–geometry transmission experiment was performed on a well–collimated beam line at the LAMPF–WNR facility at the Los Alamos National Laboratory. Spallation neutrons emerging at 30° from the 800 MeV LAMPF proton beam provide usably high intensities up to 600 MeV. Neutrons were detected by time of flight with a small, fast plastic scintillation counter. Targets were cycled in and out of the beam every 20 seconds in order to compensate for variations in intensity and spectral shape of the monitored beam. Excellent beam conditions allowed the achievement of 1% counting statistics in 1% energy bins in almost every case.

Results of the experiment are shown in Fig. 1 which is a modified Barschall plot in that the individual data sets are equally spaced along a false A axis. The most prominent feature of the data is the systematic behavior of the peaks and valleys as a function of E and A—sometimes called the nuclear Ramsauer effect. A second obvious feature is the disappearance of the sharp compound nucleus resonances with increasing E and A. Also evident is the absence of any significant change in the total cross section near the pion production threshold.

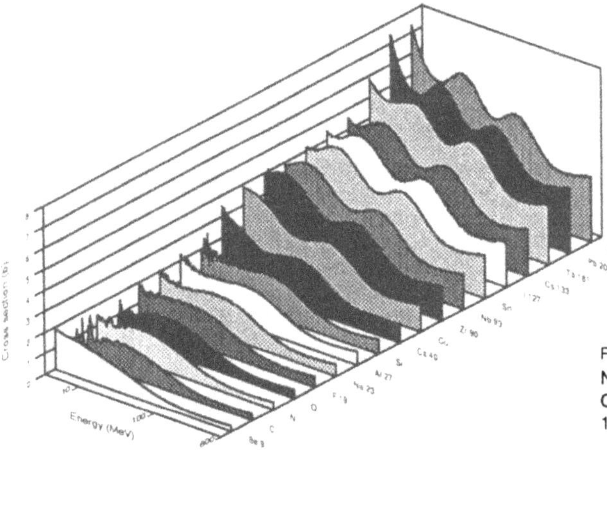

Figure 1.
Neutron total cross sections
Ohio University - Los Alamos
1990

3. Non–Relativistic Models

A) Phenomenological Potentials

In this section we explore the extension of conventional optical models [1,2] to intermediate energy. For protons, the subject has been considered in detail by Schwandt and co–workers [6,7]. In this section, a conventional optical model consists of standard local potentials with real and imaginary parts with Woods–Saxon or derivative Woods–Saxon form factors and energy–dependent potential strengths. The object of the exercise is to obtain energy–dependent potential parameters and geometrical parameters that describe the new neutron total cross section data, but we are guided by two constraints: (1) the potential should join smoothly with what we know to be valid at lower energy and (2) the potentials should be convertible to (from?) potentials that describe intermediate energy proton scattering when proper account is taken of the Coulomb term and the isovector term.

Our starting point is the very serviceable potential of Schwandt et al. [7] which describes differential proton scattering cross sections and analyzing power for 80 MeV < E < 180 MeV and 24 < A < 208. Special note should be taken of the claimed limits of validity of this potential. Any false consequences that come from extrapolating this potential outside its stated region of validity are solely this author's responsibility and are not intended as any criticism of the fine work in Ref. 7.

We start by adopting the potential parameters of Ref. 7 entirely, set the Coulomb term to zero, reverse the sign of the isovector term in real central and spin–orbit potentials, and calculate the predicted neutron total cross section for ^{90}Zr. Results are shown in Fig. 2. The agreement with the data in the region

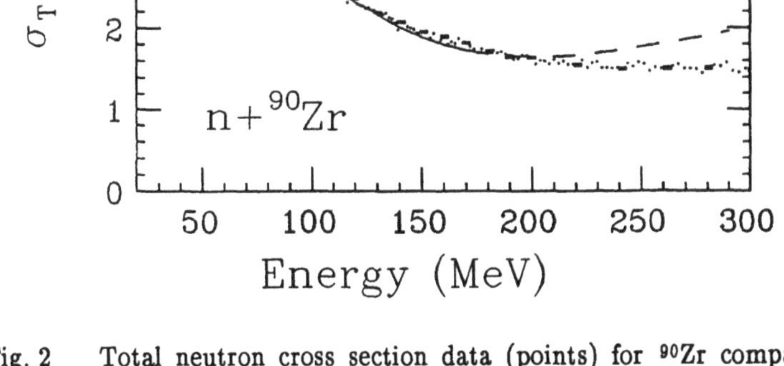

Fig. 2 Total neutron cross section data (points) for ^{90}Zr compared with the predictions of Ref. 7 from 80–180 MeV (solid) and the extrapolated potential (dashed) for 30 MeV < E < 80 MeV and E > 180 MeV.

between 80 and 180 MeV is quite reasonable. We extrapolate the potential downward to 30 MeV without any loss of general agreement. To go lower in energy would require the addition of a surface imaginary potential. Not surprisingly, when we extrapolate to higher energy (dashed line), we loose contact with the data. It is tempting at this point to begin a parameter search, but which parameters? We argue that the spin–orbit potential ought not be touched since it is based on careful measurement of proton analyzing power. The real central potential is getting quite small above 180 MeV so it might be possible to leave that term alone. In order to fit the neutron total cross sections about 200 MeV, we propose a "minimal modification" to the imaginary part of the Schwandt potential. A complication arises in that both the potential strengths and the geometrical parameters of Ref. 7 are energy dependent. Having no better guidance, we hold the geometrical parameters above 180 MeV to the values of Ref. 7 at 180 MeV and ascribe all of the modifications to the central imaginary potential strength. A proposed new potential is given in Eq. (1) and displayed in Fig. 3 where it is compared to the extrapolated Schwandt potential.

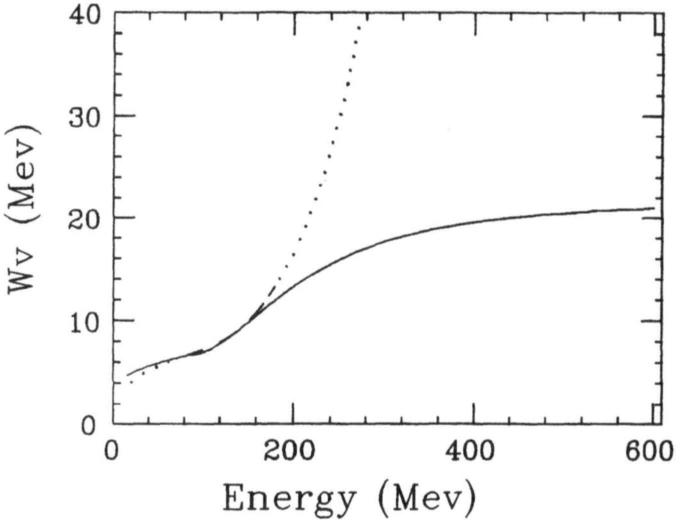

Fig. 3 The original (dashed) and extrapolated (dotted) imaginary potential of Ref. 7 compared with the recommended imaginary potential of Eq. (1).

$$W = 6.6 + 2.13 \times 10^{-2}(E-80) + 1.8 \times 10^{-6}(E-80)^3 \quad E \lessapprox 80 \text{ MeV}$$

$$= 6.6 + \frac{15.353\ (E-80)^2}{(E-80)^2 + (137.8)^2} \quad E \gtrapprox 80 \text{ MeV} \tag{1}$$

Finally (and against our better judgment) minor modifications were made to the central and spin–orbit potential to improve the fit to the data.

$$V = 104.11 - 16.715 \ln E - 16.5 \frac{(N-Z)}{A} \tag{2}$$

$$V_{so} = 19.9\,(1-0.166 \ln E) - 3.75 \frac{(N-Z)}{A} \tag{3}$$

Cross sections calculated with the potential of Eqs. (1–3) are compared with the data in Fig. 4. The new potential succeeds in describing the overall features of the total cross section data. Nonetheless, considerable caution should be used in calculating other observables with this model.

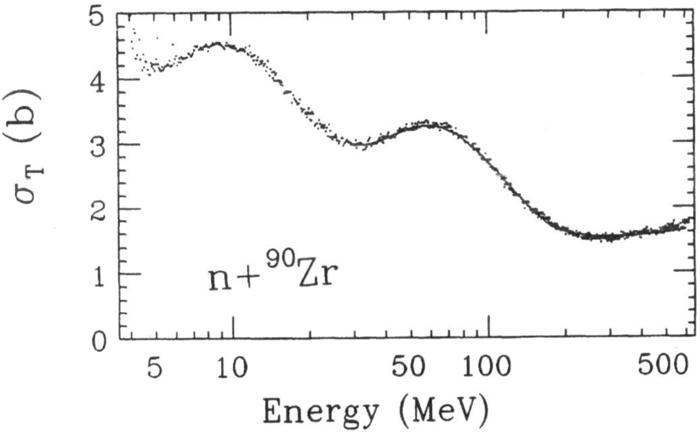

Fig. 4 Total cross section data (points) for ^{90}Zr for 5 MeV < E < 600 MeV. The solid line is a calculation based on the revised potential of Eqs. (1–3).

Returning briefly to the question of extrapolation to low energy, it could be emphasized that dispersion corrections (in addition to a surface imaginary term) will play an important role. Is it possible to develop a credible phenomenological potential that describes <u>all</u> of the neutron data from intermediate energy down to zero? An important step in this direction has been taken by R.L. Walter and co–workers [8]. The TUNL model is based on differential cross section and analyzing power measurements for 8 MeV < E < 18 MeV and on total cross section measurements for 0.5 < E < 100 MeV and includes dispersion corrections. Predictions of total cross sections for the neighboring nucleus ^{93}Nb are compared with the data over a very large energy range in Fig. 5. Agreement with the data is excellent!

B. Microscopic Models

During the last decade, much effort has gone into the development of less phenomenological, more microscopic descriptions of the nucleon–nucleus interaction. Most of these models involve a folding of an effective nucleon–nucleon

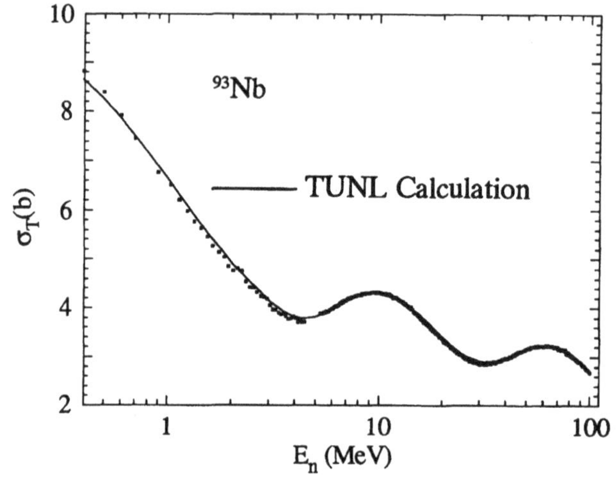

Fig. 5 Total cross section data for ^{93}Nb for 0.5 MeV < E < 100 MeV from Ref. 4 compared with the dispersive optical model analysis of Ref. 8. Data for E < 5 MeV come from several sources [8].

interaction (t–matrix) with a nuclear density, ρ. Since information on both of these quantities is available independently, one could hope that the resulting optical potentials could describe elastic nucleon–nucleus scattering in a more fundamental manner and, thus, with fewer parameters. At low energy (E < 150 MeV), the problem is complicated by the poorly–understood way in which the nucleon–nucleon interaction is modified inside the nuclear medium. Nonetheless, considerable success has been achieved in the description of scattering data. A full review appears in L. Hansen's contribution to these proceedings [9].

At higher energy, there is the hope that the problem will simplify. Medium effects should become weaker, and a simple folding or "tρ" approximation might be valid. Indeed parametrizations of the nucleon–nucleon t–matrix (e.g., Franey and Love [10]) have been a mainstay of experimentalists during this period. Many of these models had difficulty in calculating certain spin observables at intermediate energy, thus prompting great interest in relativistic models based on the Dirac equation (see Section 4).

More recently, the thrust in intermediate energy reaction theories is toward more fundamental formulation of the non–relativistic problem. The "full–folding" model approach [11,12,13] would seek to solve the first term in a multiple scattering expansion exactly, and then to begin to incorporate medium effects correctly. Therein lies the hope of describing nucleon scattering observables in between the impulse approximation region (E > 200 MeV) and the familiar phenomenology below 40 MeV. The present data may contribute to this program by providing 1% "targets" for the theorists. Already, there is some evidence [14] that the present data can provide guidance concerning the importance of inelasticities in the nucleon–nucleon interaction, correct incorporation of relativistic effects in a Schroedinger formulation, etc.

No new calculations are presented for comparison with the data. It is too early. This entire section is included to alert the reader to an area of potentially exciting developments in the future.

4. Relativistic Models

The application of the Dirac equation to nucleon–nucleus scattering has been a principal interest of B.C. Clark and collaborators for several years. This author is not qualified to review the extensive literature in the field. It is clear, however, that two main trends have emerged: the Phenomenological Dirac Potential and the Relativistic Impulse Approximation (RIA). Predictions of each of these models will be compared with the present data in turn.

A. Phenomenological Models

In this approach, the nucleon–nucleus interaction is described in terms of a large, attractive Lorentz scalar term and a large repulsive Lorentz four vector (time–like component only) term. Both terms depend explicitly on E and A. The large cancellations between these terms produce the required spin–orbit term. These cancellations also provide a natural explanation of the energy–dependent–geometry effects mentioned above. In the present work, three recent parametrizations of this model are discussed. In the first, Hama et al. [15] used a rich data base of proton scattering observables to construct an ambitious (but successful) potential model for $A > 40$ and $65 < E < 1040$ MeV. In the second, Kozack and Madland [16] used a similar data base for protons with $95 < E < 300$ MeV to which they added good quality measurements of total neutron cross sections for ^{208}Pb for $95 < E < 240$ MeV and sought to extract an isovector term in order to describe proton– and neutron–nucleus scattering in the same framework. In the third, Shen and co–workers [4] attempted to construct a neutron–nucleus optical potential using only neutron data. This is a severe handicap. As mentioned in the introduction, measurements of neutron scattering cross sections are extremely rare above 40 MeV and spin observables are non–existent.

Comparison with the present data is possible at this time only for the work of Hama et al. [15]. In Fig. 6, the present data for ^{40}Ca and ^{208}Pb are compared with predictions by Hama [15,17]. The dashed lines are the predictions of Dirac phenomenology with the Coulomb term set to zero but no other parameters were adjusted. The good agreement for ^{40}Ca (also for ^{12}C and ^{16}O, not shown) confirms the accuracy of the isoscalar part of the Ohio State potential while the significant disagreement for ^{208}Pb indicates that, for neutrons, the isovector potential is important but poorly understood.

B. Microscopic Models

Global potentials necessarily have a large number of parameters. However, at sufficiently high energy, the parameter–free relativistic impulse approximation (RIA) is quite successful at describing proton scattering. First application of the RIA to neutron scattering is shown in Fig. 7. Here the present data for ^{208}Pb above 100 MeV are shown with their error bars. The RIA calculations with the model by Horowitz, Murdock and Serot [18] are shown as a solid line. In these calculations, the ground state nuclear wave function is calculated in the Dirac–Hartree approximation, and the target density is folded with a relativistic parametrization of the free–space nucleon–nucleon interaction to obtain the

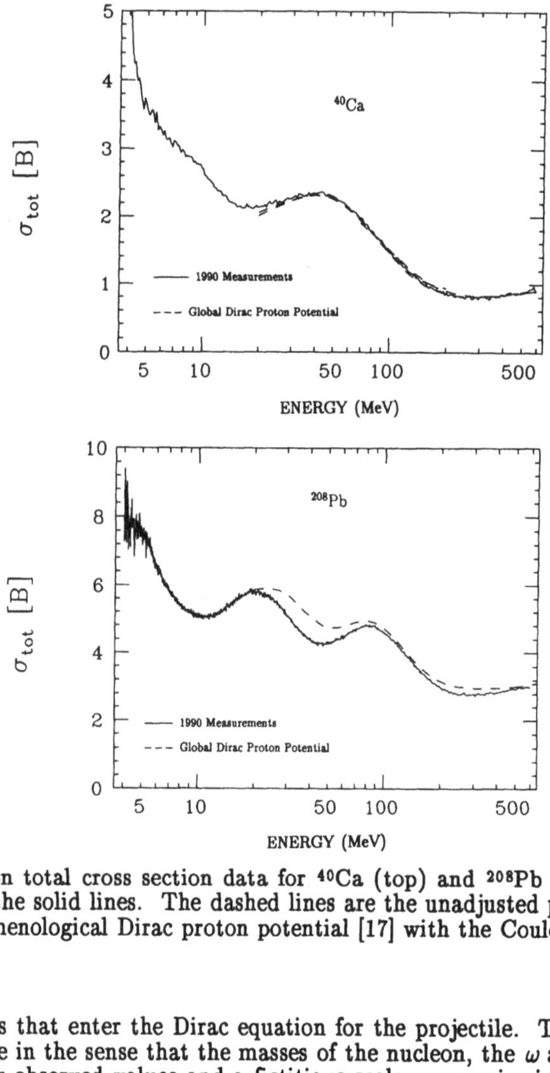

Fig. 6 Neutron total cross section data for ^{40}Ca (top) and ^{208}Pb (bottom) are shown by the solid lines. The dashed lines are the unadjusted predictions of the phenomenological Dirac proton potential [17] with the Coulomb term set to zero.

optical potentials that enter the Dirac equation for the projectile. The procedure is parameter–free in the sense that the masses of the nucleon, the ω and ρ mesons are fixed at their observed values and a fictitious scalar meson is given a mass of 520 MeV. Coupling constants are fixed by the requirement that the calculated bulk properties of infinite nuclear matter are reproduced. The relativistic Love–Franey (RLF) model [19] is used for neutron energies below 400 MeV while the McNeil, Ray, Wallace (MRW) parametrization [20] is used above 400 MeV. Agreement with the data (solid line) is remarkable for a parameter–free model. The volume integrals of the optical potentials produced by this procedure have very smooth and simple energy dependences. Very minor adjustments to the

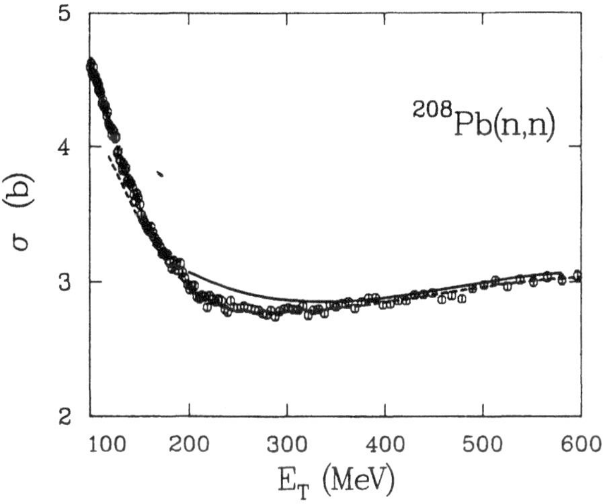

Fig. 7 Neutron total cross section data for ^{208}Pb above 100 MeV compared with RIA calculations.

potentials produce the near–perfect fit shown by the dashed line in Fig. 7. The temptation to extrapolate these results to lower energy is great but such a procedure is almost certainly not justified. Below 150 MeV, the properties of the specific target nucleus must be dealt with.

5. Conjectures and Speculations

Has anything new come out of these measurements of neutron total cross sections at intermediate energy? The answer is a somewhat speculative "perhaps". Consider the behavior of the cross section for all nuclei above their common minimum near 300 MeV. For the lightest nuclei, the rate of increase of total cross

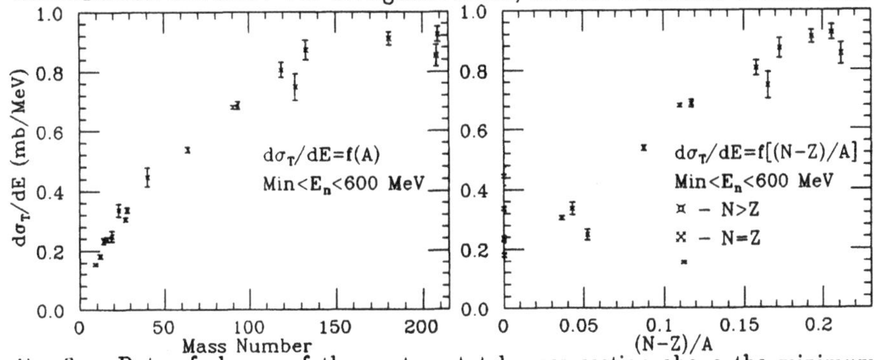

Fig. 8 Rate of change of the neutron total cross section above the minimum near 300 MeV plots vs a) mass number and b) neutron excess.

section with energy is much slower than it is for the heavier nuclei. This systematic behavior is summarized in Fig. 8a where the increase in cross section between its minimum and 600 MeV is well represented as a simple slope. These slopes when plotted against mass number have a smooth behavior with a distinct flattening for the heavy nuclei. We interpret this as a clear illustration of a "shadowing" effect in nuclei that would not be seen in experiments, e.g., (γ,n), with more penetrating probes. The persistence of this effect up to 600 MeV is somewhat surprising in view of the success of the impulse approximation in this energy region. Is this effect a manifestation of higher terms in the multiple scattering expansion mentioned in Section 3B?

Consider now the same slope, $\frac{d\sigma}{dE}$, plotted as a function of neutron excess (N–Z)/A. The points for the self–conjugate nuclei collapse to the vertical axis, but the remaining points on Fig. 8b could be described reasonably well with a straight line. (Ignore ^{9}Be whose asymmetry is the same as ^{90}Zr but must be thought of as an adolescent nucleus in these speculations.) A direct proportionality of $d\sigma/dE$ to (N–Z)/A is the signature of the formation of the Δ^- (1232) baryon resonance in nuclei.

6. Conclusions

1. Optical models are still interesting since the nucleon–nucleus interaction is a priority topic in nuclear physics.
2. Optical models require data.
3. Total cross section data are not sufficient to develop an optical model, but, by their very precision, they provide a challenging test for existing potentials.
4. There is a very great need for high–quality neutron scattering cross sections and spin observables above 50 MeV.
5. The RIA is exciting and successful above ~ 150 MeV, but its extension to lower energy is not at all obvious. Theoretical guidance is still needed in the transition region from Woods–Saxon (or Dirac) phenomenology to the high energy domain of the impulse approximation.

The author is grateful for the energetic assistance he received from Werner Abfalterer, Eric Montei, Hong Zhang and Yanhe Jin in reducing and analyzing the data. Calculations from S. Hama and R.L. Walter are also gratefully acknowledged.

References

[1] R.L. Walter and P.P. Guss, in Nuclear Data in Basic and Applied Sciences, edited by P.G. Young et al. (Gordon and Breach, New York, 1986), Vol. 2, p. 1079.

[2] J. Rapaport, Phys. Rep. 87 (1982) 25.

[3] C.H. Johnson, D.J. Horen and C. Mahaux, Phys. Rev. C36 (1987) 2252.

[4] Shen Qing–biao, Feng Da–chun and Zhuo Yi–zhong, Phys. Rev. C43 (1991) 2773.

[5] R.W. Finlay, W.P. Abfalterer, G. Fink, E. Montei, P. Lisowski, G.L. Morgan and R.C. Haight, Nucl. Instr. and Meth., to be submitted.

[6] P. Schwandt, *Proc. Workshop on the Interaction Between Medium Energy Nucleons in Nuclei,* AIP Conf. Proc. No. 97, edited by H.O. Meyer (AIP, New York, 1982), p. 89.

[7] P. Schwandt, H.O. Meyer, W.W. Jacobs, A.D. Bacher, S.E. Vigdor and M.D. Kaitchuk, Phys. Rev. C26 (1982) 55.

[8] R.L. Walter, contribution to these proceedings and private communication (1991).

[9] L.F. Hansen, contribution to these proceedings.

[10] M.A. Franey and W.G. Love, Phys. Rev. C31 (1985) 488.

[11] H.F. Arellano, F.A. Brieva, W.G. Love, Phys. Rev. Lett. 63 (1989) 605; Phys. Rev. C41 (1990) 2188.

[12] Ch. Elster, T. Cheon, E.F. Redish and P.C. Tandy, Phys. Rev. C41 (1990) 814.

[13] R. Crespo, R.C. Johnson and J.A. Tostevin, Phys. Rev. C41 (1990) 2257.

[14] Ch. Elster, private communication (1991).

[15] S. Hama, B.C. Clark, E.D. Cooper, H.S. Sherif and R.L. Mercer, Phys. Rev. C41 (1990) 2737.

[16] R. Kozack and D.G. Madland, Phys. Rev. C39 (1989) 1461; Nucl. Phys. A509 (1990) 664.

[17] S. Hama, private communication (1991).

[18] C.J. Horowitz, D.P. Murdock and B.D. Serot, "The Relativistic Impulse Approximation" in Computational Nuclear Physics, edited by K. Langanke, J.A. Maruhn and S.E. Koonin (Springer Verlag, Berlin, 1991), p. 129.

[19] C.J. Horowitz, Phys. Rev. C31 (1985) 1340.

[20] J.A. McNeil, L. Ray and S.J. Wallace, Phys. Rev. C27 (1983) 2123.

NUCLEON RADIATIVE CAPTURE AND THE INVERSE REACTION AT INTERMEDIATE ENERGIES

I. Halpern, Physics Department
University of Washington
Seattle, WA 98195, USA

Abstract

The processes which can lead to the radiative capture of fast nucleons include direct transitions in the nuclear potential, transitions in which coherent multipole resonances are excited, transitions by nucleons which are excited in early intranuclear collisions, bremsstrahlung from nucleon-nucleon collisions and photon 'evaporation' from a thermally equilibrated nucleus. Corresponding processes occur when an energetic photon ejects fast nucleons from a nucleus. As experimental information from capture and photoreactions has become more detailed, inconsistencies and uncertainties have appeared which reflect difficulties in identifying and separating the responsible processes. This has led to more sophisticated and more complicated theoretical treatments which in turn have prompted new and more demanding experiments.

1. Introduction

The study of the radiative capture of fast nucleons and of their ejection in photoreactions have, over the years, shed considerable light both on nuclear reaction mechanisms and on nuclear structure. Among the special features that photons bring to reaction studies is the relatively simple and well established form of the electromagnetic interaction. Moreover photon transitions commonly involve only small changes in angular momentum, limiting, in this way, the possiblities of what can happen in an interaction.

Even so there are enough possible mechanisms for photoemission when a fast nucleon enters a nucleus that problems in sorting them out have been an annoying source of uncertainties and inconsistencies in photonuclear studies.

The main mechanisms for the production of high energy photons when an intermediate energy nucleon enters a nucleus can be classified as follows:

1. The simplest process is the so-called *direct* process in which the nucleon drops from its incoming state in the nuclear potential to a lower state with the emission of radiation.

2. There is a two-step process which competes with the direct one in which the incident nucleon makes the same transition without radiating. Instead it excites a characteristic resonance in the nucleus through its dynamical coupling to the nuclear potential. These resonances are characterized by spin, parity and isospin. Their excitation, in this way, is called the *semi-direct* process and it can be sizeable. For

example, when the incident nucleon excites the giant dipole resonance at its peak, the cross-section for the semi-direct process is about ten times larger than that for the direct process.

3. The incoming nucleon can also interact with individual nucleons and this two-body interaction can sometimes give rise to *bremsstrahlung*. In the last few years this process has received alot of experimental and theoretical attention.[1–4] The foregoing three processes are illustrated, from a shell model point of view, in Fig. 1.

4. Most often the initial interaction of the incoming nucleon with the nucleus does not give rise to radiation but only to the excitation of other nuclear degree of freedom. It still remains possible for the degraded incident nucleon or for any nucleons that it excites to radiate by any of the foregoing processes. If such radiation occurs before the nucleus reaches thermal equilibrium, but after the intial interaction, we speak of it as *pre-equilibrium* radiation.

5. Finally there can be the emission of energetic photons from a fully thermally equilibrated nucleus. These emissions are akin to particle *evaporations*.

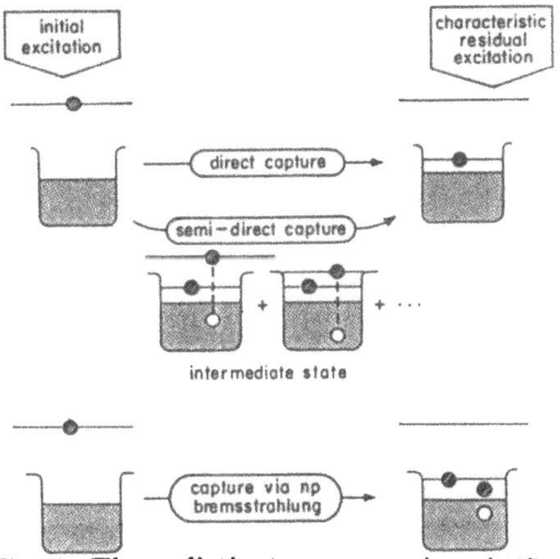

Fig. 1. Three distinct processes by which an intermediate energy nucleon, incident on a nucleus, can produce hard photons.

The inverses of these processes can all give rise to fast nucleon emission when high energy photons are absorbed by a nucleus. In the inverse of process (3) the incident photon is absorbed by a pair of nucleons. Such absorption is possible only because nucleons in nuclei exhibit a mutual (residual) interaction beyond their interaction with the overall nuclear potential. It is much more probable for *n-p* pairs to absorb a photon than for *p-p* or *n-n* pairs since only *n-p* pairs can absorb dipole radiation. For this reason and because of the way it has been modelled, photoabsorption by nucleon pairs has been called quasi-deuteron absorption since it was first invoked many years ago.[5] In this model the high energy dependence of the deuteron photocross-section is blamed entirely on the short-range (high momentum) part of the deuteron wave function which is assumed to be rather independent of the initial *n-p* relative energy. This leads to a quasideuteron photocross-section which is just $L(ZN/A)$ times that for the deuteron. Here $L(N/A)$ is simply the ratio of the mean neutron density at a typical proton in a heavier nucleus to the corresponding density in the deuteron.

From our description of these various processes, it is clear that they are not really independent or orthogonal. The internucleon forces in the nucleus are responsible for the shell model potential as well as for the various residual interactions. To treat these processes as independent will certainly correspond to some form of double counting. Even where this does not present a serious concern, one must be careful when analyzing reactions in which more than of these processes is acting since their amplitudes may interfere.

Because each of the foregoing processes is easy to picture and is correspondingly easy to model, they have generally provided the framework for the interpretation of photonuclear reactions. In those often-occurring cases where one is focussed on one of the processes, but others are also at play, these other processes are regarded as 'backgrounds.' They are unfortunately *unavoidable* backgrounds to be distinguished from those backgrounds which need not necessarily be there, for example the backgrounds of room-scattered thermalized neutrons which led, many years ago, to mistaken results for the radiative capture cross-sections of 14 MeV neutrons.[6]

In reviewing recent developments in the field of radiative capture and its inverse, our orientation in each of the following sections will be in terms of overlaps between pairs of the listed processes or mechanisms and the sorts of problems that these overlaps have given rise to. Much of the recent literature in the field has been concerned with such problems.

2. Direct and semi-direct

The direct and semi-direct amplitudes couple the same pair of initial and final states and therefore can interfere. For example, when an incoming photon drives a nuclear proton into oscillation, this proton, as a consequence of its residual couplings to the other nucleons, attracts the neutrons and repels the protons thus inducing a dipole oscillation. It follows that at low frequencies the direct and semi-direct dipole moments are out of phase and their interference is destructive. Above the resonance it is constructive. The implied asymmetry of (γ,p) and (p,γ) excitation functions does not, however, appear to be as strong as expected[7] and its absence has been attributed to a phase shift in the semi-direct amplitude arising from couplings of this oscillation to other modes.[8] The required coupling is rather large and its origin has not yet been credibly accounted for. Since a direct amplitude, D, is always accompanied by a semi-direct amplitude, SD, we will in what follows, consider them as a unit which we write DSD.

3. Overlapping DSD excitations in one reaction

When the energy of an incident nucleon is only slightly higher than that of a collective excitation to which it gives rise, the final state of this nucleon is bound and the DSD process leads to radiative capture. If however the incident energy is significantly higher than the resonance energy of the collective excitation, the final state of the nucleon is unbound and the degraded nucleon has a good chance to escape (Fig. 2). In the latter events the interaction process obviously differs in no essential way from those which lead to capture. The escaping nucleon can however be observed as an inelastic scattering. Studies of the inelastic scattering spectra of nucleons and of other projectiles have been a very important complement to capture studies of DSD excitation of multipole resonances.

In the earliest experiments of this kind the association of any broad bump in the inelastic spectrum with some particular multipole resonance was based on the angular distribution of that bump. To determine this distribution when the bump sits on a substantial background one must first subtract this background. The easiest way to proceed is to draw it in by eye. This may be a risky procedure even when the 'background' level is comparable on both sides of the bump since it is unlikely that the background under a collective bump is smooth. Such bumps are generally wide enough that there can easily be overlapping multipole excitations. The study of multipole resonance excitations by inelastic scattering has turned out to be most informative for bumps which are narrow and therefore of low excitation (*e.g.* the $E2$ and $E0$ isoscalar resonances).[9]

In general the most fruitful studies have been those at forward angles, where the possible multipolarities of excitations are restricted. It has also been useful for disentangling multipolarities to study the decay modes[10] of the excitations in coincidence with the scattered projectile. Sometimes the overlap of different multipole modes can be used to advantage when it is possible to exploit their interference (see Section 8).

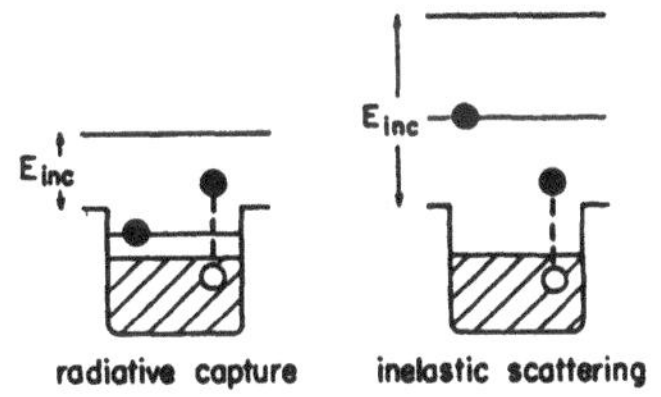

Fig. 2. The DSD process at different incident energies. At higher energies the processes can lead to inelastic scattering.

4. DSD and evaporation (photoreactions)

When incident photon energies are only a few MeV above the threshold for nucleon emission, the predominant emission mode is statistical, even to the low-lying $1h$ states in the residual nucleus. At these low energies, particle escape from the $1p1h$ states excited in a DSD process is inhibited by the poor transmission at the nuclear surface. At energies not many MeV higher, the situation reverses and the nucleon emission to low-lying $1h$ states are primarily from the DSD process rather than from evaporations. Thus there is not a very long energy stretch for which both processes are of comparable strength. The reason for this fairly sudden change of the decay mechanism is primarily the rapid fall-off of the evaporation rate to any particular state as the excitation energy E increases. This rate goes essentially as $\exp(-E/T)$ where T is the nuclear temperature. It reflects the steep rate of increase of nuclear level density with excitation.

The transition from the evaporation mode to the DSD mode has recently been examined in ^{208}Pb in the region of the giant dipole resonance (11–16 MeV) in a pair of complementary studies. In one the (γ, n) reaction was examined using tagged photons and neutron time-of-flight;[11] the other study used inelastically scattered

^{17}O with a 4π solid-angle photon detector to determine the energies of the residual states after neutron emission.[12] Both measurements compared their observed yields to those of Hauser-Feshbach calculations for statistical emission and found that below excitations of about 13 MeV emissions were mainly evaporations and above 13 MeV they were mainly *DSD*. Since 13 MeV is the peak of the giant dipole resonance in Pb, we see that the nature of the photonuclear process can change substantially across a single multipole resonance. We also see that for multipole resonances in heavy nuclei that lie higher than the dipole one need not be concerned about any background from evaporation.

The relative yields to the different low lying states in ^{207}Pb appear to differ for the two measurements, as might be expected. The (γ, n) reaction proceeds primarily through dipole absorption and mainly populates $1h$ residual states whereas the ^{17}O scattering populates a broader assortment of states. Studies of this type in which individual final states can be resolved show promise of providing new information about the structure of these states and about the mechanisms which feed them.

A rough schematic sketch (Fig. 3) for the (γ, N) reaction shows the domains in which different possible processes dominate in terms of the incident photon energy (up to about 100 MeV) and the outgoing nucleon energy. The evaporation region has been drawn in with a definite boundary to reflect the strong, exponential cutoff of the emission chance for a nucleon as a function of its energy. The *DSD* process mainly feeds the resonance region in the lower left corner. The region in which quasi-deuteron yields dominate is fuzzy. The degree of its overlap into the other regions is not yet clear. It is probably kept from reaching significantly into the lower corner of the diagram by Pauli blocking. The mechanisms which feed the very lowest-lying residual levels are of particular interest because these levels can often be experimentally resolved and their shell model structure is often reasonably well understood.

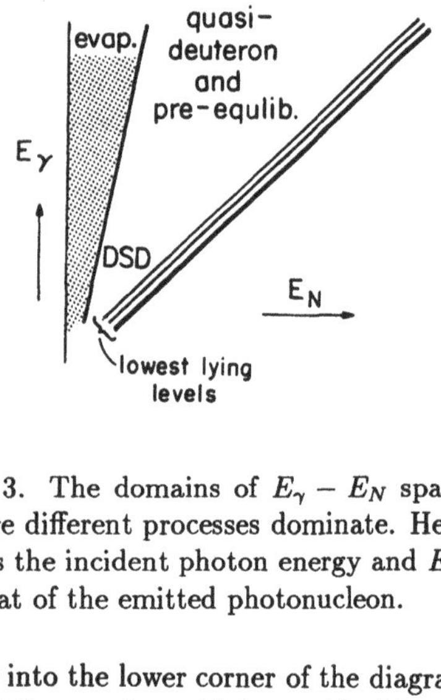

Fig. 3. The domains of $E_\gamma - E_N$ space where different processes dominate. Here E_γ is the incident photon energy and E_N is that of the emitted photonucleon.

5. *DSD* and evaporation (incident fast nucleons)

A diagram and discussion much like that in Fig. 3 applies to the inverse reaction, the ejection of hard photons in fast nucleon bombardments. One need only switch the labels on the axes and relabel the intermediate region in the figure as 'bremsstrahlung'. The discrete states with high yields in the *DSD* region are mainly $1p$

states for the (N,γ) process in place of the $1h$ states for the (γ,N) process. For incident protons as low as 10 MeV the yield to low-lying residual levels is already mainly by *DSD* through the dipole resonance, rather than by evaporation. A simple evidence for this is the much greater yield in this region for the (p,γ) reaction which can excite the dipole resonance, than for the (α,γ) reaction, which cannot.[13,14] (Fig. 4)

It is important to appreciate that the photon *evaporation* spectrum at excitations in the range of 100 MeV exhibits particular strength in the region of the collective resonances.[15] This strength does not however indicate the involvement of a semi-direct process at the start of the interaction; it is due instead to a semi-direct process at the emission stage. This can be appreciated from the required proportionality of the statistical emission probability to the inverse (absorption) cross-section of the emitted particle and the fact that photon absorption peaks dramatically at the dipole resonance because of the *DSD* process. Perhaps a less formal way to say the same thing is to note that if an incident nucleon can excite a collective oscillation in the nucleus there is no reason that the same energy nucleon, when it is about to be evaporated, would not be able to do it too. When such an excitation occurs, there is a good chance for it to decay by photon emission.

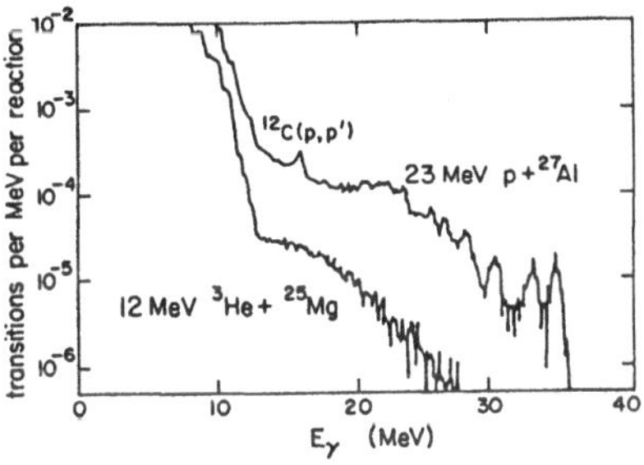

Fig. 4. Photon spectra from two reactions leading to the same excitation energy in the same compound system.[14] There are many more hard photons from the proton bombardment.

There is however an important difference between excitations produced on the way in and on the way out of nuclei. In the latter case the rest of the nucleus is generally at some considerable excitation energy rather than in its ground state. Although this difference should have only a small effect on the chance for exciting a resonance, the character of the resonance, *e.g.* its width, may strongly depend on excitation energy. The appearance of *DSD* effects in *evaporation* spectra has made possible the study of the widths of known collective resonances as a function of nuclear excitation energy and angular momentum. The observed widths of the dipole resonance, for example, are known to be sensitive to the departure from sphericity of the nuclear shape and since the shape distribution of excited nuclei depends on their excitation energy and on angular momentum, the study of the structure of the dipole resonance at high temperature has been useful for exploring the shapes of hot rotating nuclei.[16]

6. Bremsstrahlung and evaporation (incident fast nucleons)

Many more high energy photons are produced in nuclei by beams of high energy protons than can be accounted for either by evaporation or by the *DSD* process. It was suggested many years ago that the mechanism for their production may be *n-p* bremsstrahlung.[4] Interest in this problem was revived in recent experiments where again there appeared to be an excess of high energy photons in nuclear bombardments with heavy ions of tens of MeV per nucleon. It is in fact easier to study the *n-p* bremsstrahlung process with heavy ions than with incident nucleons since all *A* nucleons in a projectile can contribute to the observed radiation. The radiation cross-section should consequently be much larger with heavy ions.

For preliminary bremsstrahlung studies, however, the (N,γ) reactions has the advantage that only the target nucleons are in motion with respect to their center of mass. With heavy ions where both target and projectile nucleons are in motion, any calculation of the expected photon distribution must fold two sets of moving nucleons. A recent (p,γ) study on Au with 72 MeV protons[3] encourages the conclusion that the dominant production process for higher energy photons is indeed *p-n* bremsstrahlung. The spectra observed at different laboratory angles become identical when they are transformed to a frame moving at half the speed of the incident proton. This feature is expected since the bremsstrahlung angular distribution from *pn* pairs should be rather isotropic in the frame of the *np* center of mass. It is also encouraging in these measurements that the observed spectral shapes agree with those based on a bremsstrahlung model. The calculated bremsstrahlung cross-section is however about two and a half times smaller than the observed one when it is assumed that only the first *np* encounters radiate. This assumption may seem to be unreasonable, but the authors argue that the inclusion of a substantial amount of additional pre-equilibrium bremsstrahlung would degrade the calculated spectrum to the point that it would no longer have the same shape as the observed spectrum. The authors end up questioning the elementary *np* bremsstrahlung cross-section which they used in their calculation. Clearly there is more to be done on this basic problem.

The situation with incident heavy ions is even less clear than with protons, although here too the average radiating system appears to be moving at half the projectile speed except perhaps at the lower incident energies.[17,18] (Fig. 5) This observation tells us that the radiating system cannot be the compound system formed in the bombardment, but that it likely consists of *np* pairs, one nucleon from the projectile, one from the target.

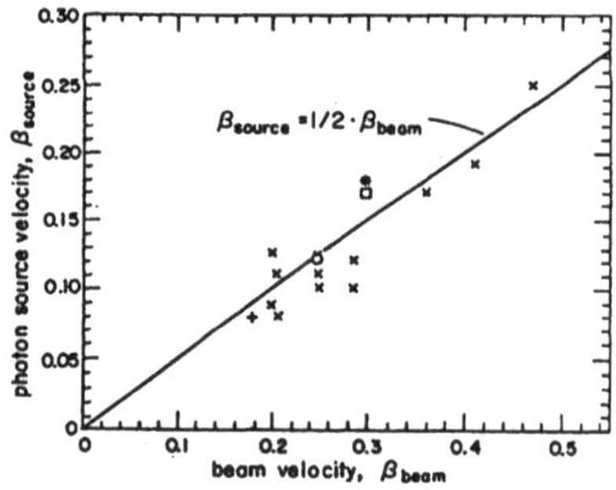

Fig. 5. Evidence for the importance of *pn* collisions for the production of high energy photons in various heavy ion bombardments.[17,18] The average speed of the radiating system appears to be half the projectile's.

At the same time that this conclusion was being generally accepted it was possible to give an account of the photon spectrum from a bombardment of Mo with a 19.5A MeV Mo beam solely in terms of statistical radiation.[19] The evaporations here are presumably from the many deep-inelastically scattered fragments which appear in this reaction. It was subsequently pointed out,[20] however, that some of the parameter values used in the statistical calculation may not have been altogether reasonable. In any event, the situation in this experiment and in the (p, γ) experiment mentioned earlier suggests that comparisons of experimental observations in this field with multi-parameter calculations of particular processes may involve too many uncertainties to be very definitive. What makes the interpretation of the observations difficult is the possible involvement of more than one process and the fact that consistent sets of parameters for any of these processes have not yet been solidly established.

7. Quasideuteron Process and Evaporation (Incident Photons)

In the inverse of the preceding process, high energy photons are incident and emitted fast nucleons are observed. Here too there may be a substantial overlap region of incident and outgoing energies in which both evaporation and the quasideuteron process contribute significantly.

The quasideuteron process was first clearly established in experiments about 40 years ago when the outgoing n and p were detected in coincidence.[21] One might therefore suppose that by now all of the parameters of this process would have been well established. This is however not the case,[22] even though it has long been clear that for photon energies from above the resonance region to photomeson threshold, ($\sim$ 30 to 140 MeV), the dominant photoabsorption process must be the quasideuteron process. Absorption by single nucleons cannot account for the magnitude of the observed cross-sections,[23] almost a full classical sum rule's worth across this energy region. (Fig. 6) Above 140 MeV, photabsorption can once again occur on a single nucleon, exciting it to its Δ resonance.

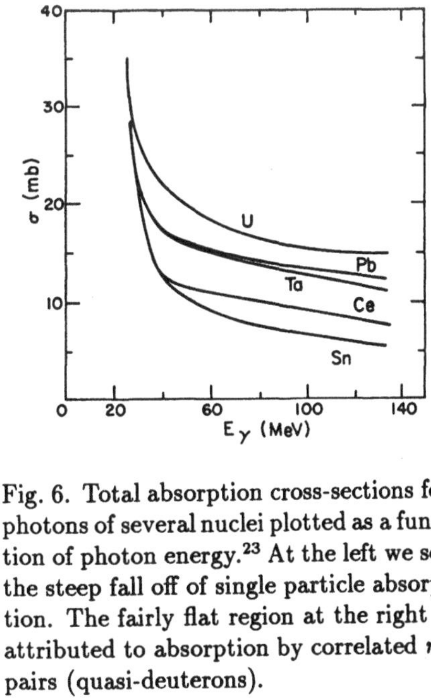

Fig. 6. Total absorption cross-sections for photons of several nuclei plotted as a function of photon energy.[23] At the left we see the steep fall off of single particle absorption. The fairly flat region at the right is attributed to absorption by correlated np pairs (quasi-deuterons).

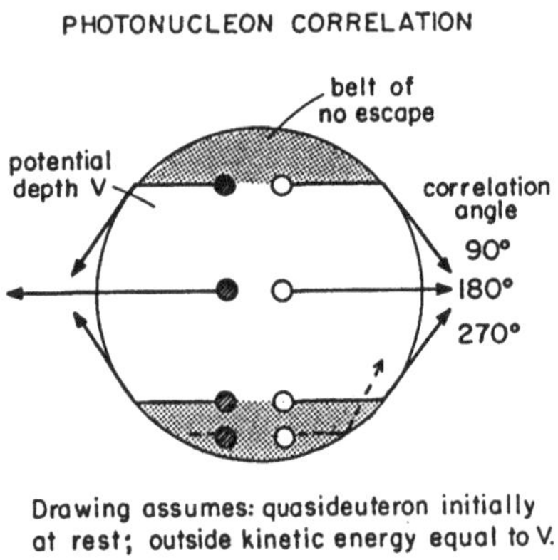

Fig. 7. Effects of refraction at the nuclear surface on the observed correlation between np pairs in photoabsorption by quasideuterons.

The possibility of definitive coincidence studies of the quasideuteron process is hindered by a number of circumstances: (1) First, the gathering of adequate statistics is difficult in coincidence experiments given the available fluxes of monochromatic photon beams. (2) Second, the quasideuteron (unlike the deuteron itself) is a moving target. The *rms* speed of a quasideuteron in a nucleus is $1/\sqrt{2}$ times that of the average nucleon and this speed is large enough to substantially spread the angular (180°) correlation characteristic of deuteron photodisintegration. (3) The nucleon mean free path inside the nucleus in this energy domain is not long compared with 3/2 the nuclear radius, the sum of the average escape paths of the pair of nucleons. In a large nucleus, at least one of the nucleons will generally change its energy and direction as the result of a collision on the way out. Finally (4), even if there are no internal collisions, the angular correlation between the escaping photonucleons is greatly smeared by refraction at the edge of the nuclear potential. Fig. 7 shows an example in which two photonucleons are each excited to kinetic energies of $\sim$ 100 MeV inside the nucleus, *i.e.* to twice the potential depth. Upon emergence these nucleons therefore have 50 MeV each and, depending on the site of the quasideuteron absorption, they will be observed in the lab with angles between them lying anywhere from 90° to 270°. These various disruptions of the basic two body correlation make it difficult to distinguish quasi-deuteron events from coincidences between pre-equilibrium nucleon pairs or even from pairs in which one of the nucleons is evaporated.

8. Quasideuteron and *DSD* processes (Incident Photons)

As we have noted, the quasideuteron photoabsorption mechanism overshadows the *DSD* mechanism for energies from about 40 to 140 MeV. Most of these quasideuteron events lead to the emission of nucleons with an equitable division of the available energy. However the motion of the quasi-deuterons in the nucleus makes it possible for some very uneven energy divisions between the neutron and proton. In that case one of these particles can emerge with energy close to the maximum possible, that is with energy close to that in a *DSD* process leading to low-lying $1h$ states. At high photon energy the *DSD* process has a very small cross-section, but since it selects $1p1h$ intermediate states whereas the quasideuteron process involves

$2p2h$ states, one can easily imagine that only a small fraction of the latter process would manage to intrude upon the former and feed low-lying $1h$ states.

It was observed however that at 60 MeV some (γ, n_0) cross-sections in light nuclei were very nearly equal to the corresponding (γ, p_0) cross-sections and it was therefore suggested that quasi-deuteron events, in which one of the photonucleons remains bound in the residual ground state, may dominate the yields even to the simple low lying states.[24] The observed forwardness of the angular distribution in these reactions was blamed on the coherence of forward emissions from the entire volume of the nucleus and a modified quasideuteron model recognizing this was developed.[24] In any event, simple *DSD* calculations of (γ, p_0) yields which use ordinary shell-model wave functions were consistently found to underestimate the observed cross-sections.[25] More ambitious and self-consistent theoretical treatments in the late 70's took into account correlations between nucleon pairs arising from residual forces, including exchange forces[26] and these calculations were found to reproduce the details of photoyields to low lying states in light nuclei reasonably well. It appears then that it is necessary to incorporate two-nucleon aspects of the photoabsorption process in analyses of photonuclear processes at intermediate energies.

Recent experiments on the (γ, p) cross-sections to *different* low levels in the same nucleus, ^{11}B, show however that the yields to residual states do not all behave consistently.[27] This presumably reflects the different particle-hole characters of these states and perhaps the different admixtures of quasideuteron and direct components that are involved in feeding them.

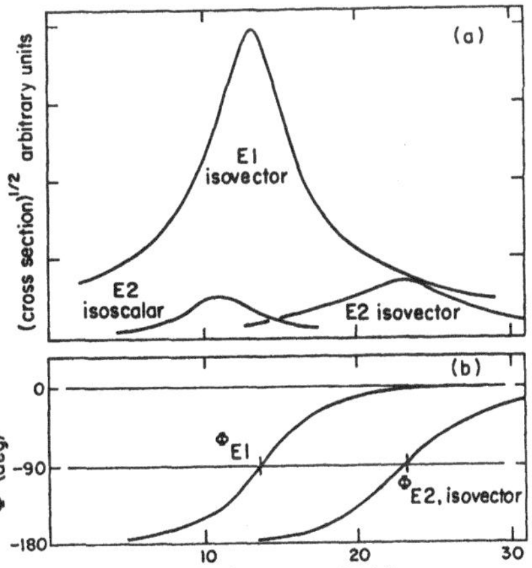

Fig. 8. A schematic drawing of relative multipole amplitudes and phases in the giant resonance region.

Just above the giant dipole resonance but much weaker in strength, there should lie the corresponding resonances of higher multipolarity. Efforts to find these resonances and to establish their basic properties (location, strength and width) may also have suffered from the intrusion of quasideuteron effects upon the basic *DSD* process which is involved in their excitation. Since these resonances occur at energies where the tail of the dipole resonance remains substantial, (Fig. 8) one way to study them has been through their interference with the dipole excitation. A strong forward folding in the (γ, N) or (N, γ) angular distributions is observed at about one oscillator spacing above the peak of the giant dipole resonance and one attributes this asymmetry to interference between an odd parity $E1$ amplitude and the even parity amplitude from the isovector quadrupole resonance.[28]

The analyses of these forward asymmetries using *DSD* calculations with $E1$ and $E2$ amplitudes give values for the $E2$ resonance energies close to expectations, but the strengths are more difficult to establish and so are the still more interesting parameters, the widths. It is significant that the expected concentration of $E2$ strength which is implied by the (γ, N) and (N, γ) experiments fails to be seen in certain reactions, *e.g.* pion charge exchange scattering[29] and $(^3He, ^3H)$ reactions.[30] It does however appear in the scattering of polarized photons.[31] These inconsistencies may simply reflect the considerable width of the $E2$ resonance. After all, the spreading and escape components of the width of the $E2$ giant resonance should both be substantially larger than those of the lower-lying dipole resonance.

The best fits to the curves which give the dependence of the observed asymmetry (the difference of the 55° and 125° cross-sections divided by their sum) upon incident energy in typical *DSD* analyses turn out to be only fair when one tries to fit the excitation functions at the same time.[32,33] (Fig. 9) In particular the calculated curves tend to underpredict the largest asymmetries which are observed. To obtain a really large asymmetry from any model involving just two amplitudes, these amplitudes must be nearly equal in magnitude *and* nearly opposite in phase at the back angle. One suspects, therefore, that there is some additional process at play whose forwardness is not based on just two amplitudes or that there are some additional amplitudes which should have been included in the *DSD* calculation. Recent calculations suggest that photoabsorption through $2p2h$ states (quasideuteron mechanism) at energies near that of the $E2$ isovector resonance are comparable in magnitude to the $1p1h$ absorptions.[34] But since the final states involved in these asymmetry experiments are $1h$ states, they are not likely to be fed nearly so strongly from the quasideuteron as from the *DSD* component. Hence it is unlikely that quasideuteron absorption is the problem here.

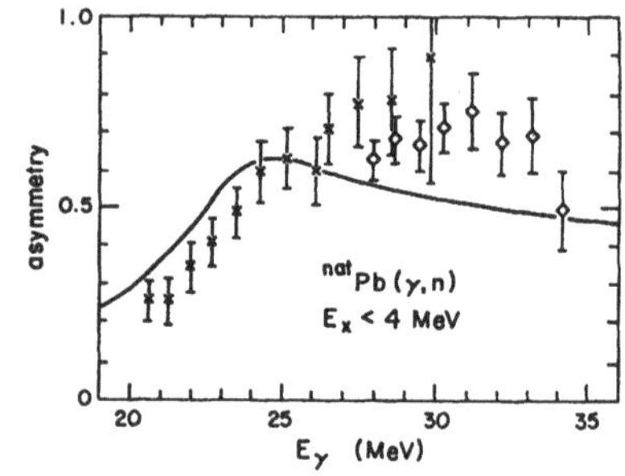

Fig. 9. The fit of a *DSD* calculation to the observed fore-aft asymmetry in (γ, n) on ^{208}Pb.[33]

Another possible omission in the conventional theoretical analysis of the asymmetries is suggested by the behavior of the observed ratio of the 90° neutron yield to the average of the yields at 55° and 125° for low-lying states in ^{208}Pb (γ, n). (Fig. 10) This ratio is observed to fall rapidly with increasing energy, and it therefore looks as though the forward peaked angular distribution is simply becoming narrower as the incident photon energy increases. This would have to mean that contributions are beginning to come in from multipolarities higher than $l = 2$. Such amplitudes would have to be collective since direct electric transitions of multipolarity >1 are negligible for neutrons. Evidence for $E3$ contributions in photoabsorption have been noted in the past,[35] and it may be that such absorption is occurring here. The fits of experiment to calculations would no doubt be improved by including this extra amplitude, if only because it introduced additional, weakly constrained fitting parameters.

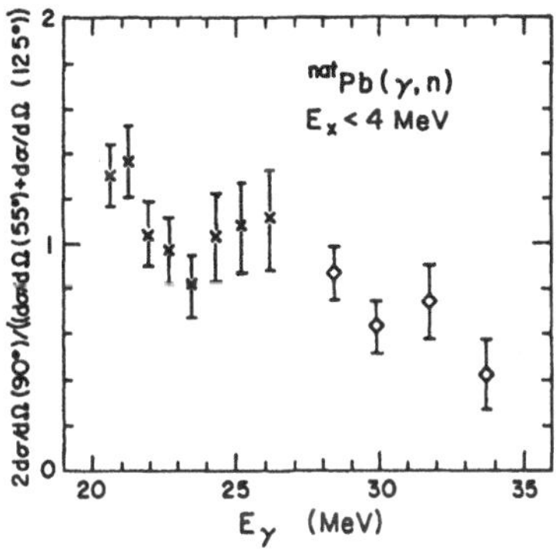

Fig. 10. The dependence of the ratio of the 90° yield to the average yield at 55° and 125° for (γ, n) on ^{208}Pb.[33]

9. Bremsstrahlung and the *DSD* process (Incident Nucleons)

Ground state capture cross-sections to a number of states in light nuclei have recently been studied for proton energies to 100 MeV.[36] There do not appear to be any estimates of the contribution of *pn* bremsstrahlung to capture into low-lying levels, but as in the inverse reactions just considered one would imagine that at high incident energy this contribution would be even smaller that the small *DSD* contributions ($\sim$ 10 to 100*nb*). Calculations for these (p, γ_0) cross-sections in light nuclei were carried out in terms of only the *DSD* process.[36] The calculations included no multipole resonance beyond dipole, and no direct amplitudes beyond quadrupole. Nonetheless they apparently track the observed cross-sections (within an order of magnitude) but were found to be quite sensitive to the uncertainties of optical model parameters.

10. Summary and Conclusion

When authors of experimental papers find only modest agreement of their results with available models, in the fields of radiative capture and photoemission of fast nucleons, they often conclude with a request for improved theoretical treatments. And many theory papers, in corresponding circumstances, plead for better and more detailed data. Substantial improvements have been occurring in both areas.

In experiment, spectrometers of large solid angle and good resolution are increasingly used to detect particles thus permitting the separation of yields to different residual states. Photon detectors of nearly 4π solid angle permit one to study the decays of specific final states at acceptable rates in coincidence experiments.[12] The measurement of analyzing powers with polarized beams[37] provides information, often critical, which is complementary to that obtained from angular distributions and cross-section magnitudes. Photon taggers have brought increasing clarity to photoreaction studies.[38] When laser taggers reach the intensities ($\sim 10^7$ photons/sec) that are now available from the bremsstrahlung taggers which have been used for (γ, n) studies, they will certainly make important contributions because of their intrinsically lower accidental rates.

On the theory side there have been steady improvements in the self-consistency of the nuclear wave functions used, for both initial and for final states and in the consistent treatment of the electromagnetic operators.[26] These changes have in part been a response to the failure of the older, simpler models to do the job. The latter, more elementary treatments involve limited degrees of freedom and are consequently easy to picture. They have played an important role in the past, helping to put in place the conceptual framework for the study of capture and photodisintegration. The main theme of this review has been that several nuclear degrees of freedom are often involved in these reactions and they must therefore be incorporated in any theoretical description which aims to make good quantitative predictions. One will still have to be careful, however, in drawing conclusions from theoretical matches with experimental results, since the fuller theory is less transparent and also involves more options and correspondingly more adjustable parameters than the simpler models.

The references which have been cited are meant to be merely representative of recent findings and problems rather than exhaustive. The review was prepared under a grant from the U.S. Department of Energy.

References

1. K. Nakayama and G. Bertsch, Phys. Rev. C **34**, 2190 (1986).
2. H. Nifenecker and J.P. Bondorf, Nucl. Phys. **A442**, 478 (1985).
3. M. Kwato Njock *et al.*, Phys. Lett. B **207**, 269 (1988).
4. J. Edgington and B. Rose, Nucl. Phys. **89**, 523 (1966).
5. J.S. Levinger, Phys. Rev. **84**, 43 (1951).
6. I. Halpern, Proc. Thousand Lakes Conf., Jyväskylä, Finland (1971).
7. D.M. Drake, S.L. Whetstone and I. Halpern, Nucl. Phys. **A203**, 257 (1973).
8. M. Potokar, Phys. Lett. **B46**, 346 (1973).
9. D. H. Youngblood *et al.*, Phys. Rev. C **23**, 1997 (1981).
10. H. Ejiri, Int. Symp. on Highly Excited States and Nucl. Struct., Orsay, France (1983); J.R. Beene *et al.*, Phys. Rev. C **41**, 920 (1990).
11. R. Alarcon, P.L. Cole, D.S. Dale, P.T. Debevec and L.J. Morford, Phys. Rev. C **43**, R2470 (1991).
12. A. Bracco *et al.*, Phys. Rev C **39**, 725 (1989).
13. D. Drake, S.L. Whetstone and I. Halpern, Phys. Lett. B **32**, 349 (1970).
14. K.A. Snover, Ann. Rev. Nucl. Part. Sci. **36**, 545 (1986).
15. J.O. Newton *et al.*, Phys. Rev. Lett. **46**, 1383 (1981).

16. M. Kicinska-Habior *et al.*, Phys. Rev. C **41**, 2075 (1990).
17. V. Metag, Nucl. Phys. A **488**, 483c (1988).
18. H. Nifenecker, Progr. Part. and Nucl. Phys. **23**, 271 (1989).
19. N. Herrmann *et al.*, Phys. Rev. Lett. **60**, 1630 (1988).
20. J. Randrup and R. Vandenbosch, Nucl. Phys. A. **490**, 418 (1988).
21. H. Myers, A. Odian, P.C. Stein and A. Wattenberg, Phys. Rev. **95**, 576 (1954); M.Q. Barton and J.H. Smith, Phys. Rev. **95**, 573 (1954).
22. S.N. Dancer *et al.*, Phys. Rev. Lett **61**, 1170 (1988).
23. A. Lepretre *et al.*, Nucl. Phys. A **367**, 237 (1981).
24. B. Schoch, Phys. Rev. Lett. **41**, 80 (1978); B. Schoch, Nucl. Phys. A **358**, 307c (1981).
25. J.L. Matthews, D.J.S. Findlay, S.N. Gardner and R.O. Owens, Nucl. Phys. A **267**, 51 (1976).
26. H. Hebach, A. Wortberg and M. Gari, Nucl. Phys. A **267**, 425 (1976); M. Gari and H. Hebach, Phys. Reports **72**, 1 (1981); M. Cavinato, M. Marangoni, P.L. Ottaviani and A.M. Sarius, Nucl. Phys. A **373**, 445 (1982).
27. S.V. Springham *et al.*, Nucl. Phys. A **517**, 93 (1990); S.A. Wender, N.R. Roberson, M. Potokar, H.R. Weller and D.R. Tilley, Phys. Rev. Lett. **41**, 1217 (1978).
28. D. M. Drake *et al.*, Phys. Rev. Lett. **47**, 1581 (1981).
29. A. Erell *et al.*, Phys. Rev. Lett. **52**, 2134 (1984).
30. S.L. Tabor *et al.*, Nucl. Phys. A **422**, 12 (1984).
31. D.S. Dale and R. Laszewski (to be published).
32. I. Bergqvist *et al.*, Nucl. Phys. A **419**, 509 (1984); R. Zorro *et al.*, Nucl. Phys. A **472**, 125 (1987); A. Håkansson *et al.*, Nucl. Phys. A **512**, 399 (1990).
33. T. Murakami *et al.*, Phys. Rev. C **35**, 479 (1987); Asmus Freytag, Thesis, Univ. of Illinois (1988).
34. P. Oblozinsky, Phys. Rev. C **40**, 1591 (1989); P. Oblozinsky, Proc. 4th Workshop on Rad. Capture, p. 329, Berkeley (1990).
35. F.S. Dietrich, D.W. Heikkinen, K.A. Snover and E. Ebisawa, Phys. Rev. Lett. **38**, 156 (1977).
36. H.J. Hausman *et al.*, Phys. Rev. C **37**, 503 (1988).
37. F. Saporetti and R. Guidotti, Nucl. Phys. A **390**, 207 (1982); M.A. Kovash *et al.*, Phys. Rev. C **40**, R1093 (1989); G. Feldman, L.H. Kramer, H.R. Weller, E. Hayward and W.R. Dodge, Phys. Rev. C **43**, 223 (1991).
38. L. Federici *et al.*, Nuovo Cimento **598**, 247 (1980); M. Preger, B. Spataro, R. Bernabei, M.P. de Pascale and C. Schaerf, Nucl. Inst. and Meth A **249**, 299 (1986).

PROGRESS OF NUCLEAR DATA WORK OF CNDC

CAI DUNJIU, ZHOU DELIN and SUN ZUXUN

Chinese Nuclear Data Centre
China Institute of Atomic Energy
P.O.Box 275(41), Beijing 102413
P. R. China

ABSTRACT

The progress of nuclear data work performed, organized and coordinated by CNDC is reviewed.

1. Introduction

Nuclear data compilation and evaluation in China was started systematically in 1975, soon after that the Chinese Nuclear Data Center (CNDC) was set up in order to create evaluated nuclear data library to meet the requirements for development of nuclear science and technology. Then China was not an IAEA member, so we only have evaluated data from publications and had to do our own evaluations and to generate group constant, etc. Therefore, many experimental and theoretical nuclear scientists in China were involved in the related activities.

This situation has changed since the exchanges were established between CNDC and IAEA / NDS, BNL / NNDC and JAERI / NDC respectively in 1980. Especially in 1984, when the People's Republic of China became an IAEA member, international cooperation and exchange between CNDC and other Nuclear Data Centers greatly increased. We participated in the international cooperation in the evaluation and compilation of neutron nuclear data, nuclear structure and decay data, fission product yields, and charged particle reaction data, etc. coordinated by IAEA / NDS or BNL / NNDC. We obtained some evaluated neutron data sets from IAEA, JAERI / NDC and CJD, and some programs from ORNL / RSIC, OECD / NEA–DB and IAEA / NDS. Of course, this is very helpful to our nuclear data work.

2. Evaluation of Neutron Nuclear Data for General Purposes

2.1. Evaluations related to the finalization of CENDL–2

In recent years, efforts have been spent to the evaluations related to the finalization of CENDL–2. 54 nuclides are included in this library and all evaluations are performed by Chinese evaluators but a few of them were completed for other evaluated nuclear data Libraries via international cooperation. About 2 / 3 of the nuclides included in CENDL–2 are from CENDL–1 after extensive revision or reevaluation. The rest are newly evaluated. Most of the evaluations have been complemented with nuclear model theoretical calculations by using the codes (MUP2, FUP1, etc.) developed by ourselves. Firstly the evaluations were reviewed by the assigned reviewer and then, reviewed and approved by the Subcommittee of Neutron Nuclear Data Evaluation. Finally, they were checked by using the codes of CHECKR, PSYCHE, FIZCON, and then accepted by CENDL–2.

Generally, most of the data measured recently in the smooth energy region were taken into account in the evaluations. Our evaluations were compared with the other newly available evaluations. For some evaluations efforts have been made to update them significantly. The evaluated resonance parameters including those for the negative level(s) have been checked and adjusted, if necessary, comparing the calculated cross sections with resolved parameters with Doppler–broadened to the available point–wise data measured at thermal and resonance energy region. To ensure the 'continually connection' of the group–averaged cross sections with the unresolved energy region physically at the boundaries, the parameters of the resonances located above the boundary of the 'resolved energy region' are added, if possible. More often, the 'background cross sections' are given properly to the corresponding interval at the boundary, similarly to other boundaries.

The unresolved resonance parameters are estimated based on the resolved parameters, then checked and adjusted to fit the newly measured point–wise data in the unresolved resonance energy region.

So CENDL–2 is greatly improved comparing with CENDL–1. And it seems that most of the evaluations included in CENDL–2 are comparable to the corresponding files of the available newest versions of other Libraries.

2.2. Preparations for further improvement of CENDL

Only files 1–5 are included in CENDL–2. So in recent years much efforts have been made to the preparations for further improvement of CENDL–2, mainly to include the files of γ–production data, double differential cross section (DDX), covariance file, and the files for special purposes etc.. Correspondingly, a code UNF which is able to calculate the DDX of neutron and charged particles emission as well as recoil nuclei, to calculate the data of γ–production concerned, etc. based on a better nuclear model theoretical framework; and a code for combining the correlated data sets in order to obtain combined (evaluated) values and their covariances correctly and reasonably have been developed. These codes have been successfully used for some calculations and evaluations. Furthermore, for improving the evaluations of averaged resonance parameters, especially the average level spacing and the level density, a method based on Bayesian approach for correcting the energy level missing related to Porter–Thomas distribution as well as Wigner distribution has been established.

In the meantime, we have taken part in the international cooperation of FENDL improvement and the related activities. We believe these activities will be valuable to the further improvement of CENDL.

2.3. Model theories and codes developed for nuclear data calculation

Various model theories have been developed since the early stage of the establishement of CNDC, and then various codes have been made correspondingly for data calculation for light, medium–heavy and fission nuclides. In brief, the codes and the related theories are summarized as following.

TSD (Chu Lianyuan et al.)

Based on Faddeev–AGS theory. It has been used to calculate the n+D reaction.

RAC (Chen Zhenpeng et al., Tsinghua University)

A comprehensive R–matrix analysis code based on multichannel and multilevel R–matrix theory. It has been used to analyse the ^{7}Li and ^{17}O systems to evaluate the data of n + ^{6}Li (E_n < 1.72 MeV) and n + ^{16}O (E_n <6.2 MeV) reactions for CENDL and ENDF / B–VI respectively.

ROP, CROP, DRM (Zhu Yaoyin et al., Jilin University)

ROP is a unified code of optical model and R matrix theory. The results obtained by means of the optical model are used as background of R–matrix analysis in the resonance region. CROP is able to fit simultaneously the experimental data of both a reaction and its inverse reaction in a system. DRM is a unified code for calculating direct inelastic scattering cross section for 1p shell nuclei with optical model and DWBA.

These codes have been used to the analyses of ^{7,8}Li system.

MUP2 (Cai Chonghai et al., Nankai University)

Second version of a unified program for theoretical calculation of fast neutron data except gamma production for medium–heavy nuclei by using the optical model, H–F theory with width fluctuation correction and pre–equilibrium statistical theory based on the exciton model and evaporation model.

MUP2 is the main code which has been used in the evaluations forCENDL–2 and it has been compared with other available programs.

FUP1 (Cai Chonghai et al.)

An extended code of MUP2 for fission nuclide calculations. With CCOM, a code mentioned below, FUP1 has been used in the evaluations of several fission nuclides for CENDL–2.

CMUP2 (Cai Chonghai, Shen Qingbiao)

A code based on MUP2, for calculation of charged particle as well as neutron induced reaction at the energy region of 4–50 MeV, including 1–4 particles emission. It has been used to calculate the data sets of p + ^{89}Y and n + ^{56}Fe, the incident energy from 4 to 50 MeV.

UNF (Zhang Jingshang et al.)

Developed recently for structural material calculation with the unified treatment of the pre–equilibrium and equilibrium reaction processes with angular momentum and parity conservations. The pick–up mechanism of composite particle (d,t, ^{3}He, ^{4}He) emission and the discrete level effects are taken into account for improving the calculations of reaction cross sections concerned. It is able to calculate DDX of particles emission as well as recoil nuclei, and γ production data. With other codes for optimal optical parameters and direct components complement, UNF has been used to calculate the data sets of n + ^{56}Fe, n + ^{209}Bi and n + ^{60}Co reactions at 14 MeV.

CCOM (Yang Zesen et al., Peking University)

A code of coupling channel optical model for deformed nucleus to calculate the direct reaction components of inelastic scattering.

2.4. Typical results of the evaluations of neutron nuclear data for general purposes

i) Calculation of n+D reaction (Chu Lianyuan et al.)

Angular distribution of elastic scattering and DDX of neutron and proton emission from D(n,2n)P reaction are calculated by using the code TSD. Some results are shown in Fig. 1.

ii) R–matrix analysis of n + ^{16}O system (Cheng Zhenpeng et al., Tsinghua University)

R–matrix analysis of n + ^{16}O system for the neutron energy region of $E_n < 6.2$ MeV has been carried out in cooperation with LASL for ENDF / B–VI. The available measured data (shown in Table 1) related to ^{17}O system are included in the analysis. The calculations have been accepted as part of the evaluation for n+ ^{16}O reaction data which has been included in ENDF / B–VI and CENDL–2.

iii) Neutron nuclear data evaluation for ^{19}F (Zhao Zhixiang et al.)

The evaluation of ^{19}F for ENDF / B–V did not satisfy the energy balance. The new evaluation for ENDF / B–VI and CENDL–2 are performed by CNDC in cooperation with ORNL and based on a consistent theory calculation by using TNG to insure energy balance. Much effort have been spent on adjusting parameters so that a good simultaneous fit to all of information, including measured cross sections, angular distributions, DDX and γ–production data is obtaned. The covariance matrix of model theory parameters are obtained based on the scatter of measured data around the calculations as the external error so the covariance matrix of the calculated cross sections estimated from the covariance matrix of the parameters looks reasonable.

The data set for ^{19}F has been accepted by ENDF / B–VI and included in CENDL–2.

iv) Neutron nuclear data evaluation for ^{197}Au (Yuan Hanrong, Liu Tong; Wang Shunuan, Yan Shiwei)

A typical evaluation for which the measured data are evaluated and complemented with theoretical calculation by using MUP2. Some results are shown in Fig. 2.

The evaluated data set for Au has been included in CENDL–2.

v) Evaluations of $^{241}Am(n,\gamma)$ reaction cross section and resonance parameters (Zhou Delin)

The experimental information of thermal capture cross section are shown in Table 2.1. From which one can see that all measured values are in agreement with each other except the sole differntial measurement which gives a lower value of 582 b and a small uncertainty of 2 %. For compromising this 'discrepancy', different evaluators obtained different evaluations based on the same data body, as shown in Table 2.2. Actually the uncertainty of Weston and Todd's measurement is underestimated and a correlated error of 7 % mentioned by the authors should be added, i.e., 582 ± 45 should be quoted instead of 582 ± 12. Then a weighted mean of the thermal value of capture cross section and the resonance parameters are obtained without any compromise. And the problem which puzzled evaluators for years are solved.

The data sets for ^{241}Am, and for ^{249}Bk and ^{249}Cf as well have been accepted in ENDF / B–VI and included in CENDL–2.

vi) Neutron nuclear data evaluation for Fe at 14.1 MeV

(Zhou Delin, Zhao Zhixiang et al.

Zhang Jingshong, Shen Qingbiao, et al.)

The new evaluations for Fe and its isotopes will be carried out. These evaluations aim to improve the evaluations and model theory calculations; to improve the covariance evaluations and to generate the covariance for file 4 and file 5. As a first step, the evaluation and calculation for 14.1 MeV have been carried out. The results of the reaction cross section evaluations are shown in Table 3. The calculations of n, p and α emission etc. are illustrated in the figures presented in Zhang Jingshong's paper in this proceedings.

3. Underlying Research Related to Nuclear Data Evaluation and Model Theory Calculation

3.1. Semi–classical theory of multi–step nuclear reaction processes

(Zhang Jinshang et al.)

(Please refer to the related paper in this proceedings.)

3.2. Non–relativistic and relativistic phenomenological and microscopic optical potentials for neutron induced reactions at low and medium energies (Shen Qingbiao, Zhuo Yizhong et al.)

(Please refer to the related paper in this proceedings)

3.3. Dispersion optical potential (Su Zhongdi et al.)

A review of optical model analyses in A = 40–60 mass region is given and the variation of the total cross section, the elastic scattering angular distribution and the absorption cross section with the optical potential parameters is studied. The comparative analyses of different parameterizations of the spherical optical potential have shown that a deep minimum in the total cross section at low energies cannot be reproduced by any existing parameterizations. These optical potentials which fit the higher energy data very accurately tend to give total cross sections at low energies that are clearly greater than the mean of the fluctuating experimental values. It has been shown that a consistent analyses using the dispersion relations can remove this 'anomaly' of the total cross sections at low energy.

3.4. Parameters of nuclear level density (Su Zongdi, Zhuang Youxiang et al.; Huang Zhongfu et al., Guangxi University)

Although the Gilbert–Cameron (GC) formula is rather universal in the application of the statistical theory, its constant temperature formula at low energy range is from an empirical approach without theoretical guidance, and the spin cut–off factor for the constant temperature formula is not well–defined and disagree with the experimental data either. Compared to GC formula, the problem mentioned above does not exist in the back–shifted Fermi gas model (BS formula), but its parameter form is not as universal as GC formula. The reasonable parts have been extracted from both formulas and combined them together. First of all, the level density parameters of BS formula for 318 nuclei have been obtained by fitting the evaluated experimental data. The spin cut–off factor at low energy region for BS formula has been improved. The systematics study of the level density parameters (a) has been carried out. The systematics results of S(Z) and S(N) have been obtained.

3.5. Estimation of averaged level spacing $<D>$ (Zhao Zhixiang, Huang Zhongfu)

Most of the methods to estimate the average level spacing $<D>$ are based on a truncated Porter–Thomas distribution of reduced neutron width and have not used the information in level position because of the fact that the level missing distorts the whole range of level spacing distribution. The using of Bayesian approach makes it possible to simultaneously use the information involved in both neutron width and level position, therefore it could improve the estimation of $<D>$.

The method has been used to estimate average level spacing of S–wave resonance for many even–even targets and also generalized to odd–A targets. Generally the results obtained are consistent with those of BNL–325 (Vol. 1 , 4–th edition), and the uncertainties of the estimated values are reduced.

4. Evaluation of Nuclear Data for Special Purposes

4.1. Activation cross section and nuclear data for dosimetry

(i) Evaluation of Activation cross section and nuclear data for dosimetry (Lu Hanlin, Zhao Wenrong et al.; Cai Dunjiu et al.)

(Please refer to the related paper in this proceedings.)

(ii) Systematics of excitation functions for neutron induced reactions (Zhao Zhixiang, Zhang Jin, Zhou Delin et al.)

Systematics studies on the excitation functions of the (n,γ), (n,2n), (n,3n) as well as (n,x) (x=p, d, t and ^{3,4}He) reactions have been performed. The parameterized formulae and all the parameters concerned have been obtained on the basis of the evaporation model with the preequilibrium mechanism and the collected data for A=25–200 in the neutron energy range from threshold to 25 MeV.

With the formulae and parameters, the excitation functions could be predicted simply for the energy regions or nuclides not measured heretofore and the formulae have been used for curve fitting in case the measured data are scarce.

4.2. Fission yields and calculation of decay heat (Wang Dao)

The Chinese Evaluated Fission–Product Yield Library has been established. The 1987 version of this library contains data in ENDF / B–V format for 10 fissioning systems, they are U235T, U235F, U235HE, U238F, U238HE, Pu239T, Pu239F, Pu241T, U233T, Th232F. (T–thermal neutron, F–fast (fission) neutron, HE–14 MeV neutron). The data library has been sent to IAEA / NDS.

At present, the effort is now being paid to obtain an improved and more complete data base of experimental data in EXFOR format and study the methods to derive a best recommended set of fission yield data, which aim at updating the 1987 version data.

Besides, the "decay heat production" research has been developed in recent years.

4.3. Nuclear structure and decay data (Zhou Chunmei et al.)

All useful ENSDF codes have been transplanted on MICRO–VAX–II of CNDC. The data for A = 51–55, 195–198, 170, 172 (total of 12 mass chains) have been evaluated and renew of some mass chains started in 1990.

4.4. Charged Particle Data (Zhuang Youxiang et al.; Shen Qingbiao et al.)

Twenty sets of data measured in China have been compiled in EXFOR format and sent to IAEA. The 220 sets of measured data (50–1000 MeV) for the reactions induced by proton or heavy charged particles were compiled in EXFOR format in cooperation with BNL / NNDC. Some activation cross sections induced by charged particles are being evaluated.

4.5. Medium Energy Nuclear Data (Shen Qingbiao, Li Zhuxia et al.)

Several studies on medium energy nuclear data calculation and evaluation have been done recently. The subjects are: 1. Calculation of 5–50 MeV neutron induced reaction data of ^{56}Fe; 2. The study of neutron emissions in 585 MeV proton on ^{56}Fe with QMD; 3. Neutron relativistic phenomenological and microscopic optical potential; 4. Systematics of intermediate energy proton nonelastic and neutron total cross section.

4.6. Atomic and Molecular (A+M) Data (Yao Jinzhang et al.)

The sputtering data for collisions of C, Al, Fe atoms and their ions with H, D, He atoms are being compiled and evaluated. The excitation cross sections of collision for He, Ne, Ar atoms with their ions, the electron impact ionization cross sections for the atoms and ions with high Z and the cross section for K–shell ionization by electron impact are also being compiled.

5. Correlated Data Combination

5.1. Codes for correlated data combination

(i) GAFC (Zhang Jin) and CPV (Shi Zhaomin), codes for group average of multi–measurements.

Usually this method is used to combine the correlated data to obtain evaluated values and their covariance in the evaluation and in a few cases in the simultaneous evaluations as well in case the measured data to be combined are distributed densely and uniformly enough.

(ii) TPFC, code for curve fitting with parameterized theoretical formulae (Zhang Jin, Zhao Zhixiang)

It has been recommended to use the parameterized formulae to fit the measured excitation functions correspondingly to obtain the whole excitation function and the covariance in case the measured data are scarce or / and distributed ununiformly, or alternatively, to fit or to normalize to such data with model theory calculations. Some typical results of curve fitting by using parameterized theoretical formulae are shown in Fig. 4.

(iii) Curve fitting of excitation function measurements with some mathematical polynomials

The least squares codes based on polynomials, orthogonal polynomials, Legendre polynomials (TPFC, by Zhang Jin, Liu Tong) and spline function (SP, by Zhou Hongmo, Liu Tingjin et al.) have been developed for curve fitting and used to combine the multi–measurements to obtain the evaluated excitation functions.

5.2. PPP in correlated data combination and its solution (Zhao Zhixiang et el.)

When the least squares method are used to combine the correlated data, sometimes the combined values are outside all of the measured values. Obviously it is unresonable and unacceptable. This phenomenon is proposed by R. W. Peelle as "Peelle's Pertinent Puzzle (PPP)". The detailed description of PPP and its solution are included in some papers to be published.

The approximate methods have been adopted in the codes GAFC, TPFC, and MPFC for correlated data combination.

6. Data Processing (Liu Guisheng, Wang Yaoqing et al.)

To meet the needs in nuclear engineering, since 1986 a large amount of nuclear data for nuclear power, nuclear safety analysis and nuclear engineering design have been processed. Based on some codes from ORNL / RSIC and NEA Data Bank, a few of program systems used for data processing and reactor calculations have been established.

1) Three programs have been implemented on computer CYBER–825 at CIAE. These codes are AMPX–II, NJOY and MINX which are used in data processing for nuclear science and technology. In cooperation with ECN (Holland), the ECNJEF–1 Library was set up by using NJOY–87.0, CRECTJ5, MILER, NPTXS and XLACS–2. The ECNJEF–1 Library is a 219 neutron group library from JEF–1 in the AMPX master format. 16 files are included in this Library with 434 elements. Some problems in JEF–1 and related processing codes were found. Most of them were solved.

2) PASC–1 was installed based on AMPX–II in cooperation with ECN. It is composed of AMPX / SCALE group constant module, several transport calculation codes and 1–3 dimension diffusion code CITATION. This code can be used in thermal and fast reactor calculations and shielding analysis. A library including 45 and 52 group constants of 37 nuclides has been processed by PASC–1 for the fast reactor calculations. A multigroup Monte –Carlo code KENO–IV from RSIC has been linked to the code system PASC–1. Through a large amount of benchmark calculations, it has been proved that the AMPX–KENO–IV is reliable for the critical safety analysis. Therefore it was applied to the critical safety calculations of the high–density spent–fuel storage racks at the Qinshan nuclear power plant.

3) Based on BINX and CINX, the CCCCPS code system, which is automatized, has been made for processing the interface file of CCCC format library. By using the code system, the 50 group constants of 49 nuclides from LIB–IV CCCC library were processed into 46 group ABBN library in the Bondarenko format for the input of the code 1DX for the fast reactor design.

4) The FRBT code system is composed of three codes EXDL, BRKP and NDP, and an integral experimental data library included 19 fast benchmark assemblies. The material compositions, geometric conditions and the other controlled variables about 19 assemblies recommended by American CSEWG have been included in the library. The EXDL is a code to retrieve an interface file for calculating a reactor by NDP. The BRKP is able to retrieve an interface file of the multigroup constants in the Bondarenko format from the ABBN library. NDP is an one–dimension diffusion code used in critical calculations of reactors. The user's input of the FRBT is very simple. Using the FRBT code system with the ABBN library of 46 groups, the calculations about 14 fast critical benchmark assemblies have been completed. The results are consistent with

those given by CSEWG.

5) The theoretical analysis for Be sphere shell neutron multiplication factor has been made. This is a cooperation project with U.S.A. and Japan.

Table 1. Channel configuration and data summary for the ^{17}O system by R–matrix analysis

Channel	L(max)	Radii (fm)
n + ^{16}O	4	4.44
α + ^{13}C	4	6.01

Reaction	Energy range (MeV)	Observable type	Data points
$^{16}O(n,n)$	0–6.5	Total, Diff., Polar.	2421
$^{16}O(n,\alpha)$	0–6.0	Integ.,Diff., Polar.	904
$^{13}C(\alpha,\alpha)$	0–4.6	Diff.	207

Table 2.1 Experimental information of thermal capture cross section of ^{241}Am

Values (b)	Authors		Comments
650 ± 104	Dovbenko et el.	(1971)	Integral meas.
612 ± 35	Harbour et al	(1973)	〃
625.4 ± 35	Pomerance	(1975)	〃
582 ± 12	Weston & Todd	(1976)	Absorp. meas.
625 ± 20	Kalebin et al.	(1976)	Transm. meas. $\sigma_t = 640$ b
639 ± 37	Wisshak et al.	(1982)	$\sigma_\gamma(^{241g}Am) / \sigma_\gamma(Au)$ meas.

Table 2.2 Evaluated values of thermal capture cross section of ^{241}Am

Values (b)	Authors		Comments
625 ± 20	Goel	(1978)	$\sigma_t - \sigma_f - \sigma_s$
600 ± 20	Lynn et al.	(1980)	Considering all meas.
610 ± 19	Froehner et al.	(1982)	〃
587 ± 12	Mughabhab	(1984)	〃
626 ± 14	Zhou Delin	(1988)	〃

Table 3.1 Evaluation of integral cross section for ^{56}Fe at 14.1 MeV individually and comprehensively

	(n,non)	(n,n–em)	(n,p–em)	(n,α–em)	(n,d–em.)
individually	1370 ± 13*	1610 ± 22+	155 ± 10	41.5 ± 1.5	8 ± 3
comprehensively	1377 ± 13	1632 ± 22	155 ± 10	41.5 ± 10	8 ± 3
BROND	1320	1616	166	38.8	
ENDF / B–VI	1386	1643	165	45.8	
JENDL–3	1426	1739	173	42.8	

* Evaluated by Pavlik and Vonach for Fe. + Evaluated value for Fe.

Table 3.2 Evaluation of reaction cross section for ^{56}Fe at 14.1 MeV individually and comprehensively

	(n,n′)	(n,2n)	(n,p)	(n,np)	(n,α)	(n,nα)	(n,d)
individually	788 ± 40	417 ± 17	114.5 ± 3*	40.5 ± 10	40 ± 1.5	1.5 ± 1*	8 ± 3
comprehensively	757 ± 40	417 ± 17	114.5 ± 3	40.5 ± 10	40 ± 1.5	1.5 ± 1	8 ± 3
BROND	659	447	112.4	63.1	38.4		
ENDF / B–VI	756	413	114.5	50.2	44.2	1.5	6.5
JENDL–3	743	467	113.0	60.4	41.5	1.3	

* ENDF / B–VI

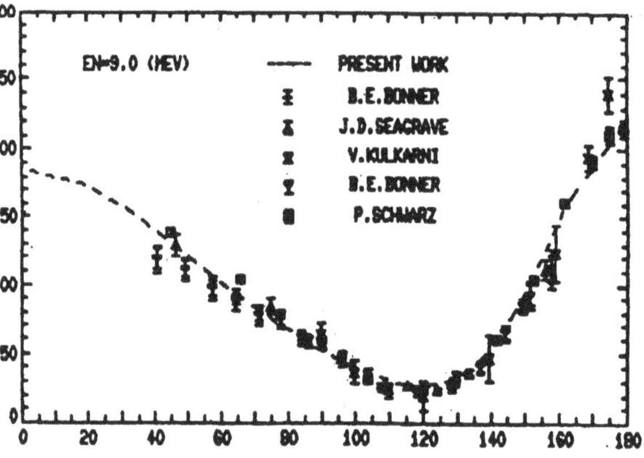

Fig. 1.3 MEASURED AND EVALUATED ELASTIC ANGULAR DISTRIBUTIONS

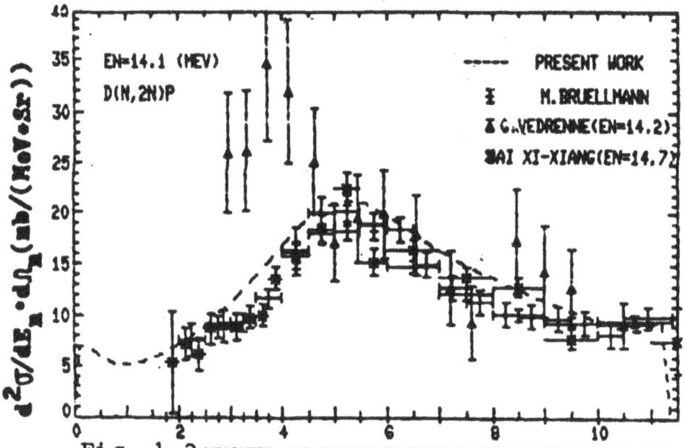

Fig. 1.2 MEASURED AND EVALUATED DOUBLE-DIFFERENTIAL E(MEV)
CROSS SECTIONS FOR THE (N,2N) REACTION AT 15°.

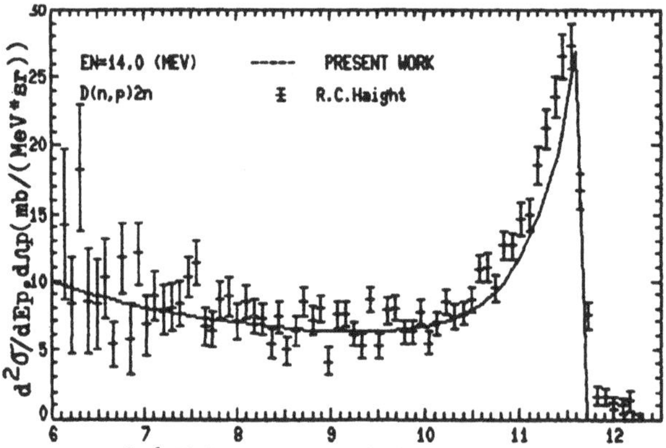

Fig. 1.1 MEASURED AND EVALUATED DOUBLE-DIFFERENTIAL E(MEV)
CROSS SECTIONS FOR THE D(n,p)2n REACTION AT 1.6 .

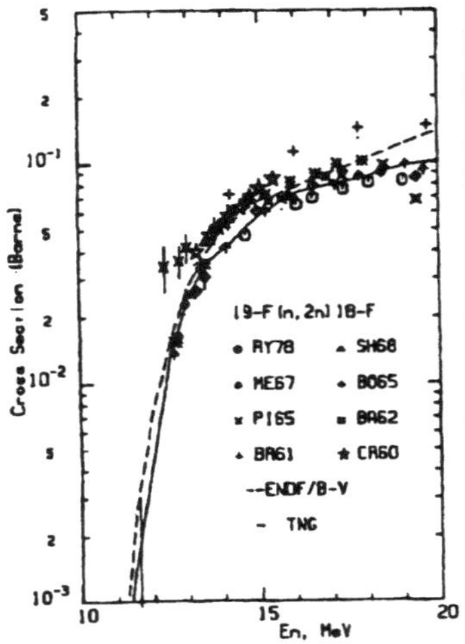

Cross sections of $^{19}F(n,2n)$ reaction.

Fig. 2.1

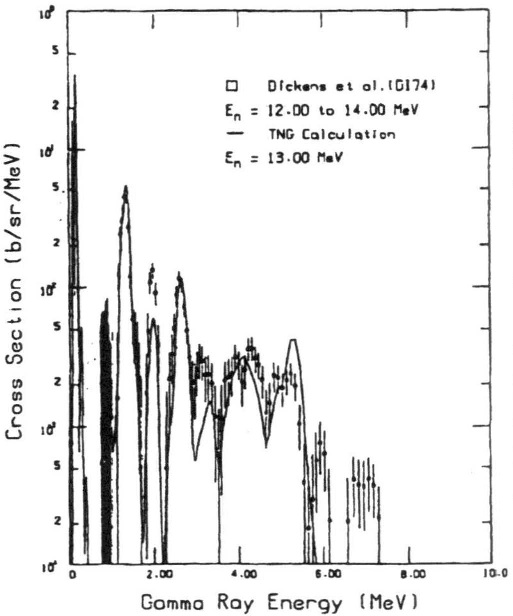

Gamma-ray spectra from $n + {}^{19}F$.

Fig. 2.2

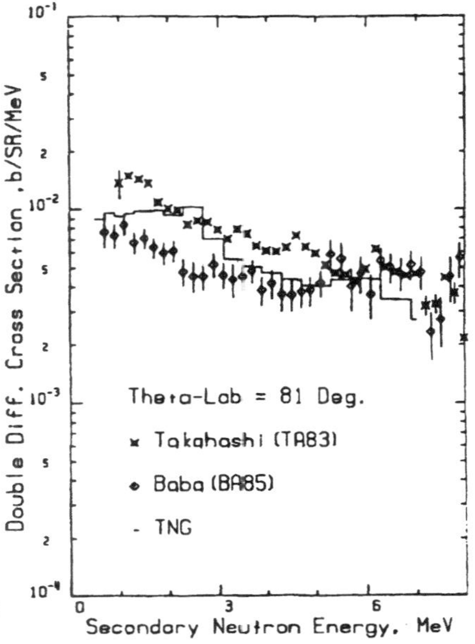

Secondary neutron spectra of ^{19}F at $E_n = 14$ MeV.

Fig. 2.3

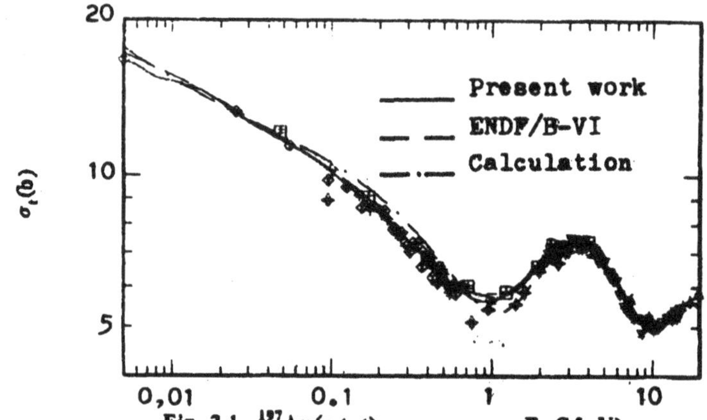

Fig. 3.1 ^{197}Au(n,tot) En(MeV)

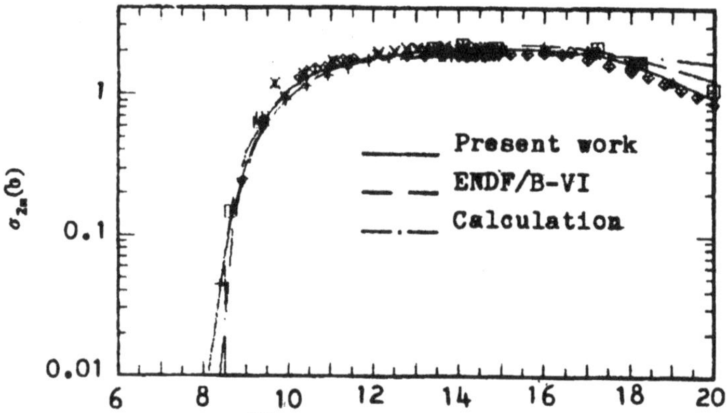

Fig. 3.2 ^{197}Au(n,2n) En(MeV)

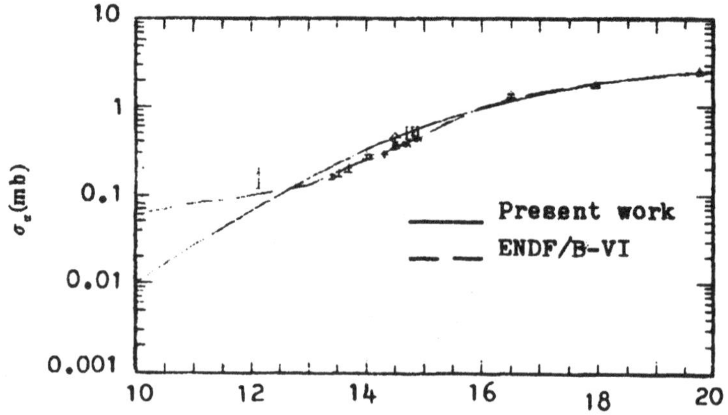

Fig. 3.3 ^{197}Au(n,α) En(MeV)

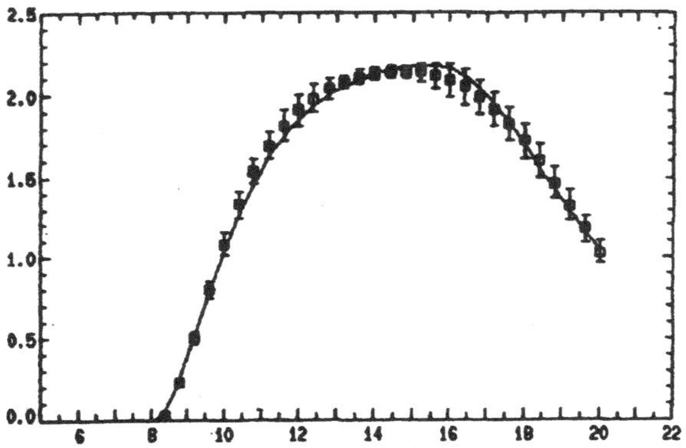

Fig. 4.1 Au-197(n,2n), _____ fitting

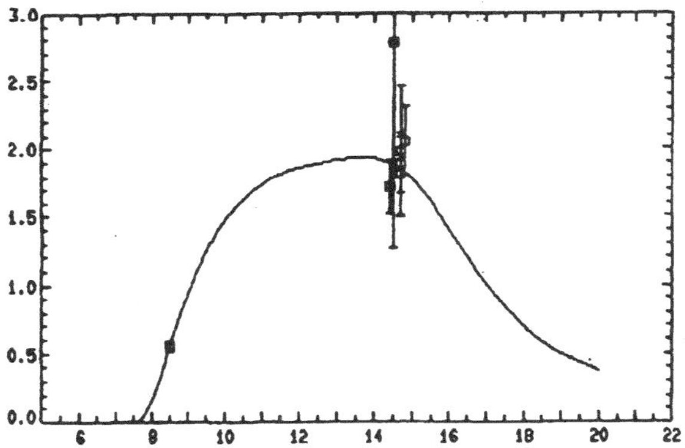

Fig. 4.2 Pt-198(n,2n), _____ fitting

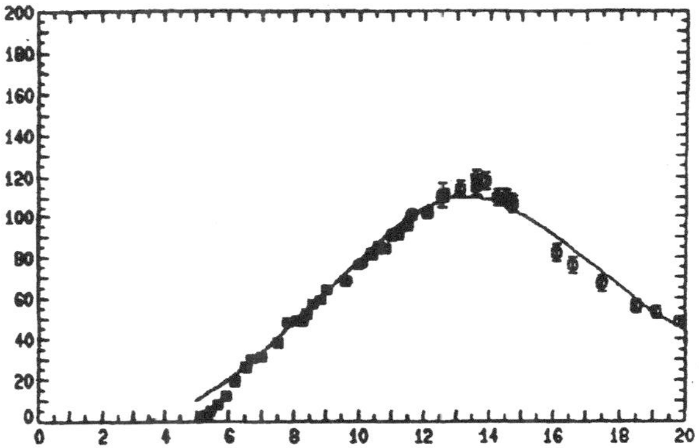

Fig. 4.3 Fe-56(n,p), _____ fitting

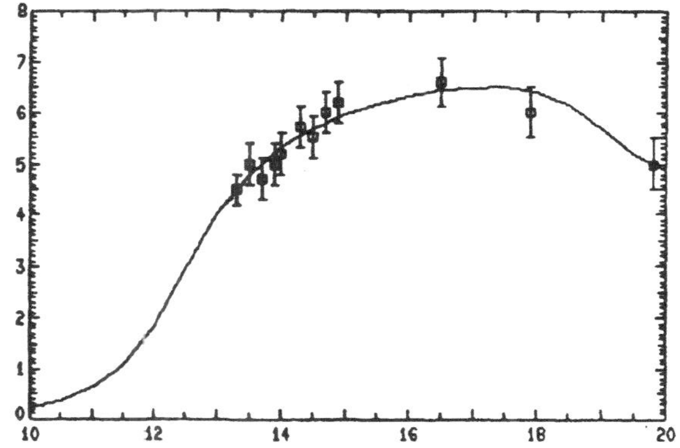

Fig. 4.4 Zr-92(n,a), _____ fitting

SUMMARY TALK

A SUMMARY of RECENT TOPICS in NEUTRON PHYSICS

M. T. McEllistrem

Physics and Astronomy, University of Kentucky,
Lexington, Kentucky 40506-0055

Abstract

Studies in fast neutron physics range from nucleon-nucleon interaction properties and the structure of the nucleon, to applications of well understood neutron induced reaction rates to problems in materials analyses, or reaction rate problems in nuclear astrophysics. They range also from studies of two-body systems to the many body behavior in complex, deformed nuclei, from simple scattering problems in a single-particle mean field to complex, multi-mechanism reactions. Reaction rates are of great interest at energies above one GeV as a means of testing nucleon structure ideas and at eV energies as tests of reaction rates in stars. In short, neutron physics is a mature field, where complex phenomena can be studied using neutrons as rather well understood probes of nuclear structure, and in practical applications of potentially great importance for developing safe and reliable means of delivering electrical energy, and aiding society in other ways. The present conference focused principally upon problems in understanding low and moderate energy neutron reaction rates, in fundamental problems of nucleon-nucleon reactions, in probing nuclear structure, and in describing accurately the complex reactions relevant to the development of fusion energy systems.

1 Introduction

Fast neutron physics is a mature field, at energies well below 100 MeV, so that probing the frontier well below that energy usually means studying complex reactions whose accurate representation requires a complicated mix of reaction mechanisms, or using well understood reactions to probe nuclear structure or to study other problems.

The frontier in neutron physics is one usually dependent upon precise and highly accurate measurements. Confidence in precision and accuracy is best established through collaborations amongst many different groups; neutron physics lends itself well to strong international collaborations. Thus an international conference of this sort is one of the very best ways to test where we are in development, and which open questions are most likely to yield to concerted efforts in the future.

There are several key topics in this varied field which were singled out by speakers to illustrate the present status of knowledge, and programs under development. The general themes of the conference could be represented by the following list:

- Double Differential (and Integrated) Emission Cross Sections
  - applications to fusion and fission technology
  - international collaborations for level density studies
  - the mixture of reaction mechanisms
- Fission Studies – from μeV to 400 MeV
  - open problems – related to neutron emission
  - spontaneous and accelerator produced
  - power of international collaborations
- Neutron Capture (and Photonuclear Processes)
  - importance in technology
  - studies of complex nuclei at high excitation energies
  - lifetimes and collective strengths of levels
- Scattering Studies – Mean Field Approaches
  - nucleon-nucleon problems and spin observables
  - both bound and scattering descriptions for complex nuclei
  - mean fields from microscopic, effective nucleon-nucleon interactions
  - inter-university collaborations

- Nuclear Data – Compilations and Evaluations at CIAE
  - tests of reaction models and level densities
  - correspondence with major evaluations from other centers
  - The Data Center – a nuclear resource for all China
- Progress at China Institute of Atomic Energy (CIAE) in Fast Neutron Physics

This is by no means a complete list of subfields in which neutron physics plays a major role; but it does exhaust the topics touched in the presentations offered here in mid-September, 1991.

2 Double Differential Neutron Emission

One of the subfields most extensively concentrated on at CIAE, as well as at several other nuclear reactions study centers, is that of the measurement and interpretation of double differential cross sections. This refers to the measurement of the energy spectrum of secondary neutrons emitted from an entire set of non-elastic reactions, such as (n,n'), (n,2n), (n,pn), and others. The neutron energy spectra are measured as a function of emission angles; hence the name "double differential". A primary motivation for these measurements comes from the need to understand energy deposition in shielding and blanket walls of proposed fusion reactor systems for neutrons of energies between 8 and 14 MeV striking the walls or shielding.

An exhaustive and well measured set of these cross sections has been developed by A. Takahashi's group at Osaka University. They have provided results for nuclei as targets ranging from Lithium to Uranium at 14 MeV neutron energy. Takahashi has shown that this extensive data set can be reasonably well characterized by a combination of pre-equilibrium and equilibrium statistical models, together with direct reactions for the discrete groups observed at high neutron energy. His work indicates that these complex neutron spectra are rather well characterized. This data set served and serves as a comparison standard for other groups initiating work on these complex neutron spectra.

Neutron Source Problems – 8 to 14 MeV Obtaining these neutron emission spectra at 14 MeV incident is feasible with low energy deuteron accelerators, since a clean neutron source is available from the $T(d,n)^4He$ reaction at very low deuteron energies. However, obtaining clean sources at other incident neutron energies, through accelerating incident ions to higher energies, has been a serious problem. The strong source for the important neutron energy region between 8 and 14 MeV is the $D(d,n)^3He$ reaction with deuteron energies above 5 MeV, or alternatively the $T(p,n)^3He$ reaction with proton energies above 9 MeV incident energy. But three-body breakup yields seriously contaminated sources at energies just a few MeV below the desired source energy. Thus this important neutron energy region has been extremely difficult to examine with conventional source techniques.

Tang Hongqing and his colleagues at CIAE have circumvented this contaminated source problem with a novel but difficult and time-demanding technique. They have departed from the usual method of placing a neutron scattering sample some 7 to 15 cm from the source, and then a neutron detector between 4 and 8 m distant from the scattering sample, as used by Takahashi's group, for example. Instead Tang placed the sample a full 2.2 m from the source, but brought the neutron detector in to a flight path of only 0.7 m from the scattering sample. This resulted in a spectrum, for 10 MeV neutrons incident, in which the highest energy breakup neutrons arrived at the detector later than the lowest energy emitted neutrons from primary neutron interactions in the scattering sample. Thus the breakup neutrons are removed from the region of interest.

However, the costs of this method are that the incident neutron flux at the sample is about 400 times smaller than those with the conventional method, and normally backgrounds would swamp the signals of interest. Tang placed a massive iron collimator, massive enough to completely shield the neutron detector at its 0.7 m flight path from the direct source neutrons, to collimate the neutrons reaching the scattering sample. This shielding allowed a good signal to background ratio to be achieved.

Adequate counting statistics required counting times of nearly five times longer than with the conventional geometry, and scattered neutron velocity resolution at a flight path as short as 0.7 m was poor. Further, large samples must be used. But the method worked well, in that the secondary neutron spectrum from double differential cross sections was clear of breakup neutrons all the way down to 1.5 MeV neutron energy. However, even with the flight distance between neutron source and sample extended to 2.2 m, this method can only work up to, at most, 12 MeV incident.

Nonetheless, it promises to cover part of the important energy region between 8 and 14 MeV incident. Detailed tests of the quality of measurements made with this arrangement are still being made, and comparisons of results with those from conventional time of flight spectrometry have been developed at an incident energy of 10 MeV. The comparison is promising.

Activation methods for charged particle emission channels Measurement of secondary neutron spectra of course cannot account for the rather weakly excited charge particle emission channels. These have been conventionally obtained by activation methods, which provide results integrated over angles and states of the residual nuclei. The advance of chemical separation techniques reported by Qaim from Jülich enables clear identification of particular reaction channels, corresponding to D, or T, or ^{3}He emission. Thus the absorption channels which would be missed in neutron detection are included.

Interpretation and Evaluation The proper description of the double differential cross sections in terms of realistic reaction models is one of the most complex problems that nuclear theorists and analysts confront. Accurate knowledge of the energies in all reaction channels is of course essential, but also central are the energy densities of levels, particularly in those energy regions where the experimental resolution is completely inadequate to separate final levels. Zhang Jingshang has taken on, and quite ingeniously completed, analyses for several sets of results from these neutron induced reactions.

The critical and central role played by level densities led to his collaborations with G. Reffo of Bologna, whose two-fluid level densities enable analysts to account for differences imposed by the fact that in heavy nuclei neutrons and protons occupy very different shells. These models, far more realistic than the earlier single fluid models with which we have all been familiar, were used with great success by Zhang in his evaluations; Zhang Jingshang's evaluations have been done with great thoroughness, and have provided really excellent characterizations of a large body of measurements. This example of the power of a strong collaboration between two nuclear centers, the ENEA of Bologna and CIAE of Beijing, shows how important cooperation can be to permit both centers to advance rapidly through the synergism achieved by collaboration. It also creates confidence in the results of evaluations, since their success is shared and released by both centers.

3 Fission

Low energy neutron emission Fission studies also reflect the maturity of an advanced science, and also the power of collaboration between centers. M. V. Blinov from St. Petersburg addressed especially the serious problem of excess low energy neutron emission; far more low energy neutrons are emitted than can be explained. It is indeed interesting that fission studies are so well advanced, with so much known, and the neutron economy of reactors is critically important; and yet a key problem exists across the decades of studies in that the neutron emission spectrum is yet to be understood.

Blinov's emphases about mysteries remaining in fission were nicely confirmed in the experiments of Bao Shanglian of Peking University and the CIAE. Thus we have another example of an important point being confirmed, generating confidence that both what is known and what is not known is clarified. The clarification comes from the successful collaboration between Blinov of the Radium Institute of St. Petersburg and Bao of P. U. Experiments correctly determining the fission neutron spectrum at low emission energies are quite difficult to develop; that is undoubtedly the reason that such a serious discrepancy could remain unresolved for so long.

The critical need for accuracy, and the careful attention to detail required to make a useful contribution in either fission studies or neutron capture was reviewed in a meticulously organized paper by A. J. Deruytter of Geel. He pointed out that we are dealing with processes over an extraordinary incident neutron energy range, from μeV to 400 MeV. This is a dynamic range spanning more than fourteen orders of magnitude! Characterizing processes accurately over such an incredible range is bound to be associated with a large set of problems.

Quite a large set of fast neutron induced fission studies has been embarked upon by the group at CIAE. The strong connections between that developing group and the established groups, like Blinov's at the Radium Institute and other groups from different countries, will be immensely valuable to the CIAE program.

4 Capture Studies

An important competing process for fission at quite low neutron energies, and an important process at all energies in many nuclei, is neutron capture. S. Raman's review of low energy capture mechanisms showed that, although the principal mechanisms are rather well understood, we are far from complete descriptions of capture cross sections. Partly the ambiguities result from interferences between different capture amplitudes. Raman's review suggested that one of the more exciting areas for future studies would be those of the weak transitions associated with M1 capture. These should show some sensitivity to nuclear currents. Very few other methods probe nuclear currents, other than high energy inelastic electron scattering. Thus information would be provided for properties of nuclear levels very little examined or understood up to the present time.

Fast neutron capture cross sections were also measured by Lo of Sichuan University using a simple, clean technique with Moxon-Rae detectors. This is a nice demonstration of the distribution of fast neutron interests in China, extending from Beijing in the Northeast to Sichuan in the Southwest and, by virtue of including Xie Kuanzhong's work on (n,2n) reactions, to Shanghai in the Southeast.

One of the special fruits of capture studies, besides examining cross sections, results from the sensitivity to nuclear structure. Through examination of primary and secondary γ-ray transitions following capture one gets a vast amount of information about level and decay schemes. No other reaction provides such a mass of information on low spin levels of nuclei. For several decades level schemes near the neutron separation energy, or at excitation energies typically of 7 to 8 MeV, have been provided, and these, when combined with levels at low excitation energies are now giving us a good description of level densities as a function of excitation energy. There are some surprises, in that the normal backshifted Fermi gas models, which have been used to characterize these densities since the late 1930s, are now seen to be inadequate.

Lifetime Determinations Another and important dimension of nuclear structure studies through capture comes from the ability to make lifetime measurements. We have been used to thinking that nucleon induced reactions provide such small nuclear recoil velocities in medium weight or heavy nuclei that it was not very practical to expect to obtain lifetime information. The improved resolution of modern Ge γ-ray detectors and the stability of electronics have both improved to the point that the nucleon-induced recoils of a few percent of the speed of light are quite routinely adequate to permit lifetime determinations in the sub-picosecond regime, or lifetimes of collective levels of many nuclei. Thus collective transition speeds can be measured in nucleon induced reactions.

But now the development of the Gamma Ray Induced Doppler (GRID) technique at the ILL in Grenoble enables lifetime studies just from the recoils induced by γ-ray emission! The GRID methods and results reported by Peter Lieb show the immense power of this method to study not only nuclear lifetimes but also the mechanisms of energy transfer of very slow heavy ions in materials. Such powerful results come from recoils $\sim 10^{-4}$ c, or one hundred times slower than those from nucleon inelastic scattering. The power of the dispersive crystal diffraction methods as developed by Borner and colleagues at the ILL will mean, when combined with lifetimes determined also in neutron inelastic scattering, that soon lifetimes will be as commonly available as level and decay schemes have been.

Lieb illustrated the value of these lifetime measurements by citing the GRID results for decays in ^{54}Cr, where strong M1 or "scissors mode" decays were observed from both 1^+ and 2^+ levels. These strong decays, which almost exhaust the expected strength for such modes, support the finding that valence neutrons and protons are excited differently from each other even in fairly light nuclei, where the neutron excess is not especially large. Both of these scissors mode states are predicted in boson models, which propose separate excitation energies and electric quadrupole strengths for proton and neutron collective excitations.

5 Neutron-Nucleus Interactions and Scattering

R. L. Walters group at Duke showed the value, for fundamental studies, of concentrating on spin observables. This work showed two important properties required of basic studies. 1) One is the need to concentrate on sensitive observables, like spin properties, and 2) second is the need even then, to work at the edge of what is possible in precision and accuracy. Neutron-proton analyzing powers are as low as 2 %, and need to be measured to an accuracy of ± 0.2 %. Such accuracies can be achieved with an intense polarized beam and careful methods.

Analyzing powers for neutron scattering in complex nuclei are important ingredients in Luisa Hansen's microscopic models of scattering from complex nuclei, using potentials constructed by folding models based on an effective nucleon-nucleon interaction. Hansen's method of constructing scattering fields microscopically allow us to replace complex phenomenological representations of scattering, which many of us use, employing more than twenty energy dependent parameters, with models having just a few fixed parameters. The success of these microscopic models confirms the validity of the whole mean-field approach to scattering, and with quite surprising accuracy.

Qi Huiqnan's very nice work on small angle scattering provided just the extra set of information that would help test Luisa Hansen's folding models. Cross section measurements below 20° are very difficult to obtain, requiring a special physical arrangement. These data are usually missing. Thus Qi's special measurements fill an important gap, and also have value in showing that collaborators from universities near the institute can execute difficult experiments working in collaboration with scientists from the CIAE.

6 Intermediate Energy Results

Another demonstration of the power of collaborations to resolve complex problems efficiently is shown in the use of Roger Finlay's total cross section measurements to extend knowledge of the energy dependence of scattering from a few MeV to 600 MeV. These data, combined with measurements at lower energies from TUNL, are being shared with Luisa Hansen, to test the predictive power of microscopic models to characterize neutron-nucleus interactions over a greatly extended energy region. Thus four components of neutron scattering studies augment each other's importance, and form a picture whose completeness could not be achieved without the cooperation which leads to their union in a single picture. The four components are the extended energy total cross sections offered by Roger Finlay, the small angle results of Qi, differential measurements similar to those of Walter's group, and the microscopic models of Luisa Hansen.

Photonuclear processes at intermediate energy explore a different dimension of nucleon-nucleus processes, and provide at present a very different view of our understanding of neutron-nucleus interactions. They promise to offer exciting insights into giant density oscillations, or coherent superpositions of giant multipole resonances.

Recent experiments reviewed by I. Halpern show, however, that the production of these highly interesting and fundamentally simple excitations is severely complicated by the large set of mechanisms which can contribute to radiative capture, especially at high nucleon energies, or to the inverse photonuclear processes. Separating these processes, and thus resolving the interferences between different multipole orders in the excitations, and the information they can offer about high energy nuclear excitations, will require many detailed experiments and considerable associated theoretical work.

The message of these studies seems almost exactly opposite that arrived at by those studying scattering, who could extend relatively simple mean field models from bound levels, with the aid of dispersion corrections to scattering potentials, to 600 MeV nucleon energy. Mean field models work well over such an extended energy region, and even yield, at least to fair accuracy, to microscopic models based on an effective nucleon-nucleon interaction. Furthermore, they do not even seem especially sensitive to the precise details of the microscopic interaction upon which the folding models are based.

In contrast, when non-elastic processes are considered, even those dominated by electromagnetic decay, we find a complicated mixture of mechanisms and processes which do not yield readily to resolution in terms of simple models. It is clear that although the mean field approaches work well, and characterize the average effects of neutron-nucleus processes easily, the extension to non-elastic processes is far from clear, particularly at intermediate energies, as I. Halpern noted in his review of photonuclear processes.

7 Nuclear Data Efforts in China

The nuclear data and reactions programs of the Institute of Atomic Energy have evolved quite rapidly over the past five years. One has to remember that although almost all of the major instrumentation of the tandem accelerator laboratory had been thoughtfully developed by 1986, at the celebration of the international symposium "Physics at Tandem", in fact the CIAE accelerator was then just barely operational.

Now, five years later almost to the day, many experimental programs are in full operation and producing results which claim the attention of the international nuclear physics community. The effectiveness of the evolving programs is reflected partly in the several international collaborations which have developed with other major nuclear centers, particularly in Russia and Japan.

The development of a nuclear data compilation and evaluation program emphasizes the large range of experimental efforts underway in China, some in many different regions of the country, but coordinated largely through the institute near Beijing. The evaluation program can be quite helpful to the world community interested in the most accurate and complete descriptions of nuclear processes, particularly those which are especially relevant to many different applications of nuclear technology. But the evaluation program is also extremely important in focusing the overall neutron physics program at CIAE, and in illustrating those discrepancies which lead to the most efficient utilization of additional measurement resources. The theoretical evaluation program is an enormous asset, both to nuclear physics in China and to the developed world.

8 Some Conclusions

Neutron physics is a mature field, in which neutron interactions at low energies, below 15 MeV, are well enough described that they can be used to fix nuclear structure properties with good accuracy. The maturity of the field enables examination of spin-observables in good accuracy, for example, and the consequent testing of the degree of success which can be obtained for an effective nucleon-nucleon approach to scattering.

Second, when characterizations of processes are extended to intermediate energies, above 100 MeV, there are striking successes for scattering, but serious and unresolved problems for non-elastic processes. Indeed, even at lower energies descriptions of capture processes is complicated by little understood competition between different reaction mechanisms. The effects of such competitions are greatly enhanced at higher energies, obviating any present simple quantitative interpretations of high energy nuclear excitations.

Third, that this is a field rich with outstanding but also with important technical problems whose resolution could be worth vast fortunes when we are able to develop practical fusion energy systems. Many laboratories are now working to resolve these practical problems, which are far less tractable than those which have largely been overcome in the neutron energy regime most important for fission energy systems.

Fission itself, however, has outstanding problems whose resolution could be very important to design of systems with enhanced long term stability and improved safety. Resolution of those problems should come about through cooperation and collaboration between laboratories in different countries, all working with extraordinary care to enhance sensitivity and accuracy in measurements.

Fourth, that each neutron physics laboratory, though often working over a substantial part of the entire range of open questions, develops a sort of unique signature, which enables it to make special contributions to the worldwide compilations of information. Thes compilations will form the base for practical applications executed efficiently and accurately. Thus the combination and comparisons of the evaluated results from all of the centers are far more valuable than just the sum of the results.

Perhaps the most important frontiers of neutron physics now are in terms of using neutron induced reactions to study outstanding structure problems, and to study the extent of validity of the notion of an effective nucleon-nucleon interaction. Open problems become more numerous and profound as one extends studies in this field into energies of several hundreds of MeV.

Acknowledgement

Those of us who have enjoyed the hospitality and encouragement of the China Institute of Atomic Energy, and particularly those who enjoy long professional associations through collaborative programs with members of the Institute, owe a debt of gratitude to President Sun of CIAE who invited us and continues to encourage collaborations and associations. We are indebted also to Mr. Tang Hongqing and Madam Zhou Zuying and the other members of the local arrangements group who solved our logistical problems with such dispatch that we were never distracted from the happy business of the conference and our enjoyable associations with each other. It was a special pleasure also to share the inspiration and leadership of Professor Wang Ganchang, honorary President of the CIAE.

Closing Address

Sun Zuxun
(President, China Institute of Atomic Energy)

Ladies and gentlemen, dear colleagues,

I have the great honour, on behalf of the organizing committee, to present some words before the close of the symposium. We have just spent five stimulating and hardworking days in participating the symposium on fast neutron physics. I think, all of you will agree with me that the symposium is a enjoyable, successful and fruitful one.

In this symposium, we have reviewed and exchanged the contemporary achievements and lastest developments of the theoretical ideas and experimental results, methods as well as techniques concerned with fast neutron induced reactions, fissions and so on, which have been so well summarized by Professor McEllistrem in his talk. I believe it was very helpful and beneficial to the participants.

As Professor Wang Ganchang mentioned in the opening address, neutron has been discovered for almost sixty years. After its discovery, especially after the discovery of fission, people quickly realized, neutron is a very useful tool. Then tremendous efforts have been made in neutron physics. It has been one of the most important part of nuclear science and nuclear physics for a long time. From this symposium, we were very pleased to see this old field is still active and vigorous. I think, the reason is the expanding applications of nuclear power and nuclear techniques need more detailed knowledge on neutron physics. And also, neutron physics can play an indispensable role in some cases in fundamental research of nuclei. To take an example, recently a new application of neutron data is the study of transmutation of long lived radiowaste. Another interesting application is the usage in space science. It demands more accurate neutron data of more wide mass region in more broad energy range. I believe the demands will never finish, so the development of this field will never stop. This has been proved by our symposium.

In this week, Chinese physicists have presented a number of theoretical and experimental results which were achieved in recent years and have learned a lot from the distinguished guests of various laboratories of Belgium, Germany, Italy, Japan, US and USSR. I hope after this symposium, the scientific exchange in this circle could be further strengthened.

Finally I would like to take this opportunity to give my sincere thanks to my colleagues of local organizing committee and secretariat. Without their excellent and enthusiastic works, the success of this symposium is impossible.

Also I would like to express our deep gratitude to the sponsors of the symposium:

China Institute of Atomic Energy,
Beijing National Tandem Accelerator Laboratory
Chinese Nuclear Physics Society

Now let me close the International Symposium on Fast Neutron Physics. Thank you and see you again.

LIST OF PARTICIPANTS

NAME	INSTITUTION	NATION
Bao Shanglian	Peking Univ., Beijing	China
Batenkov, I. O.	Radium Inst., St. Petersburg	Russia
Blinov, M. V.	Radium Inst., St. Petersburg	Russia
Cai Dunjiu	CNDC, Beijing	China
Chen Zhenpeng	Tsinghua Univ., Beijing	China
Chiba, S.	JAERI, Tokai	Japan
Deruytter, A. J.	CBNM, Geel	Belgium
Ding Dazhao	CIAE, Beijing	China
Ding Shengyao	CIAE, Beijing	China
Du Xiangwan	IAPCM, Beijing	China
Finlay,R. W.	Ohio Univ., Athens	U. S. A.
Gu Yifan	IHEP, Beijing	China
Guan Xialing	CIAE, Beijing	China
Halpern, I.	Washington Univ., Seattle	U. S. A.
Han Hongyin	CIAE, Beijing	China
Hansen, L.	LLNL, Livermore	U. S. A.
Hou Long	CIAE, Beijing	China
Hu Jimin	Peking Univ., Beijing	China
Hu Jinxiu	CNNC, Beijing	China
Huang Zhengde	CIAE, Beijing	China
Jiang Chenglie	CIAE, Beijing	China
Ke Zunjian	CIAE, Beijing	China
Klein, H.	PTB, Braunschweig	Germany
Kong Xiangzhong	Lanzhou Univ., Lanzhou	China
Li Anli	CIAE, Beijing	China
Li Shounan	CIAE, Beijing	China
Li Ze	CIAE, Beijing	China
Lieb, K. P.	Göttigen Univ., Göttigen	Germany
Liu Tingjin	CIAE, Beijing	China
Lu Hanlin	CIAE, Beijing	China
Luo Xiaobin	Sichuan Univ., Chengdu	China
McEllistrem, M. T.	Univ. Kentucky, Lexington	U. S. A.
Piao Yubo	Lanzhou Univ., Lanzhou	China
Qaim, S. M.	KFA, Jülich	Germany
Qi Huiquan	Tsinghua Univ., Beijing	China
Raman, S.	ORNL, Oak Ridge	U. S. A.
Reffo, G.	ENEA, Bologna	Italy
Shen Guanren	CIAE, Beijing	China
Shen Qingbiao	CIAE, Beijing	China

Shi Zongren	CIAE, Beijing	China
Su Zongdi	CIAE, Beijing	China
Sun Zuxun	CIAE, Beijing	China
Takahashi, A.	Osaka Univ., Osaka	Japan
Tang Hongqing	CIAE, Beijing	China
Walter, R. L.	Duke Univ., Durham	U. S. A.
Wang Ganchang	CNNC, Beijing	China
Wang Shunuan	CIAE, Beijing	China
Wang Tieshan	Inst. of Modern Phys., Lanzhou	China
Wang Xiaozhong	CIAE, Beijing	China
Wang Xuezhi	Lanzhou Univ., Lanzhou	China
Wen Chenlin	Beijing Normal Univ., Beijing	China
Wu Enjiu	Inst. of Modern Phys., Lanzhou	China
Xia Haihong	CIAE, Beijing	China
Xie Kuanzhong	Shanghai Inst. of Nucl. Res., Shanghai	China
Xu Jincheng	CIAE, Beijing	China
Yang Liming	Peking Univ., Beijing	China
Ye Bangjiao	China Univ. of Sci. & Tech., Hefei	China
Yuan Junqian	Lanzhou Univ., Lanzhou	China
Zhang Jingshang	CNDC, Beijing	China
Zhang Tianyuan	IAPCM, Beijing	China
Zhang Yumin	CNNC, Beijing	China
Zhao Jinwu	Nanjing Univ., Nanjing	China
Zhao Wenrong	CIAE, Beijing	China
Zhao Zhixiang	CIAE, Beijing	China
Zhou Delin	CNDC, Beijing	China
Zhou Zuying	CIAE, Beijing	China
Zhuang Youxiang	CIAE, Beijing	China
Zhuo Yizhong	CIAE, Beijing	China

www.ingramcontent.com/pod-product-compliance
Lightning Source LLC
LaVergne TN
LVHW082255241026
850861LV00008B/50
9789810209223